Properties of Plane Areas

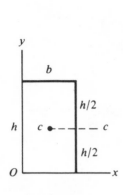

$$I_x = \frac{bh^3}{3}$$

$$I_y = \frac{hb^3}{3}$$

parallel to x,

$$I_c = \frac{bh^3}{12}$$

parallel to y,

$$I_c = \frac{hb^3}{12}$$

$$\bar{x} = b/3$$
$$\bar{y} = h/3$$

$$I_x = \frac{bh^3}{12} \qquad I_y = \frac{hb^3}{12}$$

parallel to x, $\quad I_c = \frac{bh^3}{36}$

parallel to y, $\quad I_c = \frac{hb^3}{36}$

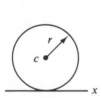

$$I_x = \frac{5\pi r^4}{4}$$

$$I_c = \frac{\pi r^4}{4} \cdot$$

$$J_c = \frac{\pi r^4}{2}$$

for $t \ll r_{\text{ave}}$

$$A = 2\pi r_{\text{ave}}\, t$$

$$I_c = \pi r_{\text{ave}}^3 t$$

$$J_c = 2\pi r_{\text{ave}}^3 t$$

semicircle

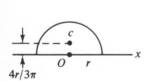

$$I_x = \frac{\pi r^4}{8}$$

parallel to x,

$$I_c = 0.035\pi r^4$$

$$J_O = \frac{\pi r^4}{4}$$

quarter ellipse

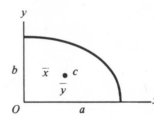

$$A = \frac{\pi a b}{4}$$

$$\bar{x} = \frac{4a}{3\pi}$$

$$\bar{y} = \frac{4b}{3\pi}$$

$$I_x = \frac{\pi a b^3}{16} \qquad I_y = \frac{\pi b a^3}{16}$$

parallel to x, $\quad I_c = 0.0175\pi a b^3$

parallel to y, $\quad I_c = 0.0175\pi b a^3$

STRENGTH
OF
MATERIALS

STRENGTH
OF
MATERIALS

BELA I. SANDOR

University of Wisconsin

PRENTICE-HALL, INC., *Englewood Cliffs, New Jersey   07632*

Library of Congress Cataloging in Publication Data

Sandor, Bela Imre.
  Strength of materials.

  Bibliography: p.
  Includes index.
  1. Strength of materials. I. Title.
TA405.S28   620.1'12   77-15506
ISBN 0-13-852418-1

© 1978 by Prentice-Hall, Inc., Englewood Cliffs, N.J.   07632

Printed in the United States of America

10   9   8   7   6   5   4   3

PRENTICE-HALL INTERNATIONAL, INC., *London*
PRENTICE-HALL OF AUSTRALIA PTY. LIMITED, *Sydney*
PRENTICE-HALL OF CANADA, LTD., *Toronto*
PRENTICE-HALL OF INDIA PRIVATE LIMITED, *New Delhi*
PRENTICE-HALL OF JAPAN, INC., *Tokyo*
PRENTICE-HALL OF SOUTHEAST ASIA PTE. LTD., *Singapore*
WHITEHALL BOOKS LIMITED, *Wellington, New Zealand*

CONTENTS

13   FAILURE CRITERIA AND DESIGN CONCEPTS   334

14   COMBINED STRESSES   356

15   PROJECT PROBLEMS   370

APPENDIXES

PREFACE

This text is intended primarily for students who are taking their first course in strength of materials. Several sections will also be of interest to more advanced students and to engineers who are seeking introductions with perspective to some of the most important recent developments in this area. The following features make this book distinguishable from others in its category.

New Material

Strength of materials is an old subject, but it is now rejuvenated with important new developments that are caused by the appearance of new materials and the need to use even the known materials in unusually severe and critical situations. Furthermore, significant advances have been made in recent years in the mechanical testing of materials. These advances not only allow more rapid and more precise tests but also have led to the practical use of entirely new concepts in strength of materials. Designers who obtained their academic education a number of years ago must now enroll in special short courses to learn about the new ideas; otherwise they must glean the information from mountains of relevant but diverse publications. It is imperative that future designers be introduced to the new as well as to the classical fundamental concepts and methods of strength of materials. Following are the most important new items (order of appearance has no significance):

1. Permutations of signs of stress and strain. Every designer should know that in many practical situations all four combinations of the signs of stress and strain are possible (+ and +, − and −, + and −, − and +). This has serious implications in using experimental stress analysis in the design process where stresses are calculated from measured strains.

2. *Cyclic stress-strain curves.* The stress-strain response of many materials depends on the prior loading history. This means that the standard tensile stress-strain curve may be inadequate for predicting a material's behavior when the loads on it vary it time, which can be expected in most cases. Sustained efforts must be made to publicize the need for obtaining the cyclic stress-strain curves besides the tensile or compressive stress-strain curves. Numerous examples of stable and unstable mechanical behavior are given in this text, and these may be of considerable interest even to experienced designers.

3. *Modern failure criteria.* The classical failure theories have limited usefulness in practice because they cannot be applied readily to deal with members that have discontinuities. Furthermore, the full significance of a flaw or machined notch is not indicated by the theoretical stress concentration factor. Certain notched members can fail in the brittle manner even if they are made of intrinsically ductile materials. The introductory mechanics analysis of such members is within the scope of what students can and should learn in a modern elementary course in strength of materials. The concept and practical aspects of fracture toughness (K_{Ic}) are also in this category.

4. *Composite materials.* Filament-reinforced composite materials such as boron-aluminum and graphite-epoxy are important in modern technology. The rule of mixtures and its limitations are presented for these. Sintered porous metals are also discussed. These can be made with wide ranges of density and strength.

5. *Modern testing equipment.* The capabilities of servo-controlled, electrohydraulic machines are discussed and demonstrated with several photographs.

Old Material in Original Presentation

There are a number of original models and analogies throughout the text.

The use of Mohr's circle is extended to deal with optimum design. There are also examples of stress analysis for nonsteady loading.

The various concepts of material toughness are clarified and presented in perspective.

The concept of the absolute maximum shear stress is given special emphasis because it has significance in modern fracture analysis and control techniques.

Mathematics

The reader should have a working knowledge of elementary calculus. The derivations of formulas are tailored to the average student.

Units

The U.S. customary and the SI units are used alternately in the text and in the problems. Note that large numbers in the U.S. system use commas, but in the SI system a space is used instead of a comma. For example, 20,000 in. = 508 000 mm.

Problems

There are approximately 160 examples worked out in the text. The solutions often show the patterns of engineering judgment in the modeling process (making assumptions) and in the evaluation of the answers.

In each chapter there are many simple, intermediate, and difficult (noted by asterisk, except in Chapter 15) problems for a wide range of experiences. Some of the difficult are open-ended, *multiple-answer problems*, which are the most common in the world. Although they are not universally liked many people consider them the most useful and the most interesting. These problems allow the student to practice the highest cognitive objectives of learning, which include analysis, synthesis, evaluation and choice, and strategies of problem-solving.

The readers should practice judgement in answering any problem. They should simply state whether the answer appears reasonable, unreasonable, or uncertain. The idea here is that anybody can make a mistake, but few people have the chance to check everything carefully. Well-educated guesses are important in guiding one to the correct answers.

Both systems of units are used in a few of the problems. This can be annoying, but real problems are this way sometimes, and perhaps can never be avoided completely.

Only a few of the project problems in Chapter 15 can be worked on in a given semester, at best. There are many ways of handling these problems, but generally they should be left for the second half or the end of the course. They are good for extra-credit work and for group projects. Students are encouraged to read Appendix B concerning the modeling process before trying to solve complex problems.

Motivation

Last, but not least, an aim of this text is to motivate the students. Most of the topics are introduced in ways that make them interesting and reasonable to explore. Some people consider this unimportant when the subject is intrinsically interesting. There is evidence from educational research, however, that affective learning (which involves interest, attitude, value, and emotion) should be nurtured to achieve the best results in educating students in any area. One could say on the basis of common sense that the most successful athletic coaches do much more for their athletes than help in muscle building and in the acquisition of techniques (and most athletes are basically more motivated than the average student in any average course). It is hoped that this book will be accepted as an enthusiastic assistant coach might be.

B.I.S.

Madison, Wisconsin

ACKNOWLEDGMENTS

At least a few of the people who have contributed to this book must be mentioned. Professor JoDean Morrow of the University of Illinois has influenced my thinking in numerous ways since about 1960. He is the grandfather-author in general and the creator of several of the most interesting problems in particular. Dr. R. W. Landgraf of the Ford Scientific Research Staff has provided guidance through his published and unpublished work in the use of recently introduced mechanical properties of materials. William Garver gave invaluable assistance in working out many examples and in critically evaluating the whole manuscript. Mrs. M. Lynch and Mrs. E. Schultz were excellent and patient typists.

STRENGTH
OF
MATERIALS

INTRODUCTION

Strength of materials has been of interest to humans since prehistoric times. All solid materials have limits to their strengths, and a qualitative understanding of these has always been essential in making useful tools, weapons, machines, vehicles, and structures. War machinery is of special historical interest because Archimedes was deeply involved with the concepts and practical problems of equipment used in warfare. He was spectacularly successful in the systematic application of a few fundamental principles of mechanics. According to an ancient historian, he developed the novel technique of dropping hooks on an attacking Roman ship, quickly raising the ship out of the water, and dropping it back. It appears that the Romans could not occupy Syracuse by force even after a long siege because Archimedes applied his knowledge and creative ability in the defense of the city (see Chapter 15, Problem A).

Among the numerous great achievements of Archimedes, one item must be mentioned in introducing this text. He formulated and communicated to later generations the concept of equilibrium conditions involving forces and moments. The concept was refined later, and now it is essential in analyzing the strength of materials.

Through most of the Middle Ages there was a gradually growing need for more knowledge in using materials, but engineers and architects relied on specific experiences and intuition instead of generally applicable analysis. To show this state of affairs and to provide a perspective for the important developments that were to follow, consider some of the efforts by Leonardo da Vinci concerning materials technology and design. One of his most ambitious projects was the design and creation of the largest equestrian statue of all time. The horse alone was to be 24 ft high. At first Leonardo boldly contemplated a rearing horse of magnificent beauty, but he was realistic in his final plans. He knew how strong bronze was, and he understood the problem of stability for a rearing horse (the total weight of the statue would have been about 200,000 lb). But he must have realized that his simple tension tests of wires did not provide enough information for the design of the most critical parts of the statue, the rear legs. Thus, he resigned himself to the idea of supporting three legs of the horse, which must have been a very painful decision because that was how all other artists made their equestrian statues. He could have gambled and won, or he could have lost disastrously. The knowledge necessary to ensure success simply was not available at the time.

The first substantial steps toward understanding the strength of materials were made by Galileo Galilei (1564–1642). He experimented extensively and made keen observations of the origins of fractures. Interestingly, the area of strength of materials was one in which Galileo made a serious error, as will be shown later. In spite of his mistake, he made such important contributions that now he is regarded as the founder of mechanics of materials as a subject for rational investigation. This little known fact is worthy of thought by both students and teachers.

Since Galileo's time, the subject has received increasing attention. It is a cornerstone of modern technology, and it will remain important in the future. There are diverse examples in this text to show the universal applicability of knowledge of strength of materials.

1

STRESS

The words *stress* and *strain* are used frequently but they denote different things to different people. The meanings are most specific to engineers who deal with the strength or deformability of materials. A number of special definitions of stress and strain are needed by engineers, and the most important of these will be presented in this text. Sometimes it is necessary to be careful and precise in using the appropriate definition in technical communication. For example, normal stress, shear stress, local stress, elastic strain, and plastic strain all have their own particular meanings.

Stress is the description of the intensity of force; strain is the description of the intensity of deformation. The two are related in most cases. It always should be remembered that all solid materials deform under the action of any force. There are no perfectly rigid bodies, even though such an assumption is often useful.

The rational basis for evaluating and comparing the mechanical performances of different materials is called *stress analysis*, which encompasses the analysis of strains. This chapter introduces the concept of stress; the deformability of solids can be generally ignored at this point.

1-1. STRESS

A good way to learn about stresses and fully appreciate the significance of several basic concepts in stress analysis is to consider a glued joint. Imagine that two pieces of flat metal are glued together with an adhesive whose strength properties are known. Assume that the finished piece is rectangular, as shown in Fig. 1-1a, but there is a choice of the angle θ of the seam.

A reasonable question is, "what is the optimum angle θ to avoid failure of

3

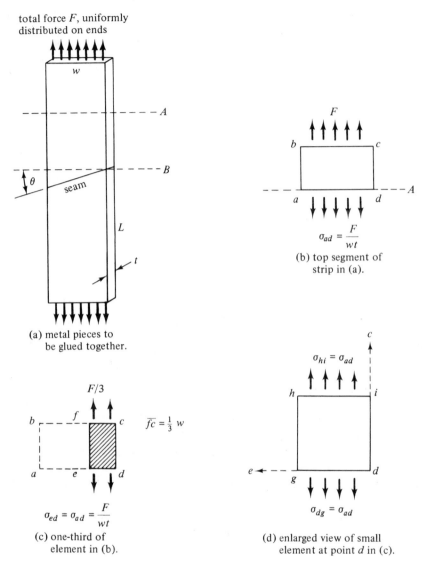

total force F, uniformly
distributed on ends

(a) metal pieces to
be glued together.

$$\sigma_{ad} = \frac{F}{wt}$$

(b) top segment of
strip in (a).

$$\sigma_{ed} = \sigma_{ad} = \frac{F}{wt}$$

$$\overline{fc} = \frac{1}{3}\,w$$

(c) one-third of
element in (b).

$$\sigma_{hi} = \sigma_{ad}$$

$$\sigma_{dg} = \sigma_{ad}$$

(d) enlarged view of small
element at point d in (c).

Fig. 1-1. (a) Metal pieces to be glued together. (b) Top segment of strip in Fig. 1-1a. (c) One-third of element in Fig. 1-1b. (d) Enlarged view of small element at point d in Fig. 1-1c.

the seam under the given conditions?" The answer to this can be found by determining the maximum values of the internal forces in the material and the directions in which these forces are acting and then relating these to the strength properties of the glue. Doing these steps for the first time will go rather slowly. Later, with a little experience, the answer can be found rapidly.

The first item of concern is the intensity of the internal forces. The intensity of a force is called *stress* and is defined as the force acting on a unit area. It is commonly denoted by the lowercase Greek letter σ. The stress is calculated by dividing the total force acting on an area by the size of that area:

$$\sigma = \frac{F}{A} \text{ pounds per square inch (psi) or pascals (Pa} = \text{N/m}^2) \qquad (1\text{-}1a)$$

Referring to the specific problem, a free-body diagram of the metal strip above line *A* (Fig. 1-1a) shows that the stresses on this piece, at the bottom surface, are downward (Fig. 1-1b). Taking a new free-body diagram, a third of the preceding one, we have only a third of the total external force that is acting upward (Fig. 1-1c). The internal stresses, σ_{ed}, are the same as for the whole cross section, σ_{ad}. This is because stress is a force on a unit area. The downward force in the last case is obtained by multiplying σ_{ed} with the area on which it acts:

$$\sigma_{ed} \frac{1}{3} wt = \frac{F}{wt} \frac{1}{3} wt = \frac{F}{3}$$

Thus, the element *cdef* is in equilibrium.

There is an important lesson to be learned from this simple exercise. Force and stress are related, but they are not the same. In fact, it is best to keep in mind that

> *The equilibrium equations of statics are valid ONLY for forces or for moments of forces.*

Return now to Fig. 1-1c. There are no stresses on side *ef* because there is no loading on side *cd*. Also, there are no stresses perpendicular to the plane *cdef*. This is called *uniaxial loading*.

In the next step let us consider an extremely small element *dghi* from section *cdef* (Fig. 1-1d). It is small, but the atoms are not distinguishable. The stresses are zero on faces *di* and *gh*; on the other faces they are equal to σ_{ad} on the basis of the preceding discussion. The correct statement of equilibrium for the element *dghi* is

$$\sigma_{ad}(\text{area } hi) = \sigma_{ad}(\text{area } dg)$$

Interestingly, the concept of stress as defined in Eq. 1-1a is most valid for such a tiny area as *dghi*. On large areas the stresses are seldom uniform, and Eq. 1-1a is only useful for calculating the average stress, even in the case of uniaxial loading. Thus, a more satisfactory definition of stress is obtained by working with the differential load dF that acts on a differential area dA:

$$\sigma = \frac{dF}{dA} \qquad (1\text{-}1b)$$

In the case of nonuniform stresses, the total force on an area is calculated by integrating Eq. 1-1b:

$$F = \int dF = \int \sigma dA \tag{1-1c}$$

The rectangular elements with all the stresses acting perpendicularly on their surfaces (as shown before) are the simplest ones in stress analysis. To analyze the stresses on the seam in Fig. 1-1a, it is necessary to consider a trapezoidal element such as the whole strip above the seam. The condition of equilibrium must be satisfied for this piece, so the total force at the seam is F (Fig. 1-2a). It is reasonable to seek elements that are as simple as possible, and one finds that for purposes of analysis the triangle in Fig. 1-2b has everything that the trapezoid in Fig. 1-2a has.

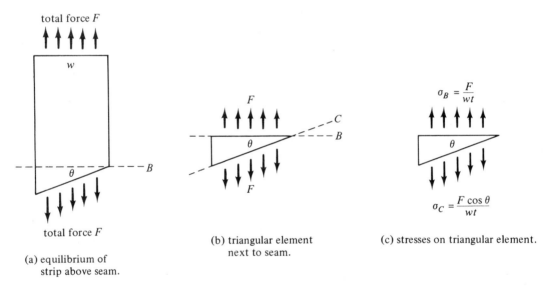

(a) equilibrium of
strip above seam.

(b) triangular element
next to seam.

(c) stresses on triangular element.

Fig. 1-2. (a) Equilibrium of strip above seam. (b) Triangular element next to seam. (c) Stresses on triangular element.

The forces on planes B and C are equal, but the stresses are not because the areas are different (Fig. 1-2c). The way of showing the stresses as in the last drawing is correct, but it is not the most convenient. The best way is to resolve the stress on each surface into two components: a stress perpendicular to the surface and a stress tangential to the surface. The first one is called a *normal* stress σ; the second is a *shear* stress τ. The latter is defined according to Eq. 1-1a, with the force being tangential to the surface. Thus,

$$\tau = \frac{\text{shearing force } S}{\text{area}} \quad \text{or, more precisely,} \quad \tau = \frac{dS}{dA} \tag{1-1d}$$

To simplify matters further, it is customary to show only one arrow for each stress, even though stress is assumed to be distributed uniformly on a given face of a small element. With these simplifications the element in Fig. 1-2c is redrawn in Fig. 1-3a. From now on, the symbol σ will always be used to denote a normal stress. The subscript θ denotes the inclination of the plane area on which the stresses are acting. Figure 1-3b shows the counterpart of the element in Fig. 1-3a on the other side of the seam. Note that the directions of the stresses on the two sides of the seam are not arbitrary. They are related according to Newton's law of action and reaction.

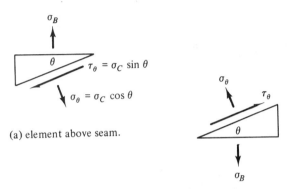

(a) element above seam.

(b) element below seam.

Fig. 1-3. Normal and shear stresses on neighboring elements.

It is reasonable to distinguish between the normal and shearing stresses in a material because they are involved in different phenomena. The normal stress may be tensile or compressive, and either of these tends to change the interatomic distances in the material. Of course in tension the atoms are farther apart than their equilibrium spacing; in compression they are closer to each other. Shear is different in that it is like friction. For a model of this, consider block A being pulled relative to block B in Fig. 1-4. Note that A will not get closer to B or farther from it for any impending or actual motion or final displacement that doesn't change the force pressing the blocks together.

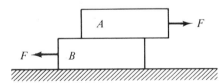

Fig. 1-4. Model for shear loading.

After these considerations the sign conventions for stresses must be clarified. Generally tensile stresses are considered positive and compressive stresses, negative. This convention is worth following, but everybody should realize that it is arbitrary. An exception to this convention is made by many people who work with the strength properties of materials that have very little strength in tension (rocks, concrete). The design loads on these materials are preferred to be compressive and these are given the positive sign. There is no physical basis for determining the signs of shear stresses. These are quite arbitrarily chosen (depending on the sense of the moment caused by τ on a small element) since there is no difference between the effects of positive and negative shear stresses (pushing block A left or right in Fig. 1-4) as far as the behavior of homogeneous materials is concerned. It is often reasonable to ignore the sign conventions for shear stress and to concentrate on consistency in using signs throughout the solution of each problem.

EXAMPLE 1-1

A bracket is to be fastened with a single bolt to a wall. The maximum force expected to pull the bracket away from the wall is 2000 N, and this is the axial load on the bolt. What is the maximum stress in the bolt if its diameter is 0.5 cm?

Solution

$$\sigma = \frac{F}{A} = \frac{2000 \text{ N}}{1.96 \times 10^{-5} \text{ m}^2} = 102 \text{ MPa}$$

EXAMPLE 1-2

A straight rod supports an axial load of 10,000 lb. There is a step change in the diameter of the rod: $d_1 = 1$ in. for one-third of its length and $d_2 = 1.2$ in. for the other two-thirds. What is the maximum stress in the rod?

Solution

The maximum stress is on the smallest of the cross-sectional areas:

$$\sigma_{\max} = \frac{F}{A_{\min}} = \frac{10,000}{\pi(1)^2/4} = \frac{40,000}{\pi} \text{ psi}$$

EXAMPLE 1-3

Two plates are riveted together with three identical aluminum rivets. The allowable normal stress in the 4-mm diameter rivets is 200 MPa. What is the maximum load that can be applied tending to pull the plates apart if the rivets share equally in resisting the external load?

Solution

The allowable load is

$$F = 3(200 \times 10^6 \text{ N/m}^2)\left(\frac{0.004^2}{4}\pi \text{ m}^2\right) = 7.54 \text{ kN}$$

EXAMPLE 1-4

A cylindrical nail of 0.1-in. diameter is driven vertically to a depth of 1 in. into a wood beam on the ceiling. A load of 50 lb is hung on the nail, and it holds. What is the shear force and the shear stress between the nail and the wood?

Solution

The shear force S is 50 lb. The shear stress is

$$\tau = \frac{S}{A} = \frac{50}{(1)(0.1\pi)} = 160 \text{ psi}$$

assuming that τ is uniform on a cylindrical area 0.1 in. in diameter and 1 in. in length.

1-2. SIMPLE PLANE STRESS

The first important milestone in stress analysis is the generalized technique of calculating unknown stresses, which is discussed here. Assume that σ_B and θ are given for an element as in Fig. 1-3. Determine the unknown stresses σ_θ and τ_θ. This can be done by solving equilibrium equations that are valid for the given element. Such an approach is often more complicated than shown in Fig. 1-2 and 1-3, but it is also more useful when the loading is complex.

Since there are two unknowns in the problem, it is necessary to write two independent equations. This is possible when all the stresses are in the same plane as here.* The obvious choices for coordinate axes are x and y or m and n, as shown in Fig. 1-5. There is a definite advantage in using the latter. Also, it is advantageous

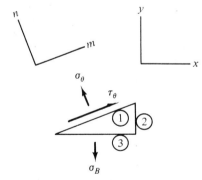

Fig. 1-5. Choice of coordinates for stress analysis.

to assume (and it is not a mistake) that the area of face 1 is equal to unity. Remembering that the equations have to be written for forces and not for stresses, they are as follows:

$$\tau_\theta(\text{area } 1) - \sigma_B(\text{area } 3) \sin \theta = 0 \quad \text{for the } m\text{-direction}$$

$$\sigma_\theta(\text{area } 1) - \sigma_B(\text{area } 3) \cos \theta = 0 \quad \text{for the } n\text{-direction}$$

*This condition is called *plane stress*.

Since (area 3) $= \cos\theta$,

$$\tau_\theta = \sigma_B (\cos\theta)(\sin\theta) \qquad\qquad\qquad (1\text{-}2)$$

$$\sigma_\theta = \sigma_B (\cos^2\theta) \qquad\qquad\qquad\qquad (1\text{-}3)$$

These are the required stresses, and they agree with the quantities shown in Fig. 1-2 and 1-3. Note that the size of the element is not important in deriving Eq. 1-2 and 1-3 as long as the stress is uniform on each plane.

Positive answers from Eq. 1-2 and 1-3 indicate that the directions of the arrows for σ_θ and τ_θ were assumed correctly in Fig. 1-5. The meaning of a negative answer is that the magnitude of that stress is correct but its real direction is opposite to that in the drawing. Thus, in setting up a problem it is not necessary to worry about the proper directions for the unknown stresses. This is particularly advantageous in more complicated problems than the present one.

The last two equations can be used to determine if there is an optimum angle θ for the seam in the metal strip (Fig. 1-1a). Of course, the optimum angle is the one that minimizes the stresses on the seam. The minimum (or maximum) values of the stresses occur where the derivatives of Eq. 1-2 and 1-3 with respect to θ equal zero. Using the double-angle trigonometric identities and equating the derivatives with zero, the results are

$$\frac{d\tau_\theta}{d\theta} = \sigma_B \cos 2\theta = 0, \qquad \frac{d\sigma_\theta}{d\theta} = -\sigma_B \sin 2\theta = 0$$

The first of these is satisfied when $\theta = 45°$ or $135°$. The second one is satisfied when $\theta = 0$ or $90°$. Substitution of these angles in Eq. 1-2 and 1-3 leads to the magnitudes of the minimum and maximum stresses. It is found that σ_θ is maximum ($\sigma_{max} = \sigma_B$) with $\theta = 0$. On the same plane, $\tau_\theta = 0$. Both σ_θ and τ_θ are zero when $\theta = 90°$. Such planes are parallel to plane 2 in Fig. 1-5, which itself has no stresses on it. The shear stress is maximum ($\tau_{max} = \sigma_{max/2}$) on two planes, $\theta = 45°$ or $135°$. There are normal stresses on these planes, but they are less than the maximum.

The most important conclusions from the foregoing are that in uniaxial loading

(a) the maximum normal stress is the external load divided by the minimum cross-sectional area of the member;

(b) the shear stress is zero on a plane where the normal stress is maximum (all planes perpendicular to the applied load);

(c) the maximum shear stress equals one-half of the maximum normal stress, and it occurs on any plane that is inclined 45° to the line of action of the applied force.

The problem of finding the optimum angle for the seam remains. This angle can be determined only if the tensile strength and the shear strength of the glue are known. Unfortunately, most manufacturers do not provide even approximately valid information in this respect (there should be a law requiring it). On the other hand, people with a little knowledge of strength of materials can find these for themselves without much trouble, as shown later. Once the strength values are known, one should select an angle θ such that σ and τ on the plane of the seam are equally below the tensile and shear strengths of the glue, respectively. If this condition cannot be satisfied, the glue is worthless and a better glue must be found or the parts redesigned to have lower stresses.

The angle that satisfies the dual conditions mentioned above can be found by trial and error on the basis of the concepts presented so far. The following discussion will lead to a superior method that can provide the answer rapidly and accurately.

EXAMPLE 1-5

Assume that the metal pieces shown in Fig. 1-1 are 2 in. wide and 0.1 in. thick. A uniformly distributed force of 500 lb is acting on the bar as shown in the figure. What are the normal and shear stresses at a section oriented such that $\theta = 30°$?

Solution

The applied stress is

$$\sigma_B = \frac{F}{A} = \frac{500}{(2)(0.1)} = 2500 \text{ psi}$$

The normal and shear stresses on the slanted plane are determined using Eq. 1-2 and 1-3:

$$\tau_{30} = 2500(0.866)(0.5) = 1080 \text{ psi}$$
$$\sigma_{30} = 2500(0.866)^2 = 1875 \text{ psi}$$

EXAMPLE 1-6

Two metal bars are joined as shown in Fig. 1-1, with $\theta = 45°$. The bars are 5 cm wide and 1 cm thick. The tensile strength of the joint is 100 MPa, and the shear strength is 50 MPa. What is the largest tensile load that can be applied to the member?

Solution

From Eq. 1-2 and 1-3,

$$\sigma_B = \frac{\tau_\theta}{\cos \theta \sin \theta} = \frac{50}{(0.707)^2} = 100 \text{ MPa}$$

or

$$\sigma_B = \frac{\sigma_\theta}{\cos^2 \theta} = \frac{100}{(0.707)^2} = 200 \text{ MPa}$$

The lower of these is the critical stress, so the external load should be less than

$$F = \sigma_B A = (10^8)(0.05)(0.01) = 50\ 000\ \text{N}$$

1-3. GENERALIZED PLANE STRESS

Situations that are more complicated than in Fig. 1-1 may occur frequently. For example, consider the following additional possibilities. There could be a second axial load, such as F_x in Fig. 1-6. Also, the external force is not necessarily uniformly distributed at the ends of the strip; there is a concentrated force F_y in Fig. 1-6. The couple $C = F_y w$ is necessary for equilibrium.

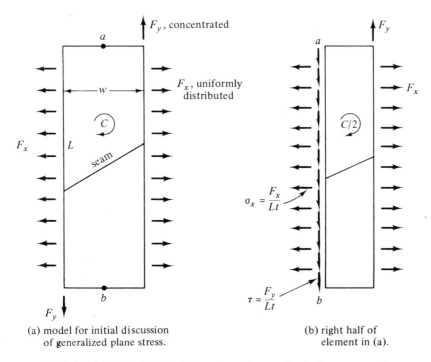

(a) model for initial discussion of generalized plane stress.

(b) right half of element in (a).

Fig. 1-6. (a) Model for initial discussion of generalized plane stress. (b) Right half of element in Fig. 1-6a.

In starting the analysis of this complex situation, consider the right half of the strip (Fig. 1-6b). The shear stresses τ acting on the face ab generate the force that opposes F_y. Assuming that these stresses are uniformly distributed, $\tau = F_y/Lt$. There are also normal stresses on face ab if the strip is in equilibrium. The magnitude of σ_x is F_x/Lt.

Next, consider a small element that includes the seam near face ab. Figure 1-7 shows the stresses on this element: σ_y is there because of the vertical tensile force F_y; τ on the left face is there because the vertical forces are applied eccentrically;

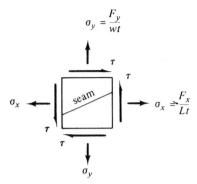

Fig. 1-7. Small element from Fig. 1-6; left face of element is part of line *ab* in Fig. 1-6b.

σ_x is caused by F_x (there is no eccentricity of this force). The element would not be in equilibrium under these stresses alone because τ on the left face generates a downward force and a counterclockwise moment. The downward force caused by τ is balanced if a shear stress of the same magnitude acts upward on the right face of the element, however, this doubles the counterclockwise moment. Equilibrium is assured only when a shear stress also acts to the right on the upper face and another one acts to the left on the lower face. The sequence of "adding" all these shear stresses is illustrated in Fig. 1-8 where only shear stresses are shown.

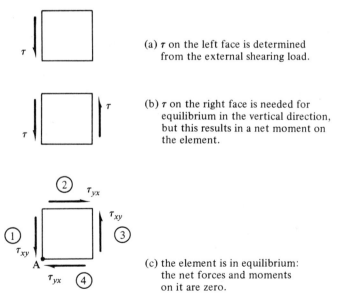

(a) τ on the left face is determined from the external shearing load.

(b) τ on the right face is needed for equilibrium in the vertical direction, but this results in a net moment on the element.

(c) the element is in equilibrium: the net forces and moments on it are zero.

Fig. 1-8. Illustration of necessity of shear stresses on all four faces of element.

The subscripts are added to the shear stresses to distinguish their directions and the planes on which they act. The first subscript denotes the direction normal to the plane, and the second indicates the direction of the stress vector. It is easy to show that the magnitudes of all four shear stresses in Fig. 1-8c are the same if the element has a constant thickness t.

For generality, assume that $a \neq b$. Assume that a shear causing a counter-clockwise moment on the element is positive. Take moments about a convenient point such as A for which two of the shears (on faces 1 and 4) have zero moment. Thus, assuming equilibrium,

$$\Sigma M_A = 0 = \tau_{xy}(bt)(a) - \tau_{yx}(at)(b), \qquad \tau_{xy} = \tau_{yx}$$

On the basis of this analysis, these subscripts will not be used for shear stresses in the rest of this text.

The stresses on the element can be completed now as shown in Fig. 1-9a. (For further considerations of the problems associated with the model shown in Fig. 1-6, see Section 1-4.) What are the normal and shear stresses on the seam in this case? These are determined in the same way as for the element in Fig. 1-3, except that the equilibrium equations contain more terms.

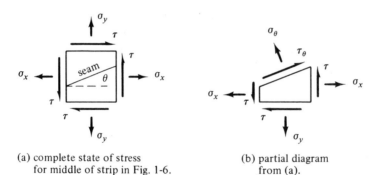

(a) complete state of stress
for middle of strip in Fig. 1-6.

(b) partial diagram
from (a).

(c) simplification of
diagram in (b).

Fig. 1-9

From the element in Fig. 1-9a a new free-body diagram is cut out to make the unknown stresses σ_θ and τ_θ external stresses as in Fig. 1-9b. A reasonable simplification of this diagram is shown in Fig. 1-9c. The faces of the triangular element* are numbered as before, and the area of face 1 is made equal to unity for convenience. The equilibrium equations for the forces on the element are

$$\tau_\theta(\text{area } 1) + \sigma_x(\text{area } 2)\cos\theta + \tau(\text{area } 2)\sin\theta$$
$$- \sigma_y(\text{area } 3)\sin\theta - \tau(\text{area } 3)\cos\theta = 0 \qquad (1\text{-}4)$$

$$\sigma_\theta(\text{area } 1) - \sigma_x(\text{area } 2)\sin\theta + \tau(\text{area } 2)\cos\theta$$
$$- \sigma_y(\text{area } 3)\cos\theta + \underbrace{\tau(\text{area } 3)\sin\theta}_{\substack{\text{force} \mid \text{to find} \\ \text{caused} \mid \text{component} \\ \text{by } \tau \mid \text{of force} \\ \mid \text{in the } n\text{-} \\ \mid \text{direction}}} = 0 \qquad (1\text{-}5)$$

The last term is explained in detail as an example for all the terms.

The unknown quantities are obtained directly from the equations above:

$$\tau_\theta = -\sigma_x(\sin\theta)(\cos\theta) - \tau(\sin^2\theta)$$
$$+ \sigma_y(\cos\theta)(\sin\theta) + \tau(\cos^2\theta) \qquad (1\text{-}6)$$

$$\sigma_\theta = \sigma_x(\sin^2\theta) - \tau(\sin\theta)(\cos\theta)$$
$$+ \sigma_y(\cos^2\theta) - \tau(\cos\theta)(\sin\theta) \qquad (1\text{-}7)$$

These equations can be simplified further with the use of the following trigonometric identities:

$$\sin\theta\cos\theta = \frac{\sin 2\theta}{2}, \qquad \sin^2\theta = \frac{1 - \cos 2\theta}{2}, \qquad \cos^2\theta = \frac{1 + \cos 2\theta}{2}$$

The resulting equations

$$\tau_\theta = \left(\frac{\sigma_y - \sigma_x}{2}\right)\sin 2\theta + \tau\cos 2\theta \qquad (1\text{-}8)$$

$$\sigma_\theta = \frac{\sigma_x + \sigma_y}{2} - \left(\frac{\sigma_x - \sigma_y}{2}\right)\cos 2\theta - \tau\sin 2\theta \qquad (1\text{-}9)$$

are called the *transformation equations for plane stress*. Note that in different books the transformation equations may appear in slightly different form (signs) from these and from one another. The reasons for such differences are the various orientations for the chosen triangular element. Any correct method should lead to the right stresses for any given plane of interest. Obviously, there is no reason to

---

*Hence the name "wedge method of analysis."

memorize the transformation equations; it is better to remember the process based on equilibrium considerations that led to these equations.

EXAMPLE 1-7

The element in Fig. 1-10a is in plane stress. Determine the stresses on a plane that is inclined 30° to plane 2. Use the wedge method of analysis.

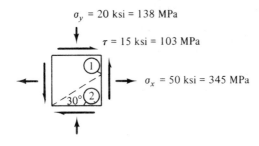

(a)

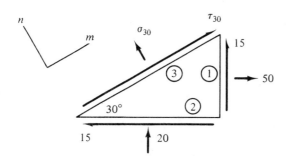

(b) wedge from (a)

Fig. 1-10. (a) Example 1-7. (b) Wedge from Fig. 1-10a.

Solution

Start by drawing a triangular free-body element taken from the rectangular element (Fig. 1-10b). The unknown stresses are σ_{30} and τ_{30}. For convenience, m-n are the chosen coordinates, and the area of face 3 is taken as unity. The equilibrium equations for forces in the m- and n-directions are

$$\tau_{30}(1) + 50 \sin 30° \cos 30° + 15 \sin 30° \sin 30°$$
$$+ 20 \cos 30° \sin 30° - 15 \cos 30° \cos 30° = 0$$
$$\sigma_{30}(1) - 50 \sin 30° \sin 30° + 15 \sin 30° \cos 30°$$
$$+ 20 \cos 30° \cos 30° + 15 \cos 30° \sin 30° = 0$$

The results are

$$\tau_{30} = -22.8 \text{ ksi} = -158 \text{ MPa}, \qquad \sigma_{30} = -15.5 \text{ ksi} = -107 \text{ MPa}$$

The negative signs of the answers indicate that the real directions of the unknown stresses in Fig. 1-10b are opposite to those assumed in the figure. In many cases it is not necessary to redraw the free-body diagram with the correct directions of the newly found stresses.

EXAMPLE 1-8

The state of stress is the same as in Example 1-7. Determine the stresses on a plane that is inclined 45° to plane 2. Use the transformation equations.

Solution

According to the sign convention that was used in the derivation of Eq. 1-8 and 1-9,

$$\tau_\theta = \frac{-20 - 50}{2} \sin 90° + 15 \cos 90° = -35 \text{ ksi} = -245 \text{ MPa}$$

$$\sigma_\theta = \frac{50 + (-20)}{2} - \frac{50 - (-20)}{2} \cos 90° - 15 \sin 90° = 0$$

Note that the signs of the stresses in the transformation equations should be considered carefully to avoid the large errors that are common in these kinds of calculations.

1-4. ST. VENANT'S PRINCIPLE

Consider some elements that are not from the center of the strip in Fig. 1-6a: for example, one at the top left corner of the strip or one at the right side of the whole strip near the seam. The stresses on such elements are quite different (Fig. 1-11) from those in Fig. 1-9a. The discrepancy is caused by having idealized point forces in Fig. 1-6 and using simplifying assumptions in determining the stresses for Fig. 1-9a. The three elements in question are not comparable directly. The material at some place on the boundary cannot "feel" at all that a point force is applied far away, perpendicular to the same boundary. On the other hand, material deep inside the strip is aware of the presence of F_y, not as a point force but as a force pulling on the strip. Taking the average vertical stress as the local stress σ_y in the

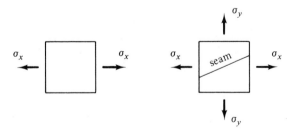

(a) top left corner
of strip in Fig. 1-6.

(b) middle region of
strip at right side.

Fig. 1-11. Elements near boundaries in Fig. 1-6. Note absence of shear stresses here.

middle region is fairly realistic if the strip is long and narrow. This is a primitive example of St. Venant's principle, which states that the stresses in a region reasonably distant from where the loads are applied on a boundary are not altered significantly if these loads are changed to other, statically equivalent loads.

St. Venant's principle is generally useful in a qualitative way. Obviously, the stress near point *a* in Fig. 1-6a depends on whether F_y is uniformly distributed or concentrated (if concentrated, the location is also important). As the distance increases from point *a*, the stress becomes less dependent on the location and distribution of F_y but still depends on the magnitude of F_y. It is often assumed that at distances larger than the largest transverse dimension of the member the stress does not depend on how or where a given force is applied at the boundary. Thus, if the distance from the upper edge of the plate in Fig. 1-6a is larger than *w* (up to *L*/2 here), one could expect σ_y to depend mainly on the magnitude of F_y. The reader is cautioned that the stresses do not change abruptly in the member according to the principle presented here. One could say that the localized effects of a force gradually disappear with increasing distance from the region where the force is applied.

Reconsiderations. Another careful look at Fig. 1-6 reveals that there are several ways to reach equilibrium. The strip could rotate until the vertical forces are collinear; horizontal forces could be applied at the corners to create the necessary counter-moment; F_x may be distributed nonuniformly to achieve the same effect. In any case, the stress distribution in the middle region of the strip will be altered. This is not important for the present discussion of the fundamental concepts of stress analysis. What should be remembered is that a given face of an element can have, at most, only one normal stress and one shear stress on it. In this sense Fig. 1-9a is as complicated as possible in plane stress (of course, the directions of the stresses may be different in other problems). The basic problem is to draw an element with the correct stresses acting on it. This may require considerable judgment in some situations.

1-5. MOHR'S CIRCLE

The transformation equations for plane stress are useful as they are, but they can be manipulated to obtain a more powerful method of stress analysis. The manipulation involves squaring Eq. 1-8 and 1-9, adding the resulting equations, and rearranging the new equation. The aim is to obtain a form of this equation that does not include any trigonometric functions of θ. Since this derivation is not entirely straightforward, the major steps are given here. Equation 1-8 is squared as it is, but Eq. 1-9 is squared with the trigonometric functions alone on the right side.

$$\tau_\theta^2 = \left(\frac{\sigma_y - \sigma_x}{2}\right)^2 \sin^2 2\theta + (\sigma_y - \sigma_x)\sin 2\theta(\tau \cos 2\theta) + \tau^2 \cos^2 2\theta$$

$$\left(\sigma_\theta - \frac{\sigma_x + \sigma_y}{2}\right)^2 = \left(\frac{\sigma_x - \sigma_y}{2}\right)^2 \cos^2 2\theta + (\sigma_x - \sigma_y)\cos 2\theta(\tau \sin 2\theta) + \tau^2 \sin^2 \theta$$

After adding these two equations and using the identity $\sin^2 2\theta + \cos^2 2\theta = 1$,

the result is

$$\left[\sigma_\theta - \left(\frac{\sigma_x + \sigma_y}{2}\right)\right]^2 + \tau_\theta^2 = \left(\frac{\sigma_x - \sigma_y}{2}\right)^2 + \tau^2 \qquad (1\text{-}10)$$

Here σ_θ and τ_θ are the unknown stresses on an arbitrarily chosen plane, and all the other stresses can be considered as known quantities, that is, constants. Otto Mohr (1835–1918) first recognized that Eq. 1-10 is the same as the equation of a circle:

$$(x - a)^2 + (y - b)^2 = c^2 \qquad (1\text{-}11)$$

Equations 1-10 and 1-11 are related as follows:

Eq. 1-11 Eq. 1-10

x	$=$	σ_θ	horizontal coordinate of a point on the circle
y	$=$	τ_θ	vertical coordinate of a point on the circle
a	$=$	$\dfrac{\sigma_x + \sigma_y}{2}$	horizontal coordinate of the center of the circle
b	$=$	0	vertical coordinate of the center of the circle
c	$=$	$\sqrt{\left(\dfrac{\sigma_x - \sigma_y}{2}\right)^2 + \tau^2}$	radius of the circle

The physical meaning of Mohr's circle of stress is difficult to see at first. Striving to understand it is worthwhile, however, because it leads to an excellent, labor-saving method of analysis.

Some qualitative understanding can be obtained by imagining that the seam in Fig. 1-9a rotates through different positions. For each position of the plane there are unique normal and shear stresses on the plane. As the angle θ varies smoothly, so do σ_θ and τ_θ. Mohr's circle is the smooth plot of the stresses for the infinite possible positions of the plane. Each point on the circle has two coordinates, the normal and the shear stress for a particular position of the plane.

EXAMPLE 1-9

The peculiarities of Mohr's circle are best illustrated by using a specific example. Figure 1-12a shows the state of stress on a small element that is similar to that in Fig. 1-9a. Two of the faces are identified by numbers, and the angle is measured as shown. This, of course, is identical in effect to having an inclined plane with the same angle θ coming out from the lower left-hand corner of the element. The plotting of Mohr's circle can be done in several equivalent ways after the σ and τ coordinates are drawn (Fig. 1-12b). One possibility is to start with face 1 of the element. The stresses on this face satisfy Eq. 1-8 and 1-9 with $\theta = 0$. These two stresses are also the coordinates of one point in the σ-τ-plane in Fig. 1-12b. The stresses on face 2 are the coordinates of another point. These stresses satisfy Eq. 1-8 and 1-9 with $\theta = 90°$.

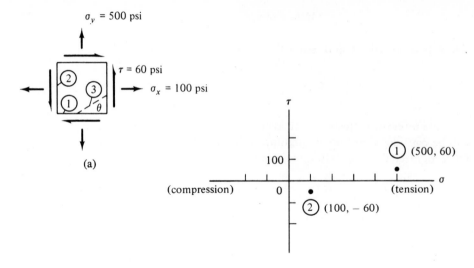

(a)

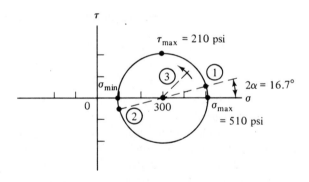

(b) coordinates and the stresses for two
planes in (a); the planes are represented
by points ① and ② in the diagram.

(c) complete Mohr's circle
for the element in (a).

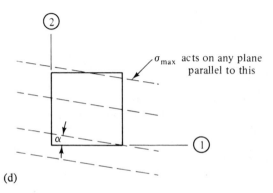

(d)

Fig. 1-12. Example 1-9.

20

Note that τ was taken as positive on face 1. This is arbitrary, but then τ on face 2 must be taken as negative since it causes an opposite moment (see remarks in Example 1-10).

At this stage two points (1 and 2) are plotted in Fig. 1-12b. They must be points of a circle because they satisfy Eq. 1-8 and 1-9, which were combined as the equation of a circle, Eq. 1-10. The next step is to construct the whole circle using the available information. This is easier than it appears at first. The center of the circle must always be on the σ-axis ($b = 0$), and its horizontal coordinate is $a = (\sigma_x + \sigma_y)/2 = 300$ psi. With this additional information the circle can be drawn as in Fig. 1-12c.

Several things are interesting in the completed circle. Since every point on the circle represents the stresses on some plane (at the same location in the material), it is obvious that there are stresses exceeding those on faces 1 and 2. The largest normal stress is slightly larger than σ_y in this case, and the maximum shear stress is always the radius of the circle. The maximum and minimum normal stresses are always on the σ-axis since the center of the circle is on this axis. These stresses act on planes where there is no shear, and they are called *principal stresses*.

Where are these planes? They can be found from Mohr's circle after thoughtful analysis. Consider plane 3 in Fig. 1-12a sweeping from plane 1 to a position parallel to plane 2. In the latter position the stresses on plane 3 are the same as those on plane 2. This means that the change from the stresses of plane 1 to the stresses of plane 2 required a rotation through 90° in the element. The same change in stresses occurs after a rotation of 180° in the Mohr's circle, in going from point 1 to point 2 in Fig. 1-12c. Continuing the counterclockwise rotations in the two related cases leads to similar results. A second 90° rotation of plane 3 in the element makes this plane again coincident with plane 1 where $\sigma = 500$ psi and $\tau = 60$ psi. The orientation of a plane determines what stresses act on it; the original and final positions of plane 3 (after a rotation through 180°) are not distinguishable. In contrast, returning to the original stresses in the Mohr's circle requires a rotation through 360° (going from point 1 through point 2 to point 1).

These are the extreme examples of what is true in general. *A rotation through an angle θ in the element is represented by a rotation through 2θ in the appropriate Mohr's circle.* The direction of rotation may be clockwise or counterclockwise, but it should be the same for the element and the circle.

The orientations of the principal planes (that have the principal stresses on them) are found using these concepts. The angle between plane 1 and the plane on which σ_{max} is acting is 16.7° in the Mohr's circle, so it must be 8.35° in the element. This principal plane is not just one plane. It represents all planes of the same orientation within the element or even outside it (but not far away), as shown in Fig. 1-12d. The principal plane with σ_{min} on it is perpendicular to the plane with σ_{max} on it.

The maximum shear stress occurs on planes that are halfway between the principal planes (at 45° from both). It appears from the Mohr's circle that there is a minimum shear stress equal in magnitude to the maximum shear stress. This has no physical significance since there is only one kind of shear. In other words, the sign of the shear stress is immaterial as far as the strength of the material is concerned. The minimum shear stress is always zero and it occurs on the principal planes (where σ is maximum or minimum).

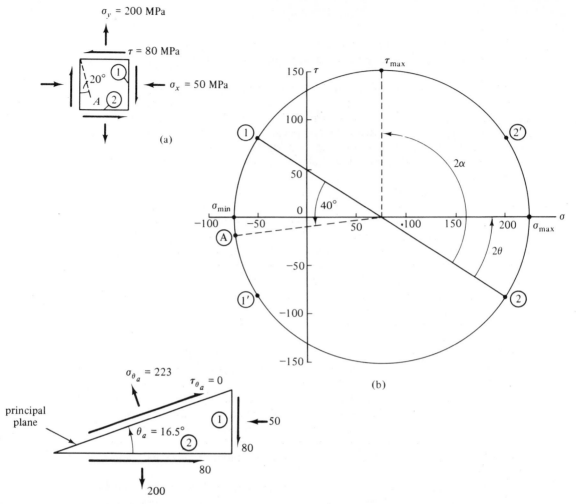

(a)

(b)

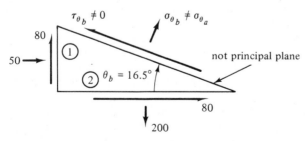

(c) wedge analysis to check direction of rotation from plane 2 to the principal plane; θ_a is in correct direction because τ_{θ_a} is zero in this element.

(d) clockwise rotation from plane 2 does not lead to the principal plane (even though $\theta_a = \theta_b$) because $\tau_{\theta_b} \neq 0$ here.

Fig. 1-13. Example 1-10.

EXAMPLE 1-10

A state of stress is given in Fig. 1-13a. Determine (a) the maximum normal and shear stresses and the orientations of the planes on which these stresses are acting and (b) the normal and shear stresses on plane A.

Solution

Assume, arbitrarily, that τ on face 1 is positive. Plot the stresses for planes 1 and 2 in Fig. 1-13b. The two points are on Mohr's circle for the given state of stress, and the center of the circle is the intersection of the σ-axis and the line connecting points 1 and 2. The required maximum stresses are

(a) $\sigma_{max} = 223$ MPa on a plane $\theta = 16.5°$ counterclockwise from plane 2.
      $\tau_{max} = 148$ MPa on a plane $\alpha = 61.4°$ counterclockwise from plane 2.

(b)   Plane A requires a rotation of $20°$ counterclockwise from plane 1 (or a $110°$ rotation from plane 2). This means a rotation of $40°$ in Mohr's circle, as shown in Fig. 1-13b. The required stresses are the coordinates of point A: $\sigma_A = -72$ MPa (the negative sign indicates compression) and $\tau_A = -20$ MPa.

Note that assuming τ on face 1 is negative gives points $1'$ and $2'$ in Fig. 1-13. Mohr's circle and the principal stresses are the same. The magnitudes of the angles to the principal planes are also the same, but it is advisable to check the direction of the rotation to the principal planes by analyzing a wedge with the given angle to see if τ_θ is zero (if not, the rotation should be in the opposite direction by the same angle).

EXAMPLE 1-11

A member is in uniaxial compression with the applied stress equal to 100 ksi. Determine (a) the principal stresses and the maximum shear stress and (b) the inclination of planes (with respect to the direction of the applied load) on which the magnitudes of the normal and shear stresses are equal.

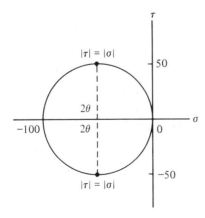

Fig. 1-14. Example 1-11.

Solution

In plotting Mohr's circle, either $\sigma_x = 0$, $\sigma_y = -100$ or $\sigma_x = -100$, $\sigma_y = 0$.

(a) $\sigma_{max} = 0$, $\sigma_{min} = -100$ ksi, $\tau_{max} = \frac{100}{2} = 50$ ksi
(b) According to Fig. 1-14, any plane inclined $\theta = 45°$ satisfies the requirement.

EXAMPLE 1-12

A small element is in biaxial tension with $\sigma_x = 1000$ MPa and $\sigma_y = 900$ MPa. What is the maximum shear stress that can be applied on the x- and y-planes if the normal stress should not exceed 1200 MPa in any direction in this material?

Solution

It is seen from Fig. 1-15 (or obtained from calculation) that the center of the circle must have coordinates of 950 MPa and 0. The radius of the circle is 250 MPa. The allowable shear stress on the x- and y-planes is slightly smaller than the radius,

namely $\tau_{allowable} = \sqrt{250^2 - 50^2} = 245$ MPa

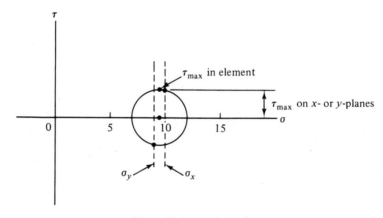

Fig. 1-15. Example 1-12.

EXAMPLE 1-13

The normal stresses on a small element are $\sigma_x = 150$ ksi and $\sigma_y = -50$ ksi. The shear stress alternates sinusoidally between the values of $+70$ and -70 ksi. What are the maximum normal and shear stresses in the element during the service loading?

Solution

The three states of stress shown in Fig. 1-16a, b, and c are the extremes that are worth considering. The stresses shown in elements (a) and (c) result in the same

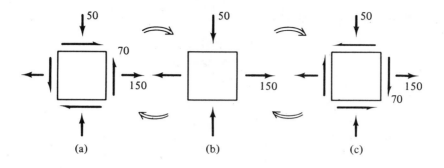

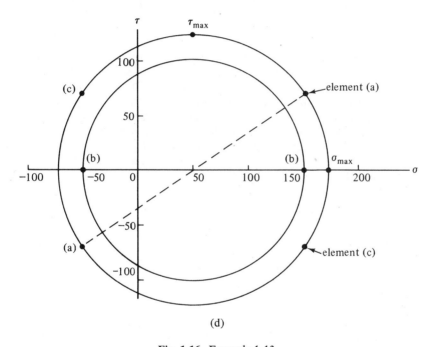

Fig. 1-16. Example 1-13.

circle, the outer circle in Fig. 1-16d. The maximum stresses in each complete cycle of loading are

$$\sigma_{max} = 172 \text{ ksi} \quad \text{(tension)}$$
$$\sigma_{min} = 72 \text{ ksi} \quad \text{(compression)}$$
$$\tau_{max} = 122 \text{ ksi}$$

EXAMPLE 1-14

The normal and shear stresses on a small element vary in a random fashion and independently of one another. Only the extreme values of each applied stress are known with any certainty. These are

$$\sigma_x : \text{maximum} \quad 500 \text{ MPa}$$
$$\text{minimum} \quad -200 \text{ MPa}$$
$$\sigma_y : \text{maximum} \quad 100 \text{ MPa}$$
$$\text{minimum} \quad -300 \text{ MPa}$$
$$\tau : \text{maximum} \quad 150 \text{ MPa}$$
$$\text{minimum} \quad 50 \text{ MPa}$$

What are the maximum normal and shear stresses that may be caused by the above loading?

Solution

The number of possible combinations is enormous even if only significantly different stresses are considered, so judgment is necessary in selecting states of stress for evaluation. It is reasonable, on the basis of Example 1-13, to consider only the maximum value from the applied shear stresses that are possible (for a given set of normal stresses, the largest applied shear stress gives the largest Mohr's circle). Of the normal stresses, those combinations are of interest that have the largest differences between σ_x and σ_y because these have the largest circles associated with them. Thus, $\sigma_x = 500$ and $\sigma_y = -300$ seem to be the critical values. From the large Mohr's circle in Fig. 1-17,

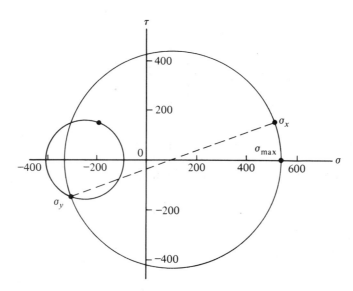

Fig. 1-17. Example 1-14.

$$\sigma_{max} = 525 \text{ MPa} \qquad \text{(tension)}$$

$$\sigma_{min} = 330 \text{ MPa} \qquad \text{(compression)}$$

$$\tau_{max} = 427 \text{ MPa}$$

It appears, however, that not all of these are absolute maximum values. For example, $\sigma_x = -200$ and $\sigma_y = -300$ with $\tau = 150$ results in a larger compressive stress (408 MPa) than the 330 MPa found in Fig. 1-17. The reader should check carefully to see if any other combinations of stresses may result in circles that are not within the large circle in Fig. 1-17.

Positions and Sizes of Mohr's Circles. It has been demonstrated that the center of Mohr's circle is on the σ-axis ($\tau = 0$). There are no exceptions to this. The horizontal position and size of a circle depend on the magnitudes and directions of the stresses on a small rectangular element. Several more examples using general notation show some of the possibilities.

EXAMPLE 1-15

Uniaxial loading. Figure 1-18a shows an element under tension. Mohr's circle for this situation is drawn in Fig. 1-18b. Note that the magnitudes of σ_{max} and τ_{max} and the orientations of the planes on which they are acting agree with the results obtained in Section 1-2 for the same problem.

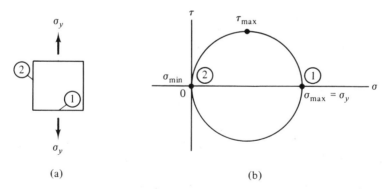

(a)   (b)

Fig. 1-18. Mohr's circle for uniaxial tension.

EXAMPLE 1-16

Biaxial loading. Figure 1-19a shows one of several possibilities, compressive loading of equal magnitude in both directions. The stresses (negative for compression) for faces 1 and 2 of the element and the calculated center for the circle are all at the same point on the σ-axis. Mohr's circle is a point circle.

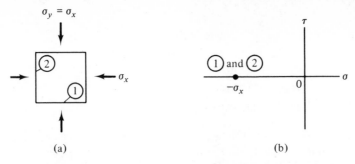

Fig. 1-19. Mohr's circle for biaxial compression.

EXAMPLE 1-17

Biaxial loading. The special case of tension in one direction with equal compression in the other direction is shown in Fig. 1-20a. The center of the circle is at the origin of the stress axes.

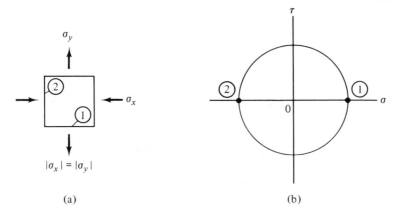

Fig. 1-20. Mohr's circle for equal tension and compression.

EXAMPLE 1-18

Pure shear. Mohr's circle is centered at the origin (Fig. 1-21b) as in Example 1-17. It is remarkable that different loadings could lead to identical Mohr's circles.

Fig. 1-21. Mohr's circle for pure shear.

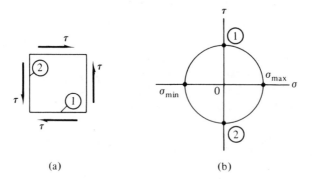

28

The tools are available now for finding the optimum angle of the seam in two joined pieces. Assume the state of stress shown in Fig. 1-22a and a tensile strength of 900 psi and a shear strength of 375 psi for the glue. The solution could start with a graphical showing of the stress levels that should not be exceeded because of the limitations of the glue. The joint should not have a shear stress that is above the horizontal dashed line in Fig. 1-22b. The normal stress should not be to the right of the vertical dashed line. These limitations are independent of the size and position of the Mohr's circle. Next, Mohr's circle is drawn in the same coordinates (Fig. 1-22c). The darker line on the circle represents a group of planes that satisfy the conditions of the imposed stresses and the limitations of strength. In theory, any plane within this segment can be used. In practice, it is best to be as far as possible from both limiting stresses, so the angle θ found from Fig. 1-22c is the ideal orientation of the seam.

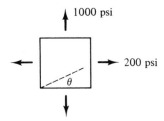

(a) element for finding optimum plane of seam

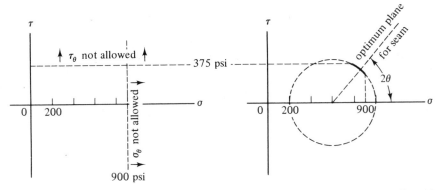

(b) limitations imposed by strength of glue   (c) dark segment of circle shows allowable orientations for plane of seam

Fig. 1-22. Finding the optimum plane of the seam from Mohr's circle.

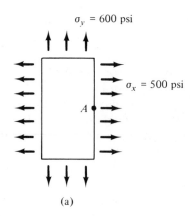

(a)

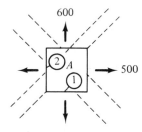

(b) dashed lines are edge
views of shear planes.

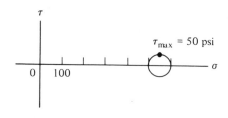

(c) Mohr's circle for
element in (b).

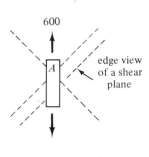

(d) element viewed
edgewise.

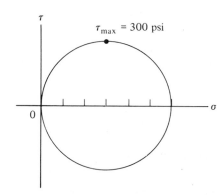

(e) Mohr's circle for element in (d).

Fig. 1-23. Stress analysis of element in biaxial tension.

There is something intriguing about shear stresses. They can be generated not only by applying shearing forces to a member (which is fairly obvious) but also by uniaxial loading, as in Fig. 1-1. This is sufficient reason to explore other problems that are apparently simple to see if there are shear stresses that one could overlook easily.

Consider a state of biaxial stress for a flat strip of material as in Fig. 1-23a. A small element at point A (Fig. 1-23b) has the same state of stress as the whole strip, and its Mohr's circle is drawn in Fig. 1-23c. The maximum shear stress is 50 psi, and it is on planes that are inclined 45° to the applied stresses (which are the principal stresses here). All of these shear planes are perpendicular to the plane of this paper.

The same element A can be viewed from other directions. For example, its view from the right is shown in Fig. 1-23d. Note that σ_x has no component in this diagram (σ_x is perpendicular to the plane of this paper). Mohr's circle based on the latter free-body diagram is drawn in Fig. 1-23e. The maximum shear stress is 300 psi, in contrast to that in Fig. 1-23c. Thus, if a homogeneous material has low shear strength, it would fail parallel to the shear planes shown in Fig. 1-23d rather than along those shown in Fig. 1-23b. It is perhaps easy to see that τ_{max} in Fig. 1-23e would be even larger if there were horizontal compression on the element in Fig. 1-23d (Mohr's circle would have a larger radius).

These observations lead to the conclusion that a three-dimensional element can have three two-dimensional free-body diagrams. For each diagram there is a Mohr's circle. The absolute maximum shear stress in an element is the radius of the largest circle. The formal statement of this is

$$\tau_{max} = \tfrac{1}{2}(\sigma_{max} - \sigma_{min}) \tag{1-12}$$

where σ_{max} and σ_{min} are the two extreme values of the three principal stresses that a three-dimensional element has.

Perhaps it is not surprising that the three Mohr's circles for the biaxial states of stress of a given three-dimensional element are closely related to one another. These relationships as well as the meaning of the absolute maximum shear stress can be seen and appreciated from the following examples.

EXAMPLE 1-19

A thin sheet of polyethylene is subjected to biaxial tensile stresses of $\sigma_x = 5$ MPa and $\sigma_y = 6$ MPa. The stress normal to the sheet is $\sigma_z = 0$. Draw all the Mohr's circles relevant to this state of stress, and determine the absolute maximum shear stress.

Solution

The three circles are drawn in Fig. 1-24b. The small circle is obtained by looking at the element in (a) from the z-direction; in this view it does not matter what σ_z

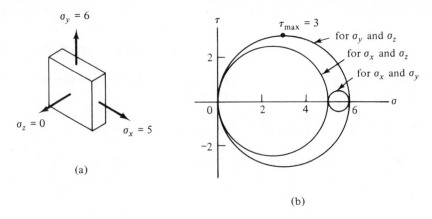

$\sigma_y = 6$

$\sigma_z = 0$

$\sigma_x = 5$

(a)

$\tau_{max} = 3$

for σ_y and σ_z

for σ_x and σ_z

for σ_x and σ_y

(b)

Fig. 1-24. Example 1-19.

is. The intermediate circle is obtained by looking at the element from the y-direction; in this view the magnitude of σ_y is irrelevant but that of σ_x and σ_z are important (even a zero magnitude). The largest circle is drawn after looking at the element from the x-direction; in this view σ_x does not appear, σ_y appears, and σ_z would if it were not zero (thus, it is significant that $\sigma_z = 0$).

The absolute maximum shear stress is 3 MPa and it acts on innumerable planes that are parallel to one another and perpendicular to the y-z-plane. Note the similarity of this problem and result to that shown in Fig. 1-23.

EXAMPLE 1-20

A rectangular block of concrete is to be subjected to $\sigma_x = -4$ ksi, $\sigma_y = 0$, and $\sigma_z = -2$ ksi. Determine the principal stresses and the absolute maximum shear stress.

Solution

The state of stress is represented by the circles in Fig. 1-25b; $\tau_{max} = 2$ ksi, and the principal stresses are the three given stresses: σ_x, σ_y, and σ_z.

Fig. 1-25. Example 1-20.

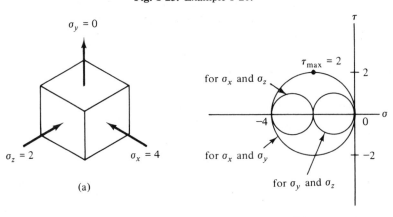

$\sigma_y = 0$

$\sigma_z = 2$

$\sigma_x = 4$

(a)

$\tau_{max} = 2$

for σ_x and σ_z

for σ_x and σ_y

for σ_y and σ_z

(b)

32

EXAMPLE 1-21

A metal plate is loaded such that $\sigma_x = 200$ MPa, $\sigma_y = -200$ MPa, and $\sigma_z = 0$. Draw the Mohr's circles for these stresses.

Solution

The circles are sketched in Fig. 1-26b.

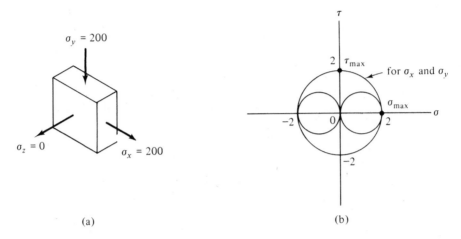

(a)   (b)

Fig. 1-26. Example 1-21.

EXAMPLE 1-22

A member is under triaxial tension with $\sigma_x = 20$ ksi, $\sigma_y = 40$ ksi, and $\sigma_z = 80$ ksi. Draw the Mohr's circles that completely describe this state of stress, and determine the absolute maximum shear stress.

Solution

The absolute maximum shear stress is 30 ksi, from Fig. 1-27b.

Fig. 1-27. Example 1-22.

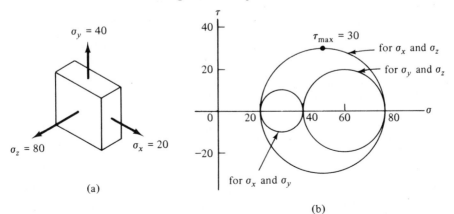

(a)

(b)

Sec. 1-1

1-1. Sketch a simple member in uniaxial tension in which the normal stresses along the given axis are of the same magnitude and yet the member is not in equilibrium.

1-2. In your opinion, what is the smallest element for which the concept of stress is valid?

1-3. Consider a large, flat piece of material that is in equilibrium under the action of external forces. If the shear stress is τ on each side of a small square $a \times a$ within the large piece, what is the shear stress on
   (a) a square $b \times b$ where $b = 2a$? The large square encloses the small one, and their respective sides are parallel to each other.
   (b) a rectangle $a \times b$? The two elements are again in the same neighborhood and have parallel sides.

1-4. The drawbar of a locomotive has a cross section 7 in. $\times$ 7 in. What is the allowable load on the drawbar if the stress in it should not exceed 100 ksi?

1-5. What is the relative load-carrying ability of two Kevlar filaments, one with a diameter of 0.01 mm and the other with 0.02 mm? The maximum stress is 3.2 GPa in both cases.

1-6. The large superconducting magnet of a proposed fusion reactor has copper strips joined with an epoxy to stainless steel. For the shear test of this bond, 1-in. wide strips of the two metals are joined. The lap joint is 1 in.² in area. What is the minimum thickness of each metal if the shear strength of the epoxy is 7000 psi or less and the tensile strength of the copper is 30 ksi and of the steel, 200 ksi? The metal strips should not fail during the test.

1-7. What is the required force in a punch press for punching a 1-cm diameter hole in a 2-mm thick plate? The shear strength of the plate is 200 MPa.

1-8. A U-shaped bracket is riveted to a beam on all three of its sides. All the rivets are 0.3 in. in diameter and have a tensile strength of 50 ksi and a shear strength of 30 ksi. What is the maximum load that can be applied to the bracket if 6 rivets are in direct tension and 16 are in shear?

Sec. 1-2

   In Prob. 1-9 to 1-16, σ is the uniaxial stress applied vertically on a rectangular element. Determine the normal and shear stresses on planes whose normals are inclined to the horizontal by the indicated angles (the planes themselves are inclined to the vertical by the indicated angles).

1-9. $\sigma = 100$ ksi
   $\theta_1 = 10°, \quad \theta_2 = 20°$

1-10. $\sigma = 500$ MPa
   $\theta_1 = 20°, \quad \theta_2 = 40°$

1-11. $\sigma = 200$ ksi
   $\theta_1 = 10°, \quad \theta_2 = 80°$

1-12. $\sigma = 50$ MPa
   $\theta_1 = 30°, \quad \theta_2 = 60°$

1-13. $\sigma = 300$ ksi
$\theta_1 = 5°, \quad \theta_2 = 45°$

1-14. $\sigma = 10$ MPa
$\theta_1 = 45°, \quad \theta_2 = 90°$

1-15. $\sigma = 150$ ksi
$\theta_1 = 90°, \quad \theta_2 = 140°$

1-16. $\sigma = 1.6$ GPa
$\theta_1 = 95°, \quad \theta_2 = 175°$

Sec. 1-3

1-17. Take moments about two different points of reference to show that the element in Fig. 1-8c is in equilibrium.

1-18. Prove that the elements in Fig. 1-9b and c are the same for all practical purposes. Which one do you prefer to use and why?

1-19. Use the concepts presented so far to approximate the optimum angle (by trial and error) for the seam in Fig. 1-1. Assume that the tensile strength of the glue is 900 psi and its shear strength is 375 psi. The applied stress is 1000 psi.

1-20 to 1-33. Use the wedge method of analysis for each of the states of plane stress shown in Fig. P1-20 to P1-33. Determine the normal stress and the shear stress acting on the plane indicated by a dashed line in each element. Note that 1 ksi = 1000 psi $\simeq$ 7 MPa if you wish to compare some of the results from different problems.

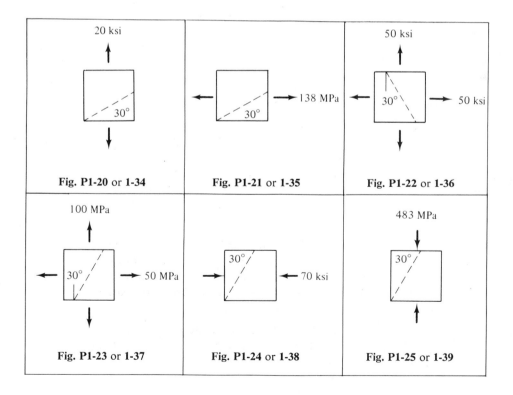

Fig. P1-20 or 1-34   Fig. P1-21 or 1-35   Fig. P1-22 or 1-36

Fig. P1-23 or 1-37   Fig. P1-24 or 1-38   Fig. P1-25 or 1-39

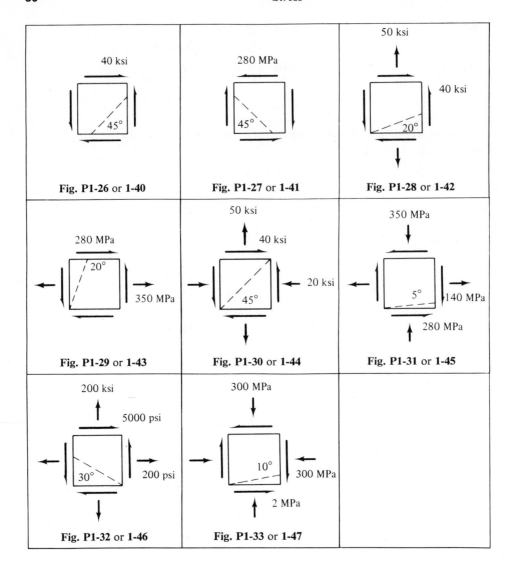

Fig. P1-26 or **1-40**

Fig. P1-27 or **1-41**

Fig. P1-28 or **1-42**

Fig. P1-29 or **1-43**

Fig. P1-30 or **1-44**

Fig. P1-31 or **1-45**

Fig. P1-32 or **1-46**

Fig. P1-33 or **1-47**

Sec. 1-5

1-34 to 1-47. Plot the Mohr's circle for the elements in Prob. 1-20 to 1-33. Determine the principal stresses and the maximum shear stress and the orientations of the planes on which these stresses are acting. For those problems that you have also analyzed by using the wedge method, check the earlier results by considering the appropriate Mohr's circle.

1-48. Draw a Mohr's circle with its center at the origin of the coordinates and with a radius of 10 ksi (70 MPa). What are the possible states of stress that could result in this circle?

Sec. 1-7

1-49. A solid metal cube rests on the ocean floor under 3000 m (10,000 ft) of water. Consider three small elements within this cube and draw the stresses on them:

(a) The element is at the center of the cube.
(b) The element is at an upper corner of the cube.
(c) The element is in the middle of the face that touches the ocean floor (assume sand).

What are the maximum shear stresses in the three elements?

1-50. Determine the maximum shear stress in a material where the principal stresses are $\sigma_1, \sigma_2 = 2\sigma_1, \sigma_3 = 5\sigma_1$. Show on a sketch a plane on which this shear stress is acting.

1-51. How many different kinds of states of stress can you mention in which there are no shear stresses?

Assume that in each of Prob. 1-52 to 1-65 the given stresses must be held because of design considerations of the overall structure. Determine the maximum allowable value of the externally applied unknown stress indicated. These are the general conditions that must be satisfied: All the externally applied stresses must be in the same plane; magnitude of maximum allowable tensile stress $= \sigma_t$ (given); maximum allowable compressive stress $= \sigma_c = -2\sigma_t$; absolute maximum shear stress $= \tau_{max} = \sigma_t/2$.

*1-52. $\sigma_x = 10{,}000$ psi　　　　　***1-53.** $\sigma_x = 100$ MPa
$\sigma_y = 0$　　　　　　　　　　　　　$\sigma_y = 100$ MPa
$\sigma_t = 12{,}000$ psi　　　　　　　　$\sigma_t = 150$ MPa
$\tau = \,?$　　　　　　　　　　　　　$\tau = \,?$

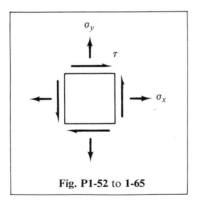

Fig. P1-52 to 1-65

*1-54. $\sigma_x = 70{,}000$ psi　　　　　***1-55.** $\sigma_x = -200$ MPa
$\sigma_y = -30{,}000$ psi　　　　　　　$\tau = 40$ MPa
$\sigma_t = 120{,}000$ psi　　　　　　　$\sigma_t = 150$ MPa
$\tau = \,?$　　　　　　　　　　　　　$\sigma_y = \,?$

*1-56. $\sigma_x = 0$
$\tau = 50{,}000$ psi
$\sigma_t = 150{,}000$ psi
$\sigma_y = ?$

*1-57. $\sigma_x = -80$ MPa
$\tau = 0$
$\sigma_t = 100$ MPa
$\sigma_y = ?$

*1-58. $\sigma_x = 0$
$\sigma_y = -100{,}000$ psi
$\sigma_t = 100{,}000$ psi
$\tau = ?$

*1-59. $\tau = 60$ MPa
$\sigma_t = 150$ MPa
$\sigma_x = ?$
$\sigma_y = ?$

*1-60. $\sigma_x = -6{,}000$ psi
$\sigma_t = 5{,}000$ psi
$\sigma_y = ?$
$\tau = ?$

*1-61. $\sigma_x = -50$ MPa
$\sigma_y = -50$ MPa
$\sigma_t = 40$ MPa
$\tau = ?$

*1-62. $\sigma_y = 100{,}000$ psi
$\tau = 30{,}000$ psi
$\sigma_t = 120{,}000$ psi
$\sigma_x = ?$

*1-63. $\sigma_y = 200$ MPa
$\tau = 0$
$\sigma_t = 250$ MPa
$\sigma_x = ?$

*1-64. $\sigma_y = 0$
$\tau = 50{,}000$ psi
$\sigma_t = 75{,}000$ psi
$\sigma_x = ?$

*1-65. $\sigma_x = 160$ MPa
$\sigma_t = 190$ MPa
$\sigma_y = ?$
$\tau = ?$

2

STRAIN

The extent of the deformation in a material is of interest to the designer for two reasons in general. One concerns the possibility of fracture in the region of the material where the deformation is most severe and the other with side effects or with satisfying functional requirements. For example, the wing of a large airplane may be designed adequately in every respect, except that it is quite flexible and the wing tip may touch the ground during a difficult landing. Such a wing is clearly unacceptable. At the other extreme, a fiber glass pole may be too rigid for pole vaulting. In that case there would be little reason to expect fracture; yet the pole is not desirable for vaulting.

All solid materials deform when forces are acting on them. There are no perfectly rigid bodies; even the strongest material deforms under an infinitesimal force. The deformation may be extremely small and thus difficult to measure, but this alone is not an assurance that it can be ignored. For example, the tolerances in the dimensions of many machine parts (gas turbines, high-precision fabricating and testing equipment, etc.) are small. Deformation under load is one way in which the allowable tolerances can be exceeded. This could lead to situations such as a turbine rotor rubbing into the stationary parts, which could have serious consequences. Other problems with small deformations occur when they are repeated many times; this may cause fatigue failures. Some deformations may be undesirable even if they do not result in fracture. For instance, a modern chair may be perfectly safe to sit on but may deflect and sway too much to be comfortable.

The deformation of a member depends on its geometry, on the mechanical properties of the material, and on the applied load. In some cases the deformation of greatest interest can be calculated easily. Most engineering problems are complicated, however, because different regions in a member may have different local deformations. This means that the total deformation may not be indicative of the most severe local deformation in the same member.

To deal with this problem, a concept somewhat analogous to the concept of stress has been established. The idea here is to analyze the deformation in such a region that every subregion in it has the same severity of deformation. The localized severity of deformation defined this way is called *strain* in the area of strength of materials. The concept of strain is described in more detail as follows.

2-1. AXIAL STRAIN

Consider a uniform cylinder whose original length is L_0. This is called the *gage length*. An axial force F is applied and the cylinder stretches to a new length L_1 (Fig. 2-1a). The difference between the two lengths is the elongation, e. Assuming

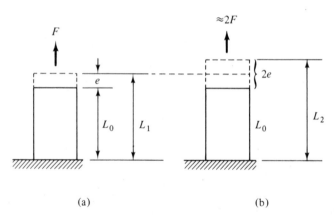

(a)   (b)

Fig. 2-1. Models for the calculation of axial strains.

that every part of the cylinder participated equally in the stretching, the severity of the axial deformation at any point is the axial strain, commonly denoted by ϵ. It is defined, with the gage length approaching zero in the limit, as

$$\epsilon = \frac{e}{L_0} \qquad (2\text{-}1)$$

The strain obtained from this formula is called the *engineering strain*. It is a useful quantity in dealing with small deformations.

Strain does not have a natural unit because e and L_0 should have the same units. It has become customary over the years, however, to identify the quantities of strain with units. Many different ways of doing this have evolved, and they are all arbitrary. Three basic schemes can be distinguished:

(a) Length divided by length, both in the same units.

EXAMPLE 2-1

$$L_0 = 1 \text{ in.} = 2.54 \text{ cm}$$
$$e = 0.01 \text{ in.} = 0.0254 \text{ cm}$$
$$\epsilon = 0.01 \text{ in./in.} = 0.01 \text{ cm/cm} = 0.01 \text{ m/m}$$

Note that the numbers in ϵ are always the same, regardless of what units are used. The units are normally written, even though they could be eliminated.

(b) Percentage.

EXAMPLE 2-2

From Example 2-1,

$$\epsilon = \frac{0.01}{1} \times 100 = 1\%$$

(c) *Microstrain.* It is frequently necessary to deal with very small strains. This has led to the custom of giving strain in millionths of the unit length for which it applies. There are several accepted ways of writing microstrain.

EXAMPLE 2-3

From Example 2-1,

$$\epsilon = 10,000 \text{ microinches per inch } (\mu \text{ in./in.})$$
$$= 10\ 000\ \mu\text{m/m} = 10,000 \text{ microstrain } (\mu\epsilon)$$

2-2. TRUE STRAIN

There are no problems with the concept of engineering strain as long as the deformation e is small compared to the gage length. A difficulty exists when the deformation becomes large. For example, a tensile strain of 100% should mean that the cylinder was stretched to a length that is twice the original length [$\epsilon = (2L_0 - L_0)/L_0 = 1.0 = 100\%$]. The same magnitude of compressive strain would require a new length of zero for the cylinder!

The cause of the difficulty can be explained with the aid of Fig. 2-1. When the cylinder is stretched from L_0 to L_2, the strain is

$$\epsilon_a = \frac{2e}{L_0}$$

The strain calculated is different from ϵ_a if the stretching to L_2 is considered in two steps: first to L_1 and then to L_2. Here are some possibilities for calculating the strain in the two-step process: the strain in the last step in the stretching,

$$\epsilon_b = \frac{e}{L_1}$$

the sum of the strains in the two steps,

$$\epsilon_c = \frac{e}{L_0} + \frac{e}{L_1}$$

or the average of the strains in the two steps,

$$\epsilon_d = \frac{1}{2}\left(\frac{e}{L_0} + \frac{e}{L_1}\right)$$

None of these is equal to ϵ_a, even though there is no reason to assume any physical difference as the material is stretched from L_0 to L_2 in one step or more. This indicates that perhaps the definition of strain should be reevaluated.

The calculation of ϵ_c above is appealing intuitively because the instantaneous severity of the deformation was considered in each step. This leads to the observation that in successive increments of identical elongations ($e + e + e + \ldots$) the true strain in each step is always smaller than the strain in the preceding step:

First step: $\epsilon_1 = \dfrac{e}{L_0}$

Second step: $\epsilon_2 = \dfrac{e}{L_0 + e}$

Third step: $\epsilon_3 = \dfrac{e}{L_0 + 2e}$, etc.

The denominator is always the instantaneous length L and the numerator is an infinitesimal change in this length. Thus, the total strain experienced by the material in deforming to a new length L_f is

$$\epsilon = \int_{L_0}^{L_f} \frac{dL}{L} = \ln\frac{L_f}{L_0} \tag{2-2}$$

Equation 2-2 is the definition of true, or natural, strain. The symbol ϵ frequently implies that the strain is true strain. Engineering strain is nearly identical to true strain in small deformations (up to a few percent of engineering strain), however, and it is easier to calculate, so many people use Eq. 2-1 and call it true strain.

2-3. SHEAR STRAIN

Angular distortions are called *shear strains*. For a model of this, consider the rectangular element in Fig. 2-2. Its lower side is fixed and a shearing force S is

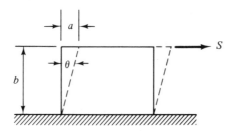

Fig. 2-2. Model for the calculation of shear strain.

applied at the top. The element deforms into a rhomboidal shape. The shear strain, commonly denoted by γ, is defined as

$$\gamma = \frac{a}{b} = \tan \theta \tag{2-3}$$

For small angles of distortion (which are most common in solids), it is also acceptable to say that

$$\gamma = \sin \theta = \theta$$

The shear strain is customarily expressed in radians.

2-4. MOHR'S CIRCLE FOR STRAINS

It was demonstrated previously that the stress transformation equations and Mohr's circle for plane stress allow one to determine stresses in any arbitrary

Fig. 2-3. Strain transformations.

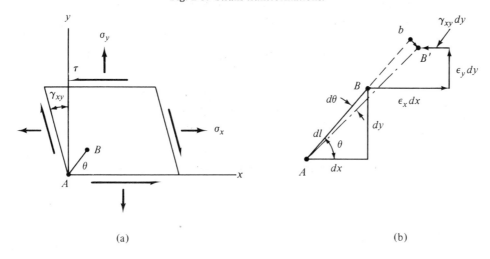

(a)   (b)

orientation quite easily. Similar opportunities exist for obtaining much informa-tion from what appears to be rather limited data when dealing with strains. The strain transformation equations are derived with the aid of Fig. 2-3. A rectangular element is loaded in biaxial tension and shear until it has a rhomboidal shape as in Fig. 2-3a. The shear strain is γ_{xy}, the angle of change from the original rectangle. The combined effect of the biaxial linear strains and the angular distortion* is analyzed by considering what happens to an arbitrary line AB in the element. Figure 2-3b is an enlarged view of this line. Assume that point A is fixed and dx and dy are the coordinates of point B. The three strains result in the displacement of point B to B' as shown by the displacement vectors in the figure. The magnitude of each of these vectors is calculated by multiplying the strain with the corresponding infinitesimal gage length.

There is an advantage from establishing the equivalent displacement vectors Bb and bB' as shown in Fig. 2-3b. It is clear that Bb is the increase in length of the line AB, and bB' causes the angular change $d\theta$ in the position of the line. The magnitude of Bb is obtained by projecting the displacement vectors $\epsilon_x dx$, $\epsilon_y dy$, and $\gamma_{xy} dy$ on a line coincident with AB:

$$Bb = \epsilon_x \, dx \cos \theta + \epsilon_y \, dy \sin \theta - \gamma_{xy} \, dy \cos \theta$$

The linear strain ϵ_l of line AB is obtained by dividing the elongation Bb with the gage length dl:

$$\epsilon_l = \frac{Bb}{dl} = \frac{\epsilon_x \, dx \cos \theta}{dl} + \frac{\epsilon_y \, dy \sin \theta}{dl} - \frac{\gamma_{xy} \, dy \cos \theta}{dl}$$

This equation can be simplified using

$$\frac{dx}{dl} = \cos \theta \quad \text{and} \quad \frac{dy}{dl} = \sin \theta$$

$$\epsilon_l = \epsilon_x \cos^2 \theta + \epsilon_y \sin^2 \theta - \gamma_{xy} \sin \theta \cos \theta$$

Another simplification results from substituting the double-angle identities that were used in the analysis of stresses, and the linear strain is

$$\epsilon_l = \frac{\epsilon_x + \epsilon_y}{2} + \frac{\epsilon_x - \epsilon_y}{2} \cos 2\theta - \frac{\gamma_{xy}}{2} \sin 2\theta \tag{2-4}$$

The angular displacement $d\theta$ is obtained by dividing the perpendicular displacement bB' by the gage length dl. Note that this is an approximation based on the assumption of small deformations. Thus, Bb is very small compared to dl (Fig.

---

*This is in the category of *plane strain*. In general, plane strain means that $\epsilon_z = \gamma_{xz} = \gamma_{yz} = 0$, while ϵ_x, ϵ_y, and γ_{xy} may be other than zero. Plane strain and plane stress are not necessarily simultaneous (consider the Poisson effect, p. 56).

2-3b is exaggerated to show the individual components of the deformation). The three parts of the displacement of point B are projected on a line perpendicular to the line AB, and

$$bB' = \epsilon_x \, dx \sin \theta - \epsilon_y \, dy \cos \theta - \gamma_{xy} \, dy \sin \theta$$

The angular displacement is

$$d\theta = \frac{bB'}{dl} = \frac{\epsilon_x \, dx \sin \theta}{dl} - \frac{\epsilon_y \, dy \cos \theta}{dl} - \frac{\gamma_{xy} \, dy \sin \theta}{dl}$$

$$= \epsilon_x \sin \theta \cos \theta - \epsilon_y \sin \theta \cos \theta - \gamma_{xy} \sin^2 \theta$$

Since shear strain is not simply the angular displacement of a line but the change in angle between two lines, it is necessary to work with another line besides AB. It is convenient to take the second line as one perpendicular to AB. The orientation of this line can be described as $\phi = \theta + 90°$. The preceding analysis is also valid for the second line, and

$$d\phi = -\epsilon_x \sin \theta \cos \theta + \epsilon_y \sin \theta \cos \theta - \gamma_{xy} \cos^2 \theta$$

because $\qquad \sin(\theta + 90°) = \cos \theta \qquad \cos(\theta + 90°) = -\sin \theta$

It is not immediately obvious that $d\theta$ and $d\phi$ must represent rotations in opposite directions. Intuitively, if both were in the same direction ($d\theta + d\phi$), they would tend to represent a rotation of coordinate axes rather than a distortion of those axes. Thus, the net change in the angle between the line AB and the line normal to it is $\gamma_\theta = d\theta - d\phi$, and this defines the shear strain for the chosen orthogonal lines. Specifically,

$$\gamma_\theta = d\theta - d\phi = \epsilon_x(2 \sin \theta \cos \theta) - \epsilon_y(2 \sin \theta \cos \theta) + \gamma_{xy}(\cos^2 \theta - \sin^2 \theta)$$

which can be simplified using trigonometric identities to

$$\frac{\gamma_\theta}{2} = \frac{\epsilon_x - \epsilon_y}{2} \sin 2\theta + \frac{\gamma_{xy}}{2} \cos 2\theta \qquad (2\text{-}5)$$

Equations 2-4 and 2-5 are the *transformation equations for strain*, and they are identical in form with the transformation equations for stress (Eq. 1-8 and 1-9). This leads to the idea that strains can be represented by a Mohr's circle with the same rules for construction as in the case of stresses. The only exception is that *half values of the shear strains are plotted* on the Mohr's circles for strain.

The equation of the circle for plane strain in rectangular coordinates is

$$\left[\epsilon_\theta - \left(\frac{\epsilon_x + \epsilon_y}{2} \right) \right]^2 + \left(\frac{\gamma_\theta}{2} - 0 \right)^2 = \left(\frac{\epsilon_x - \epsilon_y}{2} \right)^2 + \left(\frac{\gamma_{xy}}{2} \right)^2 \qquad (2\text{-}6)$$

where ϵ_x = axial strain along the x-axis

ϵ_y = axial strain along the y-axis ($\perp$ to x)

ϵ_θ = axial strain along an arbitrary line inclined by an angle θ to the x-axis

γ_{xy} = shear strain of originally orthogonal x-y-axes

γ_θ = shear strain in direction of the line inclined by the angle θ

In the plot, ϵ_θ is the abscissa and $\gamma_\theta/2$ is the ordinate. It is clear from Eq. 2-6 that the center of the circle is always on the ϵ_θ axis. Tensile strains are considered positive and compressive strains negative. Shear strains are generally considered positive when the right angle of the axes increases, and negative when the angle decreases. Of course, there is no difference in the physical effects of positive and negative shear strains of equal magnitude on the behavior of homogeneous solids. The coordinate systems for strains and stresses are slightly different from one another, but the relationship between the circle and the element from the body is the same in the two systems: *A rotation through an angle θ in the element is represented by a rotation through 2θ in the appropriate Mohr's circle.*

The maximum and minimum axial strains (called ϵ_1 and ϵ_2) are the principal strains and these occur in directions in which the shear deformation is zero. To appreciate that shear strain depends on orientation in an element, consider Fig. 2-4. The element is in pure shear, so a shear strain is expected as shown in (b). It is easy to see, however, that the coordinates x and y remain perpendicular to each other throughout the deformation ($\gamma_{xy} = 0$). There is, of course, axial strain along x and y, and these are the principal strains.

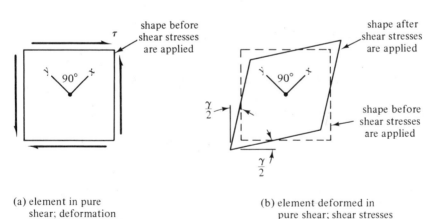

(a) element in pure
    shear; deformation
    caused by τ is not shown.

(b) element deformed in
    pure shear; shear stresses
    are not shown.

Fig. 2-4. Element in pure shear.

EXAMPLE 2-4

The following values of strain were found experimentally at a given point in

a structure: $\epsilon_x = 0.08\%,$ $\epsilon_y = -0.01\%,$ $\gamma_{xy} = 0.04°$

What are the principal strains and the maximum shear strain at the given point?

Solution

At first the given strains should be converted to have compatible units.

$$\epsilon_x = 0.0008 \text{ in./in.}, \qquad \epsilon_y = -0.0001 \text{ in./in.}, \qquad \gamma_{xy} = 0.0007 \text{ rad}$$

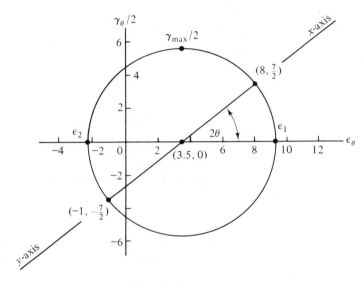

Fig. 2-5. Example 2-4.

In plotting the Mohr circle in Fig. 2-5 a strain of 10^{-4} is represented by one unit of distance for convenience. The center of the circle has coordinates 3.5 and 0, and the radius is $\sqrt{4.5^2 + 3.5^2} = 5.7$. The required strains are

$$\epsilon_1 = \text{center of circle} + \text{radius} = 0.00092$$
$$\epsilon_2 = \text{center of circle} - \text{radius} = -0.00022$$
$$\gamma_{max} = 2\,(\text{radius}) = 0.00114$$

The principal strain ϵ_1 is in a direction that is $\theta = 18.9°$ clockwise from the *x*-axis. The relative orientation of planes (clockwise or counterclockwise rotations) should be checked in some cases by considering elements and coordinates before and after deformation as in Fig. 2-4. Note that the strains given without units mean natural units of strain, in./in. or radians.

EXAMPLE 2-5

What are the maximum shear strain and the principal strains in a member where the measured strains are $\epsilon_x = 0$, $\epsilon_y = -0.1\%$, and $\gamma_{xy} = 0$?

Solution

The principal strains are the measured axial strains, ϵ_x and ϵ_y. The maximum shear strain is the diameter of Mohr's circle in Fig. 2-6, $\gamma_{max} = 0.001$ rad.

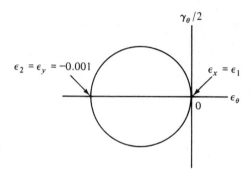

Fig. 2-6. Example 2-5.

EXAMPLE 2-6

What are the principal strains in a member where the measured strains are $\epsilon_x = \epsilon_y = 0$ and $\gamma_{xy} = 0.0005$ rad?

Solution

From the Mohr's circle in Fig. 2-7, $\epsilon_1 = 0.00025$ and $\epsilon_2 = -0.00025$. Note that the recording of zero axial strain in two directions does not preclude the presence of an axial strain at the same place.

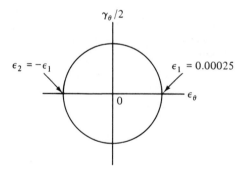

Fig. 2-7. Example 2-6.

EXAMPLE 2-7

The following information is known at a certain stage in the design process of a member: $\epsilon_y = -1$ and $\gamma_{xy} = 1.5$, where each unit represents 10^{-4}. It is also known that the principal strain in tension should not exceed $\epsilon_1 = 6$ in any direction in the member. What is the maximum allowable strain ϵ_x that satisfies these conditions?

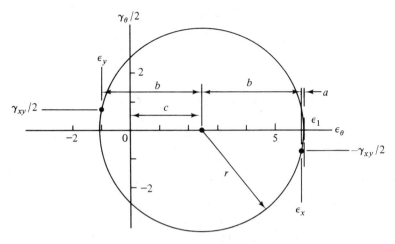

Fig. 2-8. Example 2-7.

Solution

(a) A graphical solution using trial and error is fast and may be reasonably accurate. A circle should be drawn as in Fig. 2-8, going through the given points $(\epsilon_y, \gamma_{xy}/2)$ and $(\epsilon_1, 0)$ and with its center on the ϵ_θ axis. It is evident that ϵ_x should be slightly smaller than ϵ_1; its value can be read off from a properly drawn circle.

(b) An analytical solution is especially desirable if similar calculations must be made frequently. The solution is based on the geometry of a circle (a freehand sketch is sufficient) such as in Fig. 2-8. Here ϵ_y, ϵ_1, and γ_{xy} are constants; the others are unknown.

$$\epsilon_1 = \epsilon_y + 2b + a = c + r, \qquad a = r - b$$

$$b = \sqrt{r^2 - \left(\frac{\gamma_{xy}}{2}\right)^2} = c - \epsilon_y, \qquad c = \epsilon_y + b = \epsilon_1 - r$$

The quantity sought is

$$\epsilon_x = b + c = \epsilon_1 - a = \epsilon_y + 2b$$

Subtracting $\epsilon_y = c - b$ from $\epsilon_1 = c + r$,

$$\epsilon_1 - \epsilon_y = r + b = r + \sqrt{r^2 - \left(\frac{\gamma_{xy}}{2}\right)^2}$$

Letting $m = \epsilon_1 - \epsilon_y$ and $n = (\gamma_{xy}/2)^2$,

$$r = \frac{m^2 + n}{2m},$$

$$\epsilon_x = b + c = \epsilon_1 - \epsilon_y + c - r = 2\epsilon_1 - \epsilon_y - 2r = 5.92$$

The answer is judged reasonable.

Sec. 2-1

> Determine the axial strain, or the total deformation if the strain is given, in Prob. 2-1 to 2-8.

2-1. $L_0 = 10$ in., $e = 0.2$ in.

2-2. $L_0 = 1$ m, $e = -5$ cm

2-3. $L_0 = 200$ ft, $e = 15$ in.

2-4. $L_0 = 800$ m, $e = 3$ m

2-5. $L_0 = 1$ in., $\epsilon = 30\%$

2-6. $L_0 = 10$ cm, $\epsilon = -0.002$

2-7. $L_0 = 3000$ ft, $\epsilon = 0.01$

2-8. $L_0 = 5$ mm, $\epsilon = -5\%$

2-9. A 30-ft long guy wire is tightened with a 6-in. long turnbuckle. In this process the wire elongates 1 in. and the turnbuckle elongates 0.02 in. What are the axial strains in the wire and the turnbuckle?

2-10. A thin-walled pressure vessel increases in diameter from 1 to 1.01 m when pressurized. What is the circumferential strain caused by the internal pressure?

Sec. 2-2

2-11. Prove that engineering strain e and true strain ϵ are related according to

$$\epsilon = \ln(e + 1)$$

2-12. Consider two identical cylindrical specimens. One is stretched to twice its original length; the other is compressed to half its original length. Evaluate and compare the engineering strains and the true strains for the two cases. What is the lesson from these test results?

2-13. Plot the engineering strain and the true strain on the same diagram for $L_0 = 1$ to $L_f = 2$.

2-14. Find an equation for the engineering strain that deviates from the corresponding true strain only by 10% of the true strain.

Sec. 2-4

> Determine the principal strains, the maximum shearing strain, and the directions in which these strains are in Prob. 2-15 to 2-22. Also find the strains in the direction that is 30° clockwise from the x-axis in each element.

2-15. $\epsilon_x = 0.1\%$
$\epsilon_y = 0$
$\gamma_{xy} = 0.01°$

2-16. $\epsilon_x = 5000\ \mu\text{in.}/\text{in.} = 5000\ \mu\text{m}/\text{m}$
$\epsilon_y = \epsilon_x$
$\gamma_{xy} = 0.001°$

2-17. $\epsilon_x = 0.6\%$
$\epsilon_y = -0.1\%$
$\gamma_{xy} = -0.001°$

2-18. $\epsilon_x = -10,000 \ \mu\epsilon$
$\epsilon_y = -2000 \ \mu\epsilon$
$\gamma_{xy} = 0$

2-19. $\epsilon_x = 0$
$\epsilon_y = 0$
$\gamma_{xy} = 0.007$ rad

2-20. $\epsilon_x = 0$
$\epsilon_y = 0$
$\gamma_{xy} = -0.014$ rad

2-21. $\epsilon_x = 1.2\%$
$\epsilon_y = -\epsilon_x$
$\gamma_{xy} = 0$

2-22. $\epsilon_x = -8000 \ \mu\epsilon$
$\epsilon_y = 0$
$\gamma_{xy} = 0$

2-23. What kinds of loading could result in point circles on the Mohr's plots for strains?

2-24. A balloon is barely holding its spherical shape (gage pressure $= 0$) when its diameter is D. The strain in the skin of the balloon is zero at this stage. Determine the principal strains and the maximum shear strain in the skin when the balloon's diameter becomes (a) $2D$; (b) $3D$.

2-25. A spherical balloon is made of an elastomer that has a maximum allowable strain of 600%. What is the maximum diameter of this balloon if its diameter is 2 m when the gage pressure is essentially zero (inside pressure ≈ 1 atm $= 100$ kPa)?

3

STRESS VERSUS STRAIN

In most cases, stress and strain are closely related. Analysis of this relationship can lead one to a significant understanding of the mechanical behavior of solid materials. The background for this analysis is the theory of springs that Robert Hooke discovered in 1660.

3-1. HOOKE'S LAW

The experiments that Hooke performed with springs were exceedingly simple; yet their results revolutionized the science of strength of materials. The basic experiment is with coil springs. Increasing forces are applied to a spring, as in Fig. 3-1, and the resulting deflections are plotted in Fig. 3-2. Frequently the plot is linear at least near the origin, and the spring returns to its original length after the external load is removed. The slope of the line is called k, the spring constant. It depends on the material of the spring, on the diameter of the wire, and on the diameter of the coil. Because of the two geometric factors involved, k is not a material property. Hooke's law in concise form is

$$F = kd \qquad (3\text{-}1)$$

The inverse of k is called the *compliance*; it is frequently used in some areas.

3-2. THE STRESS-STRAIN CURVE

Hooke realized that the concept of relating the force and the deformation was not limited to coil springs and was applicable to all "springy bodies." To visualize this, consider Fig. 2-1 again. The deformation in (b) is twice that in (a). It seems

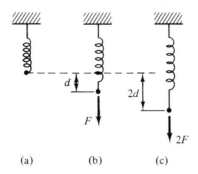

Fig. 3-1. Experiment for the theory of springs.

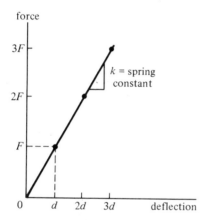

Fig. 3-2. Force versus deflection of a spring.

logical that the forces required in the two cases are proportional to the deformations. A plot of these would be a straight line similar to Fig. 3-2. The spring constant for the cylinder is expected to be quite high (that means higher stiffness) because there is no unwinding of coils in this case. k is still not a material property because it depends on the material *and* the diameter of the cylinder.

The most specific information about the "springy" behavior of materials is obtained when stresses are plotted with the corresponding strains. Such plots show material properties since they do not depend on any geometric factors. They can be obtained simply from the force versus deflection data in axial loading according to the definitions of stress and strain. The equivalent of Eq. 3-1 for axial stress and strain is

$$\sigma = E\epsilon \tag{3-2}$$

where the constant E is called the *modulus of elasticity* or *Young's modulus*. It is the slope of the stress versus strain plot (Fig. 3-3), and it is the measure of the stiffness of a material. A similar relationship applies to shear stress and strain:

$$\tau = G\gamma \tag{3-3}$$

where the constant G is called the *modulus of rigidity* or the *shear modulus*. These

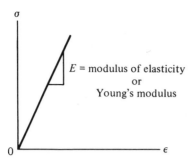

E = modulus of elasticity
or
Young's modulus

Fig. 3-3. Stress versus strain for perfectly elastic material.

moduli are material properties. Many typical stress-strain curves will be given in Chapter 4.

EXAMPLE 3-1

What are the tensile strains in aluminum (Al) and tungsten carbide (WC) of different sizes if they are subjected to identical stresses of $\sigma = 50$ ksi? $E_{Al} = 10.4 \times 10^6$ psi, $E_{WC} = 10.4 \times 10^7$ psi.

Solution

$$\epsilon_{Al} = \frac{50,000}{10.4 \times 10^6} = 4.8 \times 10^{-3}$$

$$\epsilon_{WC} = \frac{\epsilon_{Al}}{10} = 4.8 \times 10^{-4}$$

EXAMPLE 3-2

What is the shear modulus of a material for which the following limited data are available: $\tau = 100$ MPa at $\gamma = 10^{-3}$? Assume elastic behavior.

Solution

Assuming that the plot of τ versus γ is linear from 0 to $\tau = 100$ MPa,

$$G = \frac{\tau}{\gamma} = 10^5 \text{ MPa}$$

3-3. STRAIN WITHOUT STRESS. THERMAL STRAIN

Stress and strain are not always present simultaneously at a given point in a material. The simplest example of this is thermal strain. A solid bar without constraints can expand or contract freely as it is heated or cooled. The strain is calculated in the same way as when the deformation results from the application of a force. Sometimes it is instructional first to calculate the deformation caused by a certain change in temperature:

$$e = \alpha L \Delta T \tag{3-4}$$

where α = coefficient of thermal expansion (change in length per unit length per degree change in temperature)

     L = length of member

     ΔT = change in temperature

The thermal strain is

$$\epsilon = \frac{e}{L} = \alpha \Delta T \qquad (3\text{-}5)$$

EXAMPLE 3-3

     A 20-m long concrete column is designed with continuous steel reinforcing rods in it. Two kinds of rods are considered: a low carbon steel (cheap) with $\alpha = 4.9 \times 10^{-6}$ per °F and an alloy steel (expensive) with $\alpha = 6.5 \times 10^{-6}$ per °F. Describe the tendency of each steel to separate from the concrete in terms of unconstrained, relative elongations caused by a 100°F change in temperature. For the concrete, $\alpha = 4.7 \times 10^{-6}$ per °F.

Solution

     Using $e = \alpha L \Delta T$,

$$e_{\text{concrete}} = 4.7 \times 10^{-6}(20)(100) = 9.4 \times 10^{-3} \text{ m}$$
$$e_{\text{low C steel}} = 4.9 \times 10^{-6}(20)(100) = 9.8 \times 10^{-3} \text{ m}$$
$$e_{\text{alloy steel}} = 6.5 \times 10^{-6}(20)(100) = 1.3 \times 10^{-2} \text{ m}$$

     Clearly, the low carbon steel is the more compatible with the concrete. Note that in this case the units can be mixed without any complications.

EXAMPLE 3-4

     A large concrete reservoir is designed to hold liquid nitrogen. It is to be a vertical cylinder, 80 ft in diameter, with a wall thickness of 10 in. Assuming that there are no reinforcing elements in the concrete, what is the change in diameter in inches as it cools from +130° to −320°F? $\alpha = 4.7 \times 10^{-6}$ per °F.

Solution

     The wall thickness is not relevant to this particular aspect of the design. The change in diameter is obtained by calculating the change in circumference:

$$e = \alpha L \Delta T = \alpha \pi D \Delta T = 4.7 \times 10^{-6} \pi(80)(-450)(12) = -6.38 \text{ in.}$$

The corresponding new diameter is

$$d = 960 - \frac{6.38}{\pi} = 958 \text{ in.}$$

so the diameter decreases by 2 in.

3-4. STRAIN WITHOUT STRESS.
POISSON'S RATIO

A special kind of strain without stress was defined by the French mathematician and scientist S. D. Poisson (1781–1840): Materials subjected to uniaxial forces deform in the direction of the applied load *and* in directions perpendicular to the applied load. This behavior is easiest to observe with a rubber band or a thin piece of polyurethane foam. They become thinner while being stretched. The reverse is true in compression. A solid rubber cylinder (such as an eraser on a pencil) becomes fatter while it is compressed.

Compression test of rock with lateral displacement as feedback in a million-pound testing machine. *J. A. Hudson, E. T. Brown, and C. Fairhurst, "Optimizing the Control of Rock Failure,"* Closed Loop, **2**, *No. 7 (1970)*, *6–11.*

It is found that the transverse (or, lateral) strain is always smaller than the axial strain if there is no stress in the transverse direction. The ratio of the transverse strain to the axial strain is called *Poisson's ratio*, and it is frequently denoted by the symbol μ (or sometimes by v):

$$\mu = -\frac{\epsilon_{\text{transverse}}}{\epsilon_{\text{axial}}} \tag{3-6}$$

Observed values of this ratio are about $\frac{1}{4}$ to $\frac{1}{3}$ for loadings that cause linear stress-strain responses. Poisson's ratio is a material property; it is given as a positive quantity.

EXAMPLE 3-5

The cylindrical rod of a double-acting hydraulic ram is steel and has a diameter of 10 cm when there is no load on the rod. The maximum axial load on the rod is ± 500 kN. It is necessary to know the required clearances in the bearings that must guide the linear movements of the rod. What are the largest and smallest diameters of the rod during service? $E = 200$ GPa, $\mu = 0.3$.

Solution

The axial stress is

$$|\sigma| = \frac{5 \times 10^5}{\pi(0.05)^2} = 63.7 \text{ MPa}$$

The corresponding axial strain is

$$|\epsilon_{\text{axial}}| = \frac{63.7}{200\ 000} = 3.18 \times 10^{-4}$$

The lateral strain is

$$|\epsilon_{\text{lateral}}| = 0.3(3.18 \times 10^{-4}) = 9.55 \times 10^{-5}$$

The maximum changes in diameter are $\pm|\epsilon_{\text{lateral}}|$ (10 cm), so

$$D_{\text{min}} = 9.99904 \text{ cm} \quad \text{and} \quad D_{\text{max}} = 10.00096 \text{ cm}$$

3-5. STRESS WITHOUT STRAIN. THERMAL STRESS

The thermal strain described above can be used as the basis for discussing the opposite extreme: The material "feels" a stress but there is no strain. For a model of this, assume that the bar A is confined between ideally rigid walls in Fig. 3-4. Initially, the bar is barely touching the vertical walls; there is contact between them but the force is zero in the horizontal direction. Assume next that the bar alone is

heated. It wants to expand but it cannot become any longer. If there is no elongation, there is no strain.

Of course, the bar pushes against the walls as it wants to expand, so there is a
stress generated in the bar. The stress can be calculated. It is done by imagining
that the bar is not constrained as in Fig. 3-4b. The imaginary strain is calculated
using Eq. 2-1. Then the stress that would produce the same strain if the bar were
not heated is determined from Eq. 3-2: $\sigma = E\epsilon$. This stress would also be able to
push the heated bar back to its original length (Fig. 3-4c). Consequently, this must
be the stress in the bar that is prevented from extending as it is heated.

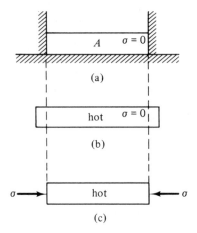

Fig. 3-4. Model for thermally
induced stress.

EXAMPLE 3-6

An aluminum rod is constrained as shown in Fig. 3-4. The rod is 10 in. long,
its modulus of elasticity is $E = 10^7$ psi, and its coefficient of thermal expansion is
$\alpha = 1.3 \times 10^{-5}$ in./in./°F. Calculate the stress in the rod caused by a 100°F increase
in the temperature.

Solution

The strain in thermal expansion of an unconstrained rod (as in Fig. 3-4b)
from Eq. 3-5 is

$$\epsilon = \alpha\Delta T = (1.3 \times 10^{-5}) \times 100 = 1.3 \times 10^{-3} \text{ in./in.}$$

From Eq. 3-2,

$$\sigma = E\epsilon = 10^7 \times (1.3 \times 10^{-3}) = 13,000 \text{ psi} \qquad \text{(compression)}$$

EXAMPLE 3-7

An aluminum wire is stretched between rigid terminals that are 10 cm apart.
The initial stress in the wire is 10 MPa at 25°C. What is the stress in the wire after it

has been cooled to liquid helium temperature (4°K)? At what temperature would the stress become zero? $\alpha = 13 \times 10^{-6}$ per °F, $E = 70$ GPa.

Solution

The temperature drop is $269 + 25 = 294°C = 561°F$. If the wire were not constrained, the strain resulting from the contraction would be

$$\epsilon = \alpha \Delta T = (13 \times 10^{-6})(-561) = -7.29 \times 10^{-3}$$

This strain can be prevented by the rigid terminals applying a stress

$$\sigma = E\epsilon = 510 \text{ MPa}$$

Since the wire is prestressed by 10 MPa, the total stress in it is 520 MPa after it is cooled to 4°K.

The stress would be zero at a temperature that would cause the same extension of the wire as that caused by the tensile stress of 10 MPa. Hence the thermal strain must be

$$\epsilon = \frac{10 \text{ MPa}}{70 \text{ GPa}} = 1.43 \times 10^{-4}$$

This can be produced by a temperature change

$$\Delta T = \frac{\epsilon}{\alpha} = 11°F = 6.1°C$$

The stress in the wire would be zero at 31.1°C.

EXAMPLE 3-8

A large concrete reservoir is designed to hold liquid oxygen. It is a vertical cylinder, 38 ft in diameter, 35 ft in height, with an 8-in. thick wall. The cylinder was prestressed by wrapping steel wires around it; the resulting circumferential compressive stress in the concrete is 650 psi at 70°F. What is the circumferential stress in the concrete that can be expected at −297°F? $E_{steel} = 30 \times 10^6$ psi, $E_{concrete} = 4 \times 10^6$ psi, $\alpha_{steel} = \alpha_{concrete} = 4.7 \times 10^{-6}$ per °F. Assume that the temperature in the steel and the concrete are always the same.

Solution

The initial stress is caused by what may be called a misfit between the steel wire jacket and the concrete cylinder. Upon cooling, both the steel and the concrete want to change dimensions equally. The temperature drop causes no additional misfit between them, so the only stress in the concrete is that caused by the original misfit. The final stress in the concrete is 650 psi.

3-6. GENERALIZED STRESS-STRAIN RELATIONS

The preceding sections describe the simplest possible relations for stress and strain. There are three kinds of complications that must be considered. One has to do with realistic constraints. A member may be neither perfectly free to change its dimension as the temperature changes nor ideally constrained. Another complication is that the stress-strain curve is not necessarily linear all the way to fracture. This is an extremely important aspect of the behavior of materials. It can be either beneficial or harmful from the practical point of view. The first detailed discussion of this appears in Chapter 4, but its real significance becomes clear only with much experience. The third kind of complication occurs because of the Poisson effect and an explanation of this follows.

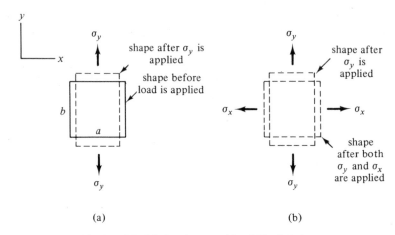

Fig. 3-5. Models for the generalized Hooke's law.

Imagine a two-dimensional element that is loaded uniaxially as in Fig. 3-5a. There are two strains resulting from this loading:

$$\epsilon_{y_1} = \frac{\sigma_y}{E} \quad \text{and} \quad \epsilon_{x_1} = -\mu\epsilon_{y_1}$$

Next, the element is loaded in the x-direction, as well. The element extends in the x-direction and contracts in the y-direction because of σ_x (Fig. 3-5b). Considering σ_x alone, the new strains are

$$\epsilon_{x_2} = \frac{\sigma_x}{E} \quad \text{and} \quad \epsilon_{y_2} = -\mu\epsilon_{x_2}$$

The total deformations in the two directions are

$$e_x = \frac{\sigma_x}{E}(a) - \mu\frac{\sigma_y}{E}(a), \qquad e_y = \frac{\sigma_y}{E}(b) - \mu\frac{\sigma_x}{E}(b)$$

where a and b are the dimensions of the element. The effective strains in the two directions are

$$\epsilon_x = \frac{e_x}{a} = \frac{\sigma_x}{E} - \mu\frac{\sigma_y}{E}, \qquad \epsilon_y = \frac{e_y}{b} = \frac{\sigma_y}{E} - \mu\frac{\sigma_x}{E}$$

An extension of this analysis leads to expressions of the effective strains under three-dimensional loading:

$$\epsilon_x = \frac{\sigma_x}{E} - \mu\frac{\sigma_y}{E} - \mu\frac{\sigma_z}{E} \tag{3-7a}$$

$$\epsilon_y = \frac{\sigma_y}{E} - \mu\frac{\sigma_x}{E} - \mu\frac{\sigma_z}{E} \tag{3-7b}$$

$$\epsilon_z = \frac{\sigma_z}{E} - \mu\frac{\sigma_x}{E} - \mu\frac{\sigma_y}{E} \tag{3-7c}$$

These equations are called the *generalized Hooke's law*. They are valid for any normal stresses that act in three mutually perpendicular directions. The modulus of elasticity and Poisson's ratio are assumed to be independent of direction in the material. Such materials are called *isotropic*.

Equations 3-7a, b, and c can be combined to express the stresses in terms of the strains:

$$\sigma_x = \frac{E}{(1 + \mu)(1 - 2\mu)}[(1 - \mu)\epsilon_x + \mu\epsilon_y + \mu\epsilon_z] \tag{3-8a}$$

$$\sigma_y = \frac{E}{(1 + \mu)(1 - 2\mu)}[(1 - \mu)\epsilon_y + \mu\epsilon_x + \mu\epsilon_z] \tag{3-8b}$$

$$\sigma_z = \frac{E}{(1 + \mu)(1 - 2\mu)}[(1 - \mu)\epsilon_z + \mu\epsilon_x + \mu\epsilon_y] \tag{3-8c}$$

For biaxial states of stress, the following simpler forms can be obtained directly from Eq. 3-7a and b, with $\sigma_z = 0$:

$$\sigma_x = \frac{E}{1 - \mu^2}(\epsilon_x + \mu\epsilon_y) \tag{3-8d}$$

$$\sigma_y = \frac{E}{1 - \mu^2}(\epsilon_y + \mu\epsilon_x) \tag{3-8e}$$

Equations 3-7 and 3-8 are valid for tensile or compressive loading if signs are used consistently. Tensile stresses and strains are positive and compressive stresses and strains are negative in most calculations.

EXAMPLE 3-9

A beryllium alloy sheet is 1 mm thick and it is loaded in biaxial tension with $\sigma_x = 500$ MPa and $\sigma_y = 200$ MPa. What is the new thickness under these loads?

How is the answer changed if a compressive stress of $\sigma_z = 300$ MPa is added? $E = 300$ GPa, $\mu = 0.1$ (this is a very strange metal).

Solution

Using Hooke's law (Eq. 3-7c),

$$\epsilon_{z_1} = \frac{1}{E}[\sigma_z - \mu(\sigma_x + \sigma_y)]$$

$$= \frac{1}{300\ 000}[0 - 0.1(500 + 200)] = -2.33 \times 10^{-4}$$

The new thickness is

$$t_1 = t_0(1 + \epsilon_{z_1}) = (1 \text{ mm})[1 - (2.33 \times 10^{-4})]$$

$$= 0.99976 \text{ mm}$$

With the addition of the compressive stress,

$$\epsilon_{z_2} = \frac{1}{300\ 000}[-300 - 0.1(500 + 200)] = -1.23 \times 10^{-3}$$

$$t_2 = 0.99876 \text{ mm}$$

EXAMPLE 3-10

A cylindrical concrete block 10 in. in diameter is located deep under water and is loaded axially. The pressure of the water is 700 psi; the axial load is 300,000 lb in compression. What are the axial and diametral strains in the block? $E = 4 \times 10^6$ psi, $\mu = 0.2$.

Solution

The axial stress from the large load is

$$\sigma_x = \frac{300,000}{\pi(5)^2} = 3820 \text{ psi} \qquad \text{(compression)}$$

The hydrostatic pressure causes

$$\sigma_x = \sigma_y = \sigma_z = 700 \text{ psi} \qquad \text{(compression)}$$

The net stresses are

$$\sigma_x = 4520 \text{ psi}, \qquad \sigma_y = \sigma_z = 700 \text{ psi} \qquad \text{(all in compression)}$$

Using Hooke's law,

$$\epsilon_x = \frac{1}{E}[\sigma_x - \mu(\sigma_y + \sigma_z)] = \frac{1}{4 \times 10^6}[-4520 - 0.2(-1400)]$$

$$= -1.06 \times 10^{-3}$$

$$\epsilon_y = \frac{1}{E}[\sigma_y - \mu(\sigma_z + \sigma_x)] = \frac{1}{4 \times 10^6}[-700 - 0.2(-5220)]$$

$$= 8.6 \times 10^{-5}$$

EXAMPLE 3-11

A test is performed on a homogeneous block of metal. The loading is uniform triaxial compression of 600 MPa, and a value of 0.6 is obtained for the Poisson's ratio. This is higher than the range of common values, so the question arises as to whether it is even possible according to theory.

Solution

The exact value of the stress is not important here.

$$-\sigma_x = -\sigma_y = -\sigma_z = \sigma, \qquad -\epsilon_x = -\epsilon_y = -\epsilon_z = \epsilon$$

A fruitful approach is to add Eq. 3-7a, b, and c and obtain

$$3\epsilon = \frac{3\sigma}{E}(1 - 2\mu)$$

Since σ and ϵ must have the same sign,

$$1 - 2\mu \geq 0 \quad \text{and} \quad \mu \leq 0.5$$

Thus, the value of 0.6 is in error according to theory.

3-7. RELATION OF ELASTIC CONSTANTS

The question arises as to whether the three elastic constants (E, G, and μ) are related to one another for a given material. An attempt to answer this question can be based on the analysis of an element for which the normal and shear stresses are related in a simple way. Such a simple situation is shown in Fig. 3-6a, which is equivalent to the pure shear shown in Fig. 3-6b. Mohr's circle is identical for these two, with the center at the origin and the radius $= \sigma_x = \tau$ (Fig. 3-6c). This equivalence is used in the analysis of the deformation of the element in Fig. 3-6b, which is shown in Fig. 3-6d.

The right angles of the element change to $90° - \gamma$ on the x-axis and to $90° + \gamma$ on the y-axis (γ = shear strain). The axial strains are, from Eq. 3-7,

$$\epsilon_x = \frac{\tau(1 + \mu)}{E} \quad \text{and} \quad \epsilon_y = -\frac{\tau(1 + \mu)}{E}$$

since $\sigma_x = -\sigma_y = \tau$.

The diagonals of the square in Fig. 3-6b are the gage lengths, so the deformation can be described by the distances

$$Oa' = Oa\left[1 + \frac{\tau(1 + \mu)}{E}\right] \quad \text{and} \quad Ob' = Ob\left[1 - \frac{\tau(1 - \mu)}{E}\right]$$

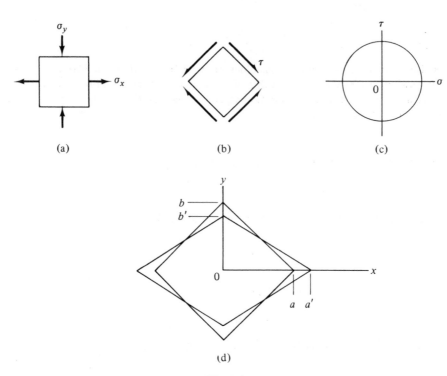

(a)   (b)   (c)

(d)

Fig. 3-6

The angular distortion can be described in terms of these distances:

$$\tan\left(45° - \frac{\gamma}{2}\right) = \frac{Ob'}{Oa'} = \frac{1 - [\tau(1 + \mu)/E]}{1 + [\tau(1 + \mu)/E]}$$

With a trigonometric identity,

$$\tan\left(45° - \frac{\gamma}{2}\right) = \frac{\tan 45° - \tan(\gamma/2)}{1 + \tan 45° \tan(\gamma/2)} = \frac{1 - \tan(\gamma/2)}{1 + \tan(\gamma/2)}$$

For elastic deformations (when the element returns to its original shape after unloading), the angle γ is always small, so

$$\tan\frac{\gamma}{2} \approx \frac{\gamma}{2}$$

Thus,
$$\frac{1 - (\gamma/2)}{1 + (\gamma/2)} = \frac{1 - [\tau(1 + \mu)/E]}{1 + [\tau(1 + \mu)/E]}$$

The last equation can be simplified to give

$$\gamma = \frac{2\tau(1+\mu)}{E} \quad \text{and} \quad \frac{\tau}{\gamma} = \frac{E}{2(1+\mu)}$$

Substituting the modulus G from Hooke's law for shear,

$$G = \frac{E}{2(1+\mu)} \tag{3-9}$$

EXAMPLE 3-12

A titanium alloy (Ti-8Al-1Mo-1V) was developed for good oxidation resistance and thermal stability up to 850°F. Tension tests were performed, and it was found that $E = 17.5 \times 10^6$ psi and $\mu = 0.32$. To save time and expense, G is to be determined from these data.

Solution

$$G = \frac{17.5 \times 10^6}{2(1 + 0.32)} = 6.63 \times 10^6 \text{ psi}$$

Shear tests performed at a later time confirmed this by giving $G = 6.7 \times 10^6$ psi.

3-8. TRANSFORMATION OF MOHR'S CIRCLES

For any given element, the Mohr's circle for stress must be related to the Mohr's circle for strain. The transformation from one to the other can be accomplished if relations are found for the centers and radii of the two circles.

The center of the circle in the stress coordinates is C_σ:

$$C_\sigma = \frac{\sigma_x + \sigma_y}{2}$$

In the strain coordinates it is C_ϵ:

$$C_\epsilon = \frac{\epsilon_x + \epsilon_y}{2}$$

The latter can be written in terms of stresses using Eq. 3-7a and b:

$$C_\epsilon = \frac{1}{2E}(\sigma_x - \mu\sigma_y + \sigma_y - \mu\sigma_x) = \frac{(1-\mu)}{2E}(\sigma_x + \sigma_y)$$

Thus,

$$C_\sigma = \left(\frac{E}{1-\mu}\right)C_\epsilon \tag{3-10}$$

The radius of the circle in the stress coordinates is R_σ:

$$R_\sigma^2 = \left(\frac{\sigma_x - \sigma_y}{2}\right)^2 + \tau^2$$

In the strain coordinates it is R_ϵ:

$$R_\epsilon^2 = \left(\frac{\epsilon_x - \epsilon_y}{2}\right)^2 + \left(\frac{\gamma_{xy}}{2}\right)^2$$

Using Hooke's law and Eq. 3-9,

$$\tau^2 = G^2\gamma^2 = \frac{E^2\gamma^2}{4(1 + \mu)^2}$$

With this, and using Eq. 3-8d and e,

$$R_\sigma^2 = \frac{1}{4}\left[\frac{(\epsilon_x + \mu\epsilon_y)E - (\epsilon_y + \mu\epsilon_x)E}{1 - \mu^2}\right]^2 + \frac{E^2\gamma^2}{4(1 + \mu)^2}$$

$$= \frac{1}{4}\left[\frac{E(1 - \mu)(\epsilon_x - \epsilon_y)}{(1 - \mu)(1 + \mu)}\right]^2 + \frac{E^2\gamma^2}{4(1 + \mu)^2}$$

$$= \left[\frac{E}{1 + \mu}\left(\frac{\epsilon_x - \epsilon_y}{2}\right)\right]^2 + \left[\frac{E}{1 + \mu}\left(\frac{\gamma}{2}\right)\right]^2$$

Thus,

$$R_\sigma = \left(\frac{E}{1 + \mu}\right)R_\epsilon \tag{3-11}$$

3-9. COMPOSITE MATERIALS

There are many important materials that are not homogeneous and isotropic (wood, reinforced concrete, and fiber glass, to name a few). The technological significance of metallic and nonmetallic composites is rapidly increasing and will continue to increase in the future. There are two reasons for this. Composites can be designed and manufactured to have certain desirable properties in specified directions in the material. For a simple illustration of this advantage, consider a member that must be loaded only in one direction. If it has equal strength in all three directions, one could say that the strengths in two directions are not fully utilized. A lighter material that has sufficient strength in the desired direction and little strength in the other directions may be far superior to the isotropic material. The second reason for the increasing role of composites in modern engineering is that the strongest known materials can only be made in the forms of whiskers, particles, or filaments of very small diameters. The strength advantage can be substantial: Some filaments can be four times as strong as the strongest steels. Another extraordinary advantage of many filaments is their relatively low density, so they have excellent specific strengths (strength-to-weight ratios). To make usable solids, however, it is necessary in most cases to embed the fibers in another material. This makes it impossible to take full advantage of the properties of the fine components, but it is still practical to do. For example, usable solids can be made that have strengths two and a half times that of the strongest steel wires.

Fractured boron-epoxy composite viewed through a scanning electron microscope. *Courtesy Professor R. E. Rowlands, University of Wisconsin, Madison, Wis.*

Composites, of course, are relatively complex to make or analyze. The simplest composites in stress analysis are those in which all fibers (most of them extending the length of the member) are parallel to one another and which are loaded in the longitudinal direction of the fibers. For such members the *rule of mixtures* gives the apparent Young's modulus:*

$$E_c = E_f V_f + E_M V_M \tag{3-12}$$

where E_c = modulus of elasticity of the composite
E_f = modulus of elasticity of the fibers
E_M = modulus of elasticity of the matrix in which the fibers are embedded
V_f = volume fraction of the fibers
V_M = volume fraction of the matrix

*For a derivation of this and other properties of composites, see Robert M. Jones, *Mechanics of Composite Materials* (New York: McGraw-Hill Book Company, 1975).

(a)

(a) Test setup for boron-epoxy composite materials; (b) schematic of test instrumentation (not all tests of composites are this complex). *Richard B. Freeman, Erwin C. Durchlaub, and B. Dale Austin, "Testing for Safelife Data with Boron/Epoxy Composite Laminates,"* Closed Loop, **4**, *No. 2 (1974), 17–21.*

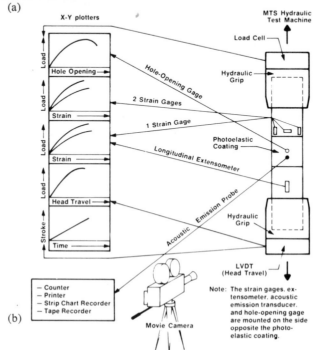

(b)

Note: The strain gages, extensometer, acoustic emission transducer, and hole-opening gage are mounted on the side opposite the photoelastic coating.

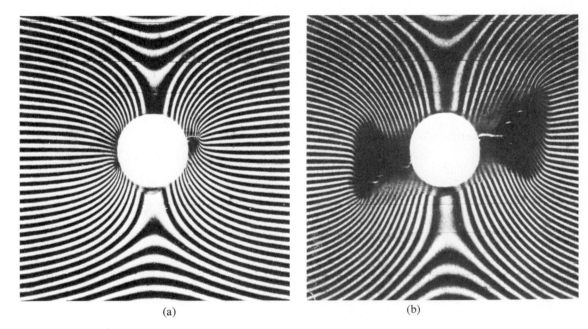

(a)   (b)

Sequence of moiré-fringe patterns in glass-epoxy plate immediately prior to total failure; the growing dark areas show delamination; (a) $\sigma = 29,850$ psi, (b) $\sigma = 30,400$ psi. *R. E. Rowlands, I. M. Daniel, and J. B. Whiteside, "Stress and Failure Analysis of a Glass-Epoxy Composite Plate with a Circular Hole,"* Experimental Mechanics, **13**, *No. 1 (1973), 31–37.*

Fibrous composites are very weak when loaded perpendicular to the fibers. For this reason, lamination is often necessary, with fibers in different layers oriented in different directions. The rule of mixtures does not apply for such constructions. The theories for them are quite complex and can be used in analysis only with the aid of computers.

EXAMPLE 3-13

Many nails are driven through a block of wood. The nails are all parallel and they are flush with the wood on both faces of the block. They take up 5% of the original volume of the block. What is the effective Young's modulus for compressive loading along the nails? $E_{wood} = 1.5 \times 10^6$ psi, $E_{steel} = 30 \times 10^6$ psi.

Solution

$$E_c = 30 \times 10^6(0.05) + 1.5 \times 10^6(0.95) = 2.92 \times 10^6 \text{ psi}$$

Thus, the nails almost double the original stiffness of the block.

EXAMPLE 3-14

A concrete block is to be reinforced with parallel steel rods. Each rod occupies 1% of the total available space and is eight times stiffer than concrete. How many of these rods are required to make the block at least three times stiffer than concrete alone? The loading is parallel to the rods.

Solution

Letting X = the number of rods required,

$$E_c = E_{concrete}(1 - 0.01X) + E_{steel}(0.01X)$$
$$3 = (1)(1 - 0.01X) + 8(0.01X)$$
$$X = 29$$

EXAMPLE 3-15

Thin filaments of tungsten coated with boron make up 70% of the volume of a high-performance composite. The elastic modulus of the fibers is 380 GPa. The matrix is an epoxy with a modulus of 35 PGa. How does the elastic modulus of the composite compare with that of a common steel for which $E = 206$ GPa?

Solution

$$E_c = 380(0.7) + 35(0.3) = 276.5 \text{ GPa}$$

This is considerably higher than the modulus of the steel, even though the matrix has an almost negligible contribution to the stiffness.

3-10. PROBLEMS

Sec. 3-1

3-1. How does the spring constant k depend on the geometry of a cylindrical test specimen?

3-2. Two springs are made of the same metal, and the coil diameters are also identical. What is different in the spring that has a lower compliance than the other?

Sec. 3-2

Determine the unknown quantity in Prob. 3-3 to 3-6. Assume linear behavior.

3-3. $\sigma = 100$ ksi, $\epsilon = 0.1\%$, $E = ?$

3-4. $\sigma = 10$ GPa, $\epsilon = 0.02$, $E = ?$

3-5. $\tau = 50$ ksi $\gamma = 10^{-3}$, $G = ?$

3-6. $\tau = 200$ MPa, $G = 300$ GPa, $\gamma = ?$

Secs. 3-3, 3-4, 3-5

3-7. What are the desirable gaps in a railroad for $\Delta T_{max} = 150°F$? Assume that the rails are 100 ft (30 m) long and should not be loaded in compression. If a new railroad is constructed there with the individual rail segments welded end-to-end, what is the maximum stress that can be expected in a rail?

3-8. What are the relative advantages of rails with gaps and those without gaps? Return to this question again later in the course when new knowledge leads you to new ideas.

3-9. Discuss the Poisson effect in thermally induced stress and strain.

Determine the unknown quantity in Prob. 3-10 to 3-15. (One-dim. stress)

3-10. $\epsilon_{axial} = 0.006$, $\epsilon_{lateral} = -0.0019$, $\mu = $?

3-11. $\epsilon_{axial} = 0.2\%$ $\mu = 0.25$, $\epsilon_{lateral} = $?

3-12. $\epsilon_{axial} = -0.01\%$, $\mu = 0.3$, $\epsilon_{lateral} = $?

3-13. $\epsilon_{lateral} = 0.001\%$, $\mu = 0.32$, $\epsilon_{axial} = $?

3-14. $\epsilon_{lateral} = -0.0008$, $\mu = 0.28$, $\epsilon_{axial} = $?

3-15. $\epsilon_{lateral} = -0.05\%$, $\mu = 0.3$, $\epsilon_{axial} = $?

Sec. 3-6

3-16. Show in sketches similar to Fig. 3-5 the deformations caused by three mutually perpendicular stresses if
(a) all the stresses are tensile.
(b) one stress is tensile and the other two are compressive.

Determine the unknown quantity in Prob. 3-17 to 3-26. $\mu = 0.3$ if not given.

3-17. $\sigma_x = 100$ ksi, $\sigma_y = 120$ ksi, $\sigma_z = 0$, $E = 30 \times 10^6$ psi, $\epsilon_z = $?

3-18. $\sigma_x = 400$ MPa, $\sigma_y = -50$ MPa, $\sigma_z = 50$ MPa, $E = 200$ GPa, $\epsilon_z = $?

3-19. $\sigma_x = 150$ ksi, $\sigma_z = -20$ ksi, $\sigma_y = 30$ ksi, $E = 30 \times 10^6$ psi, $\epsilon_y = $?

3-20. $\sigma_x = 300$ MPa, $\sigma_z = 60$ MPa, $\sigma_y = 0$, $E = 70$ GPa, $\epsilon_x = $?

3-21. $\sigma_x = \sigma_y = 50$ ksi, $\sigma_z = 70$ ksi, $E = 2 \times 10^7$ psi, $\epsilon_x = $?

3-22. $\sigma_x = \sigma_y = 100$ MPa, $\sigma_z = -50$ MPa, $E = 70$ GPa, $\epsilon_z = $?

3-23. $\epsilon_x = 0.1\%$, $\epsilon_y = 0.05\%$, $\epsilon_z = 0$, $E = 30 \times 10^6$ psi, $\mu = 0.28$, $\sigma_x = $?

3-24. $\epsilon_x = 0.1\%$, $\epsilon_y = 0.08\%$, $\epsilon_z = 0.01\%$, $E = 200$ GPa, $\mu = 0.3$, $\sigma_y = $?

3-25. $\epsilon_x = \epsilon_y = 0.2\%$, $\epsilon_z = -0.1\%$, $E = 10^7$ psi, $\mu = 0.3$, $\sigma_z = $?

3-26. $\epsilon_x = 0.15\%$, $\epsilon_y = \epsilon_z = -0.05\%$, $E = 200$ GPa, $\mu = 0.28$, $\sigma_x = $?

Sec. 3-9

   Determine the unknown quantity in Prob. 3-27 to 3-32.

3-27. $E_f = 50 \times 10^6$ psi, $E_M = 2 \times 10^6$ psi, $V_f = 60\%$, $E_c = ?$

3-28. $E_f = 200$ GPa, $E_M = 70$ GPa, $V_f = 70\%$, $E_c = ?$

3-29. $E_f = 30 \times 10^6$ psi, $E_M = 10^7$ psi, $V_f = 0.55$, $E_c = ?$

3-30. $E_f = 250$ GPa, $E_M = 30$ GPa, $V_f = 0.65$, $E_c = ?$

3-31. $E_f = 30 \times 10^6$ psi, $E_M = 3.5 \times 10^6$ psi, $E_c = 20 \times 10^6$ psi, $V_f = ?$

3-32. $E_f = 350$ GPa, $E_M = 25$ GPa, $E_c = 270$ GPa, $V_f = ?$

4

MECHANICAL PROPERTIES
OF MATERIALS

A physical property of a material could be analogous to the name of a person. It should be a constant, independent of time and place and circumstances. Unfortunately, there are no material properties that are perfectly constant (of course, some are better than others in this respect). One should realize this and strive to find and use those properties that are sufficiently constant for practical purposes. In some cases the available material properties are not entirely satisfactory, and continuing research is aimed at finding better ones.

The few material properties that were mentioned previously (E, G, μ) are among the most reliable properties. Other mechanical properties will be introduced during discussions of realistic stress-strain curves.

4-1.   LINEAR AND NONLINEAR STRESS-STRAIN CURVES

Some materials have linear or nearly linear stress-strain curves all the way to fracture. These are called *brittle* materials. Brittleness is not a material property but a qualitative term used arbitrarily for the purpose of comparing materials. It can be affected, sometimes to a great extent, by many things, as shown in Section 4-4.

Nonlinear stress-strain curves are very common. The material properties obtained from these curves include those that are relevant to the most brittle materials also.

The simplest realistic nonlinear stress-strain curve is one that can be described by a power function:

$$\sigma = K\epsilon^n \qquad (4\text{-}1)$$

73

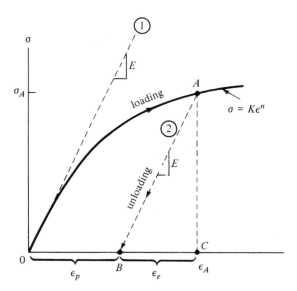

Fig. 4-1. Nonlinear stress–strain curve.

The curve shown in Fig. 4-1 is typical for fairly pure metals such as aluminum and copper in the annealed condition. These are called *ductile metals*. Ductility is the opposite of brittleness, and it is similarly qualitative in meaning. A very ductile material has little resistance to deformation, especially in the early part of the loading. The increase in deformation resistance is called *cold working* or *work hardening*. The exponent n in Eq. 4-1 is called the *strain hardening exponent*. It is determined by plotting stress versus strain on logarithmic coordinates; n is the slope of the resulting line, which is normally fairly straight. Its value for metals is in the range of 0.05 to 0.4. Low values of n indicate hard, brittle materials, while soft, ductile materials have relatively high strain hardening exponents. This does not seem right to most people at first sight, but it is correct as stated. The constant K in Eq. 4-1 is the strength coefficient. It is the stress for which the strain is 1.0. The order of magnitude for K is 10^5 psi for commonly used metals.

A common variation of the curve in Fig. 4-1 has a linear portion ($\sigma = E\epsilon$) coming from the origin and this connects smoothly to a power function at high stresses.

Both n and K are material properties, but this does not mean that chemical composition alone is what these constants depend on. As far as mechanical behavior is concerned, many "different materials" can be produced (by thermal, mechanical, or thermomechanical treatments) that have the same average chemical composition. This is an important aspect of strength of materials because

> *Some events or conditions during the normal service life of a member may alter the original mechanical properties of the material.*

The nonlinearity in Fig. 4-1 is generally referred to as inelastic behavior or plastic deformation or yielding. This behavior is always such that the curve obtained in tensile loading is below the straight line 1 which is tangent to the stress-strain curve at the origin. The straight line represents Young's modulus.

In many cases the nonlinearity indicates permanent deformation. This is shown in Fig. 4-1 also. The material is first loaded to point A on the stress-strain curve. Interestingly, unloading from the stress at A is along a line that is essentially parallel to the original line representing elastic behavior. The stress is zero at point B but there is a permanent strain. This is commonly called *plastic strain*, and it is denoted by ϵ_p.

There are several interesting and important things that can be learned from the loading-unloading process if proper measurements are made. A loading to the stress of σ_A results in the strain ϵ_A. This is the only strain that can be observed and measured directly while the material is stressed, but it is made up of discrete components. Part of ϵ_A could be recovered by unloading. It is called the *elastic strain component*, ϵ_e. This component can always be calculated in a simple way if the stress and Young's modulus are known. In Fig. 4-1,

$$\epsilon_e = \frac{\sigma_A}{E}$$

The elastic strain component is always obtained from a triangle similar to ABC in Fig. 4-1; so, in general,

$$\epsilon_e = \frac{\sigma}{E} \tag{4-2}$$

Equation 4-2 is useful in several areas of strength of materials. What should be remembered is the simple fact that it gives the effective springiness of a material after any plastic deformation.

The discussion above leads to the generalization that

$$\epsilon_t = \epsilon_e + \epsilon_p \tag{4-3}$$

where ϵ_t = total strain (it is measurable directly)
  ϵ_e = elastic strain (not measurable directly during loading)
  ϵ_p = plastic strain (not measurable directly during loading)

The total strain may have other components than the two mentioned here, but knowledge of these two is sufficient for solving many practical problems.

A few words about the nature of plastic strain are necessary. The simplest possible model of plastic deformation is two rows of identical spheres (representing atoms) shown in Fig. 4-2. In (a) there are no external forces on the spheres. Assume that shear forces are applied as shown in (b), and the two rows start slipping relative to each other. When the forces are removed, there is a permanent change in the position of the spheres as in (c).

This crude model shows that plastic deformation (which is also called *slip*, or *flow*, or *yielding*) involves internal friction in the material. Indeed, the heat gen-

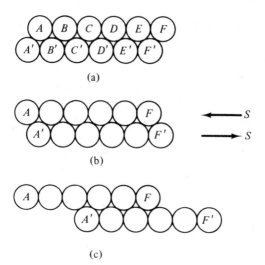

Fig. 4-2. Model for plastic deformation.

erated can be felt when a wire is bent repeatedly into new shapes. It is also seen in the model that the volume occupied by the material is not changed in the deformation. Because of this, Poisson's ratio is 0.5 in plastic deformation (this will not be proved here). Another characteristic of this kind of strain is that it is caused by shear stress. Of course, the shear stress may be either from shear or normal loading.

EXAMPLE 4-1

A metal is tested to determine its modulus of elasticity, but a mistake is made and it is plastically deformed a little. The largest measured strain is 1.1%, which is reached at the stress of 120 ksi. After unloading, there is a plastic strain of 0.25%. What is the modulus of elasticity?

Solution

The elastic strain at peak load is

$$\epsilon_e = 1.1 - 0.25 = 0.85\% = 0.0085 \text{ in./in.}$$

$$E = \frac{\sigma}{\epsilon_e} = \frac{120,000}{0.0085} = 14.1 \times 10^6 \text{ psi}$$

4-2. YIELD STRENGTH

When a material is considered for a practical application, its resistance to plastic deformation must be known. Two concepts have evolved to describe this resistance. The proportional limit is the stress at which the stress-strain curve begins to deviate from a straight line (Fig. 4-3). This is practically worthless as a material property in modern engineering because it depends too much on eyesight and judgment. It could be eliminated from the area of strength of materials. The

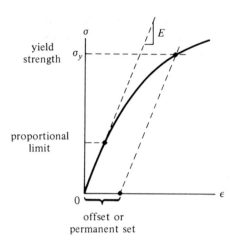

Fig. 4-3. Definition of yield strength.

other concept is also arbitrary, but it is specific in meaning and, consequently, useful in practice.

Yield strength is defined as the stress that causes a specified amount of plastic strain. The specified strain is called the *offset* or *permanent set*. Its magnitude is arbitrary and should be given with the value of the yield strength since the latter depends on the offset. The most commonly used offset is 0.2%, and this may be assumed if the offset is not given. It is important to remember that some yielding has already occurred when the offset yield strength is reached.

There are two other concepts that are related to yield strength. The first is called the *secant modulus*. It is the ratio of the stress to the total strain at any point on the stress-strain curve in the region of plasticity:

$$E_S = \frac{\sigma}{\epsilon} \tag{4-4}$$

The second is the *tangent modulus*, which is the slope of the stress-strain curve at any point in the region of plasticity:

$$E_T = \frac{d\sigma}{d\epsilon} \tag{4-5}$$

The tangent modulus is the instantaneous deformation resistance of a material. The secant and tangent moduli are not material properties. They are convenient in analyses of stress-strain curves.

Sharp Yielding. An important group of metals, low and medium carbon steels, has distinctly different stress-strain curves from that in Fig. 4-3. The curve has a straight section that ends abruptly as in Fig. 4-4. From point *A* to *B* the deformation resistance decreases. From *B* to *C* the deformation increases while the stress is essentially constant. From *C* to the right the deformation resistance increases again because of cold working. The stresses at *A* and *B* are called the *upper* and *lower yield strengths*, respectively. Point *A* is loosely called the yield point; since it has units of stress, it is more precise to call it *yield strength*.

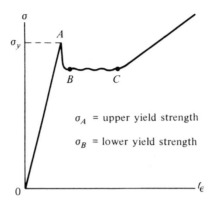

Fig. 4-4. Sharp yielding.

The yield strength for a sharply yielding metal is an important material property, but it can be changed. Important as such a change can be, few engineers realize it. For more information about this, see Section 4-4.

EXAMPLE 4-2

In Fig. 4-1, assume that $\sigma_A = 300$ MPa and $\epsilon_A = 0.02$ cm/cm. What are the secant and tangent moduli at point *A*?

Solution

$$E_S = \frac{\sigma_A}{\epsilon_A} = 15 \text{ GPa}$$

In the absence of accurate recorded data of $d\sigma/d\epsilon$ in the neighborhood of point *A*, a line should be drawn tangent to the curve at *A*. From this,

$$E_T \simeq 4 \text{ GPa}$$

4-3. STRAIN ENERGY. MODULUS OF RESILIENCE

The work done by a force is the force *F* times the distance *x* over which the force is acting:

$$W = Fx$$

When a force is deforming a spring, the force is not a constant but depends on the deformation, $F = kx$. The incremental work is

$$dW = (kx)dx$$

and the total work required to deform an ideal spring (no permanent deformation) by an amount x is

$$W = \tfrac{1}{2}kx^2$$

This is also the energy stored in the spring.

By analogy, the magnitudes of stress and strain may be used to calculate the energy in a unit volume of material. Since $\sigma = E\epsilon$,

$$dW = \sigma(d\epsilon) = \frac{\sigma}{E}\,d\sigma$$

The total energy in the unit volume is

$$W = \frac{\sigma^2}{2E}$$

which is also equal to the area under a linear stress-strain curve from the origin to the stress σ.

The maximum value of the elastic strain energy in a unit volume that has not been permanently deformed is called the *modulus of resilience*. In Fig. 4-3, this is the area under the stress-strain curve up to the proportional limit; in Fig. 4-4, it is the area under the curve up to point A. Note that the modulus of resilience is not necessarily the largest value of the stored strain energy. It is merely the largest elastic strain energy that is not accompanied by plastic deformation. Of course, there is no need for this qualification when the material is perfectly brittle.

4-4. FRACTURE PROPERTIES

Complete tension tests lead to other useful material properties. The true stress just prior to fracture is called the *fracture strength*, σ_f. The maximum plastic

area A = true toughness
area B = elastic strain energy at fracture

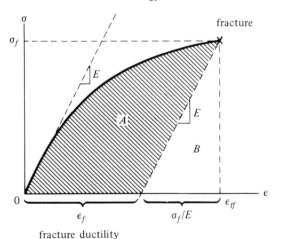

Fig. 4-5. Fracture properties.

strain is called the *fracture ductility*, ϵ_f. The latter can be calculated from the total strain just prior to fracture, ϵ_{tf}:

$$\epsilon_f = \epsilon_{tf} - \frac{\sigma_f}{E} \tag{4-6}$$

These are appreciated more easily from Fig. 4-5.

The fracture strength, σ_f, should not be confused with the ultimate strength, σ_u, which is often used by engineers. The two are the same for brittle materials, but ultimate strength has a special meaning for ductile materials that have considerable permanent deformation before fracture. This deformation is always localized as shown in Fig. 4-6. Even a well-machined, uniform cylindrical rod necks down at a place of instability that cannot be predicted easily. Fracture occurs at the minimum cross-sectional area. Ultimate strength is defined as the maximum tensile load that the specimen could resist divided by the original cross-sectional area. In a ductile material the maximum load occurs well before fracture. As the area decreases, a decreasing tensile load can keep the specimen stretching. At the same time, the true stress keeps increasing to fracture, so $\sigma_f > \sigma_u$ for a ductile material. Figure 4-7 shows the so-called engineering stress-strain curve where stress is always equal to $P_{\text{instantaneous}}/A_{\text{original}}$ in contrast with the true stress-strain curve. A generally

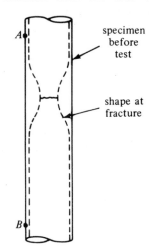

specimen
before
test

shape at
fracture

Fig. 4-6. Ductile specimen necking down.

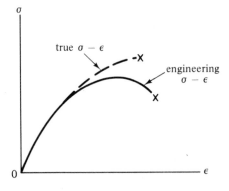

true $\sigma - \epsilon$

engineering
$\sigma - \epsilon$

Fig. 4-7. Stress–strain curves for a given material.

observed characteristic of the latter is that, aside from a possible sharp yielding and associated lower level yielding, the true stress increases monotonically* to fracture.

The fracture ductility property is normally given as the percent elongation or the percent reduction of area. Percent elongation is not a reliable measure of ductility because it makes a difference how long the original gage length is (for example, the length A-B in Fig. 4-6). Furthermore, the elongation obtained using a given gage length is not readily converted to elongation based on a different gage length. This is sometimes compounded by the human error of omitting the gage length when giving the percent elongation. All in all, percent elongation is a poor material property, and its usage should be avoided.

The concept of percent reduction of area is excellent for the quantitative measurement of ductility. It is commonly denoted by $\% \, RA$ and calculated from the original area, A_0, and the area after fracture, A_f:

$$\% \, RA = \frac{A_0 - A_f}{A_0} \times 100 \qquad (4\text{-}7)$$

A brittle material does not neck down prior to fracture, and its $\% \, RA$ is zero or nearly that. A ductile material can neck down severely (even approaching a sharp point at fracture, occasionally), resulting in a high $\% \, RA$.

The value of the $\% \, RA$ is that it depends only on the material and on the minimum specimen diameters before and after the test. Sometimes it is even possible (as it would be for the specimen in Fig. 4-6) to calculate the $\% \, RA$ years after the test if the broken specimen was saved but no test record was available.

The $\% \, RA$ can be converted to ϵ_f. This is based on a different definition of true strain. The plastic component of the deformation causes no change in the volume, so $L/L_0 = A_0/A$, and

$$\epsilon = \ln \frac{A_0}{A} = \ln \frac{L}{L_0} \qquad (4\text{-}8)$$

Thus,

$$\epsilon_f = \ln \frac{A_0}{A_f} = \ln \frac{100}{100 - \% \, RA} \qquad (4\text{-}9)$$

EXAMPLE 4-3

What is the fracture ductility of a material that failed when the stress and strain in it reached 420 ksi and 2%, respectively? $E = 30 \times 10^6$ psi.

Solution

From Eq. 4-6,

$$\epsilon_f = 0.02 - \frac{420,000}{30 \times 10^6} = 0.006 = 0.6\%$$

*Each value is larger than the ones before it.

This material has little ductility, which is not surprising in view of its high strength. Ductility and strength tend to be mutually exclusive.

EXAMPLE 4-4

A handbook gives the following properties for an aluminum alloy: $E = 70$ GPa, $\sigma_f = 350$ MPa, and $\epsilon_f = 0.9$. What is the maximum strain that can be expected in this metal?

Solution

From Eq. 4-6,

$$\epsilon_{tf} = 0.9 + \frac{350}{70\,000} = 0.905$$

Note that this metal is very ductile, and the elastic strain at fracture is negligible.

EXAMPLE 4-5

Identical cylindrical specimens are made of three different materials. The original cross-sectional area is 1 cm² for each. After tension tests, the following final minimum areas are measured: $A_1 = 0.97$ cm², $A_2 = 0.5$ cm², $A_3 = 0.05$ cm². Describe the ductility of each material by the reduction of area.

Solution

$\% \, RA_1 = 3$	This is brittle.
$\% \, RA_2 = 50$	It may be tempting to call this equally brittle and ductile or neither brittle nor ductile; in fact, it is quite ductile because the specimen has necked down to half its original area.
$\% \, RA_3 = 95$	This is very ductile.

Toughness. A concept that combines the strength and ductility properties is called *toughness*. There are several different definitions of this used in strength of materials, so one has to be careful. Basically, a material is called *tough* if it can deform plastically considerably but only if the stresses are high during this deformation. This idea has enormous technological significance because high operating stresses are desirable to reduce the weight of members, and ductility is needed to avoid brittle failures that give no warning and often have catastrophic results.

According to the simplest definition, the toughness is the area under the stress-strain curve. The definition is improved if the true stress-strain curve is specified for this calculation, but a problem remains because plastic strain is only a component of total strain.

There are two equivalent ways to calculate toughness without ambiguity. The area under the true stress versus *plastic* strain curve clearly shows how much plastic strain occurred at what stress levels (as an exercise, plot the stress versus plastic strain curve from the stress versus total strain curve in Fig. 4-5). The other method involves subtracting the area representing elastic strain energy at fracture from the total area under the stress-strain curve. In Fig. 4-5 the elastic strain energy is area $B = \frac{1}{2}(\sigma_f^2/E)$, and the true toughness is area A. Note that an ideally brittle material has zero true toughness according to this definition. It has a substantial area under its stress-strain curve (straight line), but this whole area represents elastic strain energy that is recoverable. True toughness implies irrecoverable work done on the material (the work produces heat during the plastic flow).

Other important concepts of toughness have arisen recently in the area of fracture mechanics. See Chapter 13 for an introduction to these new ideas and how they can be applied in design.

4-5. EFFECTS OF VARIABLES ON MECHANICAL PROPERTIES

The mechanical properties of a material can be affected in innumerable ways. Recently found examples, such as radiation effects on the behavior of materials, justify the idea that there are still some effects to be discovered. Thus, the variables discussed here should be considered only as a beginning for those dealing with the strength of materials. The effects of the variables are shown only qualitatively because the magnitude of the change in a mechanical property may depend considerably on the material. The aim here is to alert the reader to seek further information beyond what is available in standard tables (this may involve much effort and expense) about any chosen materials in these two general situations: (a) There are unusual conditions during fabrication or service and (b) the design requires the saving of material as much as possible. In some cases even the indicated qualitative effects are not true. A different material or one under different conditions may show an opposite tendency in mechanical behavior.

The following variables with respect to mechanical properties of metals are often important to engineers.

1. *Grain Size*

In general, resistance to deformation increases as the microstructure (sizes of grains) is made finer. Cold-working reduces the average grain size, increases

Fig. 4-8. Effect of microstructure on strength and ductility.

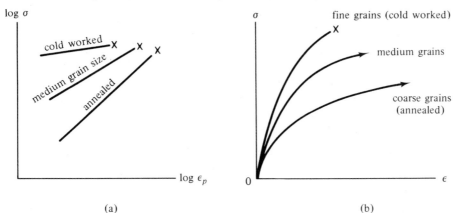

(a)　　　　　　　　　　　　　　　　　　(b)

the yield strength, reduces the fracture ductility, but leaves the fracture strength essentially unchanged, as shown in Fig. 4-8. A cold-worked metal can be softened by reannealing it any number of times. For a simple demonstration of this, bend a heavy copper wire repeatedly until it becomes stiff. Throw it in an open fire for a few minutes, let it cool, and then bend it again.

2. *Hardness*

The effect of hardness on strength and ductility is shown in Fig. 4-9. *Hardness* is a commonly used term, but it does not necessarily refer to a well-defined quantity or measurement. Here, as in many cases, it will mean indentation hardness, the resistance of a material to being indented by a harder, stiffer object. Hardness tests are simple and can be performed quickly. For this reason, much effort has been spent in trying to predict other mechanical properties from the results of hardness tests. Some of these efforts have been fairly successful, but they should be treated only as methods of approximating other mechanical properties. The validity of such approximations should always be checked carefully in the design process.

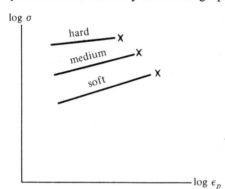

Fig. 4-9. Effect of hardness on strength and ductility.

With increasing hardness of a given material, one could expect an increase in yield strength and fracture strength and a decrease in fracture ductility.

3. *Temperature*

The common effect of temperature is that both the yield strength and the fracture strength decrease while the ductility increases as the temperature increases (Fig. 4-10). There are two important exceptions to this that must be noted here. At mildly elevated temperatures the yield strength of low carbon steel increases as shown in Fig. 4-11. The phenomenon responsible for this behavior is called *strain aging*. This strengthening mechanism depends on plastic deformations and on the diffusion rate of carbon atoms in the steel. Strain aging appears as a factor in several areas of strength of materials.

The other exception to the tendencies shown in Fig. 4-10 is the ductile-brittle transition found mainly in low carbon steels. This is shown schematically in Fig. 4-12, which applies to a metal containing flaws or machined notches. The slope and the horizontal position of the transition part of the curve depend on the metal. In most cases, the transition occurs somewhere

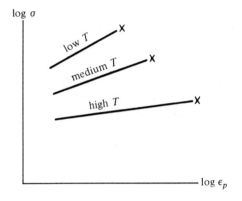

Fig. 4-10. Effect of temperature on strength and ductility.

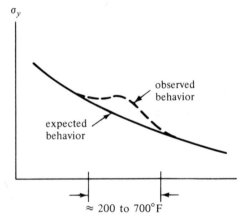

Fig. 4-11. Increase in yield strength by strain aging in mild steel.

Fig. 4-12. Ductile–brittle transition in low carbon steels.

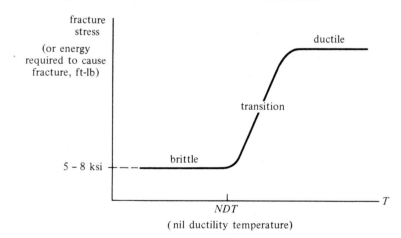

below $+50°$F. There may be less than 20°F between the upper and lower plateaus of the curve. The ductile-brittle transition has enormous practical significance because the metals affected by it are the most common and because the transition temperatures (or those below them) occur in many places in the world. Designing all members to have the average stresses in them below the lower plateau is a safe but uneconomical approach.

4. *Strain Rate*

The rate at which a material is deformed affects its strength, as shown in Fig. 4-13. Both the yield strength and the fracture strength increase and the ductility decreases at high strain rates. These effects are less pronounced in normal practice than those caused by various temperatures.

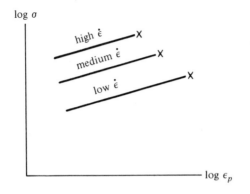

Fig. 4-13. Effect of strain rate on strength and ductility.

5. *Fatigue*

Repeatedly applied loads may cause large changes in the stress-strain curve of a metal. A given metal may cyclically soften or harden compared to its tension stress-strain curve as shown in Fig. 4-14. Which of these happens depends on the initial condition of the material. In most cases, if it was soft (annealed) initially, it will cyclically harden, and vice versa. The strain hardening exponent, *n*, which can be determined in a monotonic test, can

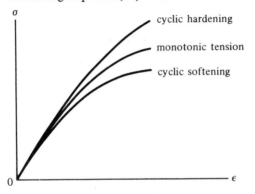

Fig. 4-14. Effect of cyclic loading on stress–strain response.

give an indication of what will happen in fatigue loading. The material will change little if $n \cong 0.15$. There is hardening for n greater than the stable value and softening for n below the stable value.

The sharp yielding of many steels that can be observed during a single tensile loading can be completely bypassed as demonstrated in the following. A single loading to stress σ_A on the tension stress-strain curve in Fig. 4-15

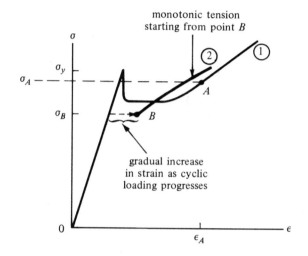

Fig. 4-15. Change in stress-strain curve after some cyclic loading. Note: curves 1 and 2 may or may not cross.

would involve sharp yielding. Assume that tensile and compressive stresses of magnitude σ_B are applied many times to a new specimen of the same metal, after which the stress is raised in a single step. A new stress-strain curve is followed after the cyclic loading as shown in Fig. 4-15.

A curve that shows the deformation resistance of a material after it has been deformed cyclically is called a *cyclic stress-strain curve*. Unfortunately, there is no unique cyclic stress-strain curve for a material even for a given temperature and other conditions. Such curves also depend on the extent of prior deformations; they may, in fact, be in constant change throughout the life of the member. The practical solution to this problem is to use a single cyclic stress-strain curve for each material that represents all possibilities fairly well. A reasonable curve can be obtained from the fatigue test results of a number of identical specimens. Each specimen is subjected to different stress or strain amplitudes that are repeated until failure. The stress amplitude versus the strain amplitude at half the life for each specimen is plotted on a single diagram. The curve through these points shows the effects of cyclic loading on the material's resistance to deformation quite well. Problems 4-30 to 4-54 show many examples of monotonic and cyclic stress-strain curves.

The important thing to remember is that in some cases the stress-strain curve obtained in a unidirectional (tension or compression) test may not be adequate to describe the stress-strain response of a material throughout its life. The cyclic stress-strain curve is generally more suitable for this purpose.

6. *Chemical Effects*

    Many chemicals are detrimental to the mechanical behavior of materials. Two broad categories of chemical effects can be established:

    (a) Continuous corrosive attack that causes an increase in stress because of the loss of material. The mechanical properties of the remaining material are not changed by this.
    (b) Embrittlement of a ductile metal without a significant loss of material. Grain boundary attack and stress-corrosion cracking are important technological problems in this category. The macroscopically measurable ductility and strength are reduced by these. Alloys of aluminum, copper, iron, magnesium, titanium, and others are susceptible to these problems.

7. *Causes of Embrittlement*

    Other causes are not classified normally among the chemical effects. Three of these are mentioned here.

    (a) *Hydrogen embrittlement* is a common and serious problem that affects steels and alloys of titanium and zirconium. The problems are caused by hydrogen introduced into the metals during electroplating or in service (drilling for oil, pipelines and containers of natural gas).
    (b) Some *liquid metals* in contact with certain solid metals may reduce the

Hydrogen-induced cracking in a welded component made of SAE 4340 steel. *A. R. Pfluger and R. E. Lewis, editors*, Weld Imperfections (*Reading, Mass.: Addison-Wesley Publishing Company, 1968*), *p. 582.*

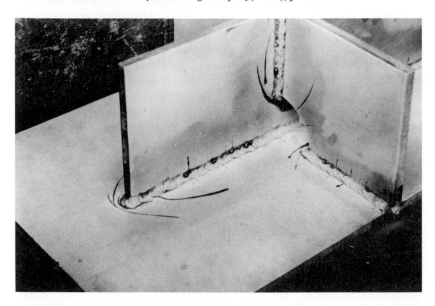

strength of the latter drastically. One should be alert to potential problems with these combinations of solid and liquid metals: steel—In, Li, Cd, or Zn; aluminum alloys—Hg, Ga, Na, In, Sn, or Zn; magnesium alloys—Na or Zn; titanium alloys—Hg; brass—Hg; Zinc—Hg. This list is not exhaustive.

(c) *Neutron irradiation* produces defects at the atomic level in materials. As a result, the yield strength increases (yes, defects can cause this) and the ductility decreases. The embrittlement caused by radiation depends on many variables, and it is always a complex problem.

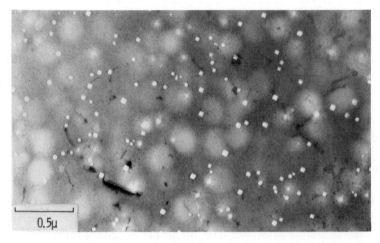

Random defect clusters in nickel irradiated by 3×10^{21} n/cm^2 at 450°C. *Courtesy Professor G. Kulcinski, University of Wisconsin, Madison, Wis.*

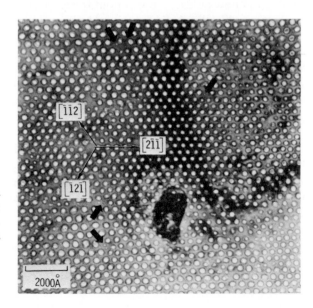

Ordered defects in niobium bombarded with 7.5-MeV Ta ions at 800°C. *Courtesy Professor G. Kulcinski, University of Wisconsin, Madison, Wis.*

8. *Time*

Time is a variable that frequently must be considered when dealing with mechanical properties. Most variables that affect the properties depend on time themselves. A few examples are mentioned here. Microstructure depends on the time that a material spends at certain temperatures (especially at elevated temperatures). Chemical effects depend on time in obvious ways. In fatigue, a certain load may be applied 1000 times to a member to cause failure within a day; if there is a day of rest between applications of the load, the life will not be 1000 cycles, most likely. It could be more than 1000 cycles if the metal is strengthened by strain aging. The cyclic life under the same loads could be reduced if the chemical effects (even corrosion caused by air) become important over the long time the member is in service.

4-6. SIMPLE AND ADVANCED MECHANICAL TESTS

As seen in Section 4-5, the mechanical properties of materials can be affected in many ways. This creates many difficult, challenging problems for engineers. Another way to look at this aspect of dealing with materials is that the number of interesting, intellectually demanding problems is practically inexhaustible and so are the job opportunities for those interested in strength of materials, because new materials are being developed all the time and old materials are frequently used in new applications.

The safe and efficient usage of a material is possible only when its appropriate mechanical properties are known. Unfortunately, these are not always available. In fact, seldom (and only in simple cases) are data on all the necessary properties at hand or sufficiently reliable. For this reason, people are forced sometimes to test prospective materials or to ask others to perform this service for them. In both cases it is desirable to know at least a little about the tests. A few ideas relevant to these are presented here, but this discussion should be considered only an introduction to the subject of mechanical testing of materials.

Simple Tests. Tension, shear, and bending tests are useful and can be performed without much difficulty. The main idea is to have small enough cross-sectional areas of the specimens so that small forces can generate large stresses.

The tests for glues are among the simplest. This is because it is easy to find materials that are stronger than any glue. For a tension test, two bars of any cross-sectional shape but equal areas should be glued together and loaded as shown in Fig. 4-16. The tensile strength of the glue is W_{max}/A. For a shear test the bars can be glued similarly but loaded as shown in Fig. 4-17. The weight should have its center of gravity on a vertical line as close to the glue joint as possible. The interface should be in a vertical plane. The shear strength is W_{max}/A obtained in this position of the bars. The surfaces to be glued must be prepared as in the service application. The results from several tests must be considered (averaged, or taking the lowest

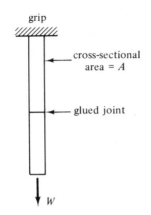

Fig. 4-16. Simple tension test.

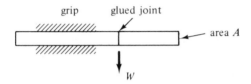

Fig. 4-17. Simple shear test.

values) because there is always some scatter in the results of mechanical tests. Six or more identical tests allow for reasonable confidence in the results.

The tensile strength of a glue can be found from a simple bending test too. The method of doing this will become obvious after studying the flexure formula.

If one wishes to determine the tensile strength of a uniform metal bar, such as those that could be used in the test shown in Fig. 4-16, there is a complication. The bar has to be gripped at two places to pull it apart, and the gripping devices indent and nick the bar. This weakens the bar and it tends to fail within the grips. There are two ways to avoid this problem.

Thin sheets and wires can be wrapped around horizontal cylindrical holders as shown in Fig. 4-18. The cylinders should have as large diameters as possible to reduce bending of the specimen, and they should be fixed to prevent rotation. Two or three turns of the specimen on each cylinder are sufficient to hold the specimen almost entirely by friction.

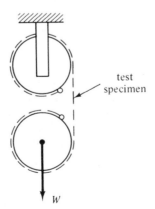

Fig. 4-18. Tension test of thin specimen (rotation of drums must be prevented).

Another method, which is essential for thick and strong specimens, is to design specimens with a reduced test section such as that shown in Fig. 4-19. The wide ends of the specimen can be gripped tightly; they are weakened by this, but the largest stresses are concentrated in the reduced section, and fracture tends to occur there. The tensile tests of most metals require the application of large forces unless the specimens are thin wires.

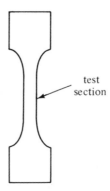

test section

Fig. 4-19. Specimen design to prevent failure of the specimen in the grips.

Advanced Tests. The tests outlined above can be performed with little equipment. Naturally, they are improved if a mechanical testing machine is available. Most laboratories have simple machines that can apply either tension or compression up to about 10,000 lb at least. These will not be discussed in detail here. Instead the equipment and methods of modern testing deserve attention because the ultimate goal should be a complete and accurate characterization of the mechanical behavior of materials.

The essential difference between a simple test and an advanced one is that, in the latter, strains can be measured and controlled if desired throughout the test. Average stresses can be measured or controlled in both kinds of tests. The difference appears trivial at first sight and, indeed, it is possible to find situations when this is the case. On the other hand, the measurement and control of strains is often difficult and requires special equipment and skills, so it serves as a basis for distinction between tests. Furthermore, the accurate determination of stresses also depends on the measurement of strains.

Any technique for the measurement of change of dimension is a technique for measuring strain. These range from the hand-held ruler to the use of X rays. The most commonly used method employs a strain gage, which is now discussed.

Strain Gages. The strain gage is a fabulous invention. It had a humble beginning in the garage of a technician in California and in a few laboratories near the beginning of World War II.* It was one of the major technological advances made in that war, and since that time it has been used universally in experimental mechanics.

*See F. G. Tatnall, *Tatnall on Testing* (Metals Park, Ohio: American Society for Metals, 1966).

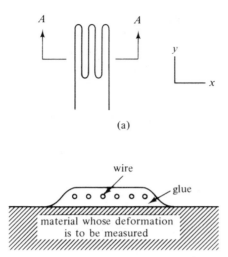

(a)

(b) section *A-A* from (a)

Fig. 4-20. Model of strain gage.

The basic idea of the strain gage is illustrated in Fig. 4-20. A thin metal wire or foil is folded back and forth several times, as in Fig. 4-20a. The wire is glued to the material whose deformation is to be measured (Fig. 4-20b). The glue deforms with the material below it and forces the wire to deform as well. The relatively large surface area of the wire ensures that a sufficiently large traction force can be transmitted to the wire. A tensile deformation in the *y*-direction (Fig. 4-20a) causes the wire to elongate and become thinner. Both of these increase the electrical resistance of the wire. A compressive deformation in the *y*-direction shortens the wire and also makes it thicker. Both of these decrease the electrical resistance. The changes in resistance can be measured quite precisely with the aid of a Wheatstone bridge. These changes are dependent on the strains in the material right below the strain gage. The material and geometry of the gage affect the resistance measurements, but calibrations are made easily to measure strains directly with strain gages.

Many different kinds of strain gages (even liquid metal) are made in sizes of gage lengths ranging from about 0.25 mm (0.01 in.) to a few centimeters. The gages are used in several ways. Many are cemented permanently to laboratory specimens or to structural members. This way the sensitivity of the gages is utilized fully, but each one is sacrificed for the measurements at one place (it is not practical to move a gage to another location or specimen). Furthermore, if there is any doubt about the measured strains after the specimen (and perhaps the gage, as well) has fractured, there is little chance to check whether the gage had been functioning properly.

Strain Rosettes. Biaxial stresses are easily determined experimentally with the aid of strain gages if the directions of the principal stresses are given. In such cases, two gages are required for each point on a member where the strains must be measured. The stresses must be calculated from the strains since there are no stress gages. When the directions of the principal stresses are not known, any unknown strains at a point can be calculated if values of ϵ_x, ϵ_y, and γ_{xy} at the same point can be measured. The problem is that it is very difficult to measure shear strains; the strain gage is not suitable for this purpose. For this reason, a technique has been developed to determine all required strains at a point by measuring linear strains in three directions at the same place.

The method of biaxial strain measurement is explained with the aid of Fig. 4-21. Three strain gages are mounted in the arbitrary directions a, b, and c with

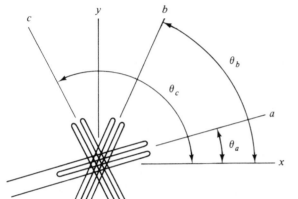

Fig. 4-21. Strain gages in rosette.

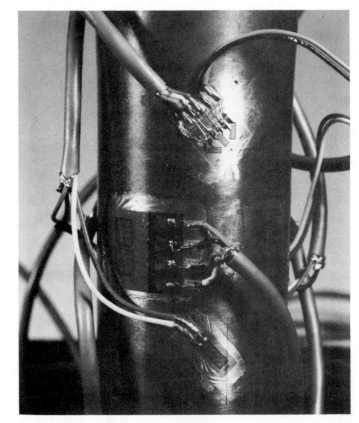

Pressure tube with strain gages and rosettes; student project in experimental mechanics. *Photograph by J. Dreger, University of Wisconsin, Madison, Wis.*

respect to the selected coordinates x and y. The strains in these directions are related according to Eq. 2-4 as follows:

$$\epsilon_a = \frac{\epsilon_x + \epsilon_y}{2} + \frac{\epsilon_x - \epsilon_y}{2} \cos 2\theta_a - \frac{\gamma_{xy}}{2} \sin 2\theta_a$$

$$\epsilon_b = \frac{\epsilon_x + \epsilon_y}{2} + \frac{\epsilon_x - \epsilon_y}{2} \cos 2\theta_b - \frac{\gamma_{xy}}{2} \sin 2\theta_b$$

$$\epsilon_c = \frac{\epsilon_x + \epsilon_y}{2} + \frac{\epsilon_x - \epsilon_y}{2} \cos 2\theta_c - \frac{\gamma_{xy}}{2} \sin 2\theta_c$$

In these equations ϵ_x, ϵ_y, and γ_{xy} are the unknown quantities, and they can be calculated by solving the equations. Further simplification results if the strain gages are mounted in regular geometric patterns. Two such arrangements are commonly used. In the *45° strain rosette*, $\theta_a = 0$, $\theta_b = 45°$, and $\theta_c = 90°$; so

$$\epsilon_x = \epsilon_a, \qquad \epsilon_y = \epsilon_c, \qquad \frac{\gamma_{xy}}{2} = \frac{\epsilon_a + \epsilon_c}{2} - \epsilon_b \qquad (4\text{-}10)$$

In the *60° strain rosette*, $\theta_a = 0$, $\theta_b = 60°$, and $\theta_c = 120°$. With these values,

$$\epsilon_x = \epsilon_a, \qquad \epsilon_y = \frac{1}{3}(2\epsilon_b + 2\epsilon_c - \epsilon_a), \qquad \frac{\gamma_{xy}}{2} = \frac{\epsilon_c - \epsilon_b}{\sqrt{3}} \qquad (4\text{-}11)$$

Note that in each rosette the gages must be electrically insulated from one another, but it is not necessary to mount them on top of each other. It is often sufficient to have them arranged in a fork or delta pattern.

Extensometers. A special technique of using strain gages is in the so-called extensometers (not all of these use strain gages). A model of an extensometer is shown in Fig. 4-22. The flexible, U-shaped beam is temporarily attached to the specimen with spring clips or rubber bands. The strain gage (or gages) is permanently

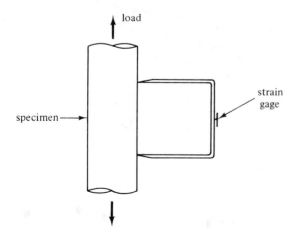

Fig. 4-22. Extensometer.

Strain measurement at elevated temperature; the specimen is heated by induction; the telescopes of the optical extensometer are aimed at two small, horizontal pegs on the specimen; the pegs define the gage length; the load cell is on top. *Harry J. Winslow, "High-Temperature Low-Cycle Fatigue Testing Using Optical Extensometer Control,"* Closed Loop, **2**, *No. 4 (1969), 14–18.*

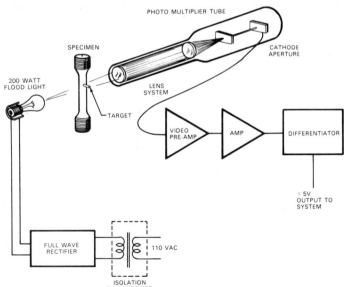

cemented to the flexible beam. The extensometer can be calibrated before and after a test, and it can be used for thousands of tests with proper care. The main problems with it are that its knife edges may slip on the specimen (this changes the gage length), and they may also scratch the specimen.

Load Cells. Another important application for strain gages is in load cells, which are force-measuring devices. In these, the electrical output of the gages is calibrated in terms of the force that deforms the material on which the gages are mounted. Load cells are always in series with the specimens to make sure that all the force acting on a member is transmitted through the load cell.

Testing Equipment. A modern materials testing facility is described schematically with the aid of Fig. 4-23. The roles of all items in this figure must be self-explanatory. Oil is delivered under high pressure from a hydraulic pump through a servo valve to the actuator (hydraulic ram). The servo valve is a high-precision device that can control the flow of oil over a wide range. The valve itself is controlled electronically. The essential feature of the system is the closed-loop feedback control. The dependent and independent variables in the test are monitored continuously, and the equipment makes automatic adjustments in the flow of oil to satisfy the predetermined test condition.

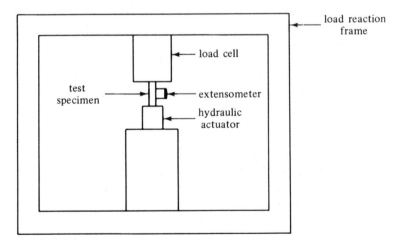

Fig. 4-23. Schematic of specimen in load frame.

For example, a constant strain rate required throughout the test means that strain must increase with time as in Fig. 4-24. For a ductile material this means that the force must increase at a decreasing rate. The equipment with feedback control can take care of such complexities. One selects the desired ramp function for strain and starts the test. The equipment samples the strains frequently, compares the measured strain with what it should be at the time of the sampling, and makes rapid corrections in the flow of oil as necessary.

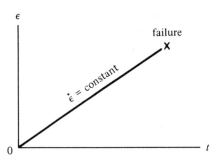

Fig. 4-24. Strain rate as a selected control function.

The items shown in Fig. 4-23 and the hydraulic pump are only part of the total test facility. Other standard equipment includes a function generator (ramp function; sine, square, triangular wave functions), amplifiers for the load cell and extensometer, a cycle counter for fatigue tests, strip-chart recorder, *X-Y* plotter, and an oscilloscope (preferably one that has memory). There is practically no end to the useful equipment that can be added to a servo-controlled machine. Digital voltmeters, amplitude-measurement units, acoustic crack detectors, optical extensometers (especially for testing at elevated temperatures), and crack-propagation measuring devices are among the numerous possibilities.

Automated systems with computer control are rather recent but they are natural extensions to the basic concept of closed-loop control. The most attractive development in this area is the interactive test system. This allows the operator to have complete control of a variety of tests at all times by simply entering statements in native language via a keyboard. Test parameters, data processing routines, and test procedures can be changed quickly. Graphic displays of the program dialog and the interim results are available to improve the communication with the machine. The system also allows additional memory for large programs and data storage, multiple command signals for synchronous or asynchronous operation, analog to digital conversion, and many other options. Figure 4-25 shows a servo-controlled test facility with computer control and three examples of graphic displays of interim or final results. Such systems are expensive, but they are expected to be fairly popular in the near future. They are extremely useful both in basic research and in routine testing for design purposes.

4-7. PROBLEMS

Sec. 4-1

4-1. Prove qualitatively by a hypothetical example that a material with high hardness has low strain hardening exponent.

4-2. How would you determine the strain hardening exponent for a
   (a) ductile material whose entire stress-strain curve is described by a power function?
   (b) low carbon steel that yields sharply?

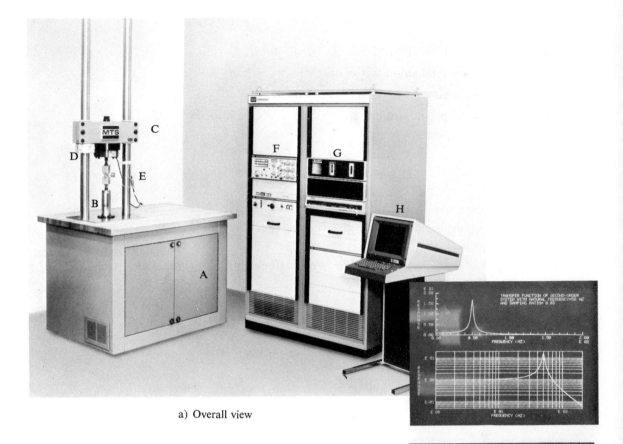

a) Overall view

A – hydraulic pump
B – hydraulic actuator
C – adjustable crosshead
D – load cell
E – extensometer
F – control electronics
G – digital processor and cassette unit
H – graphics terminal with keyboard

Fig. 4-25. Servo-controlled materials testing equipment and sample results from tests. *Herbert C. Johnson*, "*Mechanical Test Equipment in the Sixties: a Decade of Radical Change*," Closed Loop, **4**, *No. 4 (1974), 15–21.*

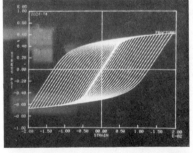

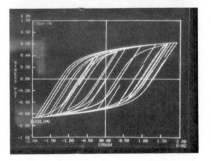

b) Examples of graphic displays

4-3. What is the total strain if $\sigma = 30$ ksi, $\epsilon_p = 2\%$, $E = 30 \times 10^6$ psi?

4-4. What is the total strain if $\sigma = 100$ MPa, $\epsilon_p = 0.08$, $E = 70$ GPa?

4-5. What is the stress if $\epsilon_t = 0.007$, $\epsilon_p = 0.002$, $E = 10^7$ psi?

4-6. What is the plastic strain if $\epsilon_t = 3\%$, $\sigma = 200$ MPa, $E = 120$ GPa?

Sec. 4-3

4-7 to 4-10. Determine the elastic strain energy in Prob. 4-3 to 4-6.

Sec. 4-4

4-11. What is the significance of the fracture ductility (ϵ_f) in manufacturing processes?

4-12. Which of the following affect the percent elongation, and to what extent?
(a) gage length
(b) diameter of specimen
(c) shape of cross section of specimen
(d) ratio of specimen length to gage length

4-13. Which is larger for a ductile metal, the true total strain at fracture or the engineering total strain at fracture?

4-14. Many people call tension or compression tests monotonic tests. Is this correct if the material tested yields sharply?

Sec. 4-5

4-15. What kinds of materials problems would you anticipate if you were designing an oil rig to be used in the Sahara?

4-16. A steel sheet is to be subjected to explosive forming. How will it behave compared to the same steel in a standard tension test?

Sec. 4-6

4-17. What kinds of errors could affect the results of a simple shear test as in Fig. 4-17? What is the most serious error?

4-18. How would you determine the shear strength of a thick, strong metal bar?

4-19. Does the strain gage indicate true strain?

4-20. What should be the physical characteristics of a good strain gage material?

4-21. What is the purpose of the "folds" of the wire or foil in most strain gages?

4-22. How does the Poisson effect influence a measurement made with a strain gage?

4-23. Are knife edges necessary on extensometers such as that shown in Fig. 4-22?

4-24. Should the U-shaped beam of an extensometer be strong or quite weak? Ideally, what kind of a cross section should be selected for the beam? You will be better prepared to answer this after reading Chapter 7; now, rely on your intuition.

4-25. What kind of error do you expect in the measured strain if a knife edge of an extensometer slips on the specimen during (a) a tension test? (b) a compression test?

4-26. Each strain gage has a maximum allowable strain specified for reliable measurements. How would you prevent a strain gage from overstraining if it is (a) mounted on a specimen directly? (b) used in an extensometer (Fig. 4-22)?

4-27. Describe the simplest load cell that you can imagine.

*4-28. Suppose you have a single specimen of a ductile metal that you want to test in tension. The reduced section is a well-machined, uniform cylinder, 4 in. long. Your extensometer has a 1-in. gage length, and you mount it at the center of the reduced section. Unfortunately, the specimen fails outside the gage length, 0.75 in. from the nearest knife edge. Draw the recorded stress-strain curve and what it might have been with the fracture occurring at the middle of the gage length (between the knife edges). In other words, try to construct the data that you hoped to obtain from the results of the test that unfortunately turned out poorly.

4-29. List as many different kinds of extensometers as you can.

General

*4-30. A quenched and tempered SAE 4340 steel ($E = 30 \times 10^6$ psi) is found to behave according to the following stress-strain relation:

$$\sigma = 380\epsilon^{0.1}$$

where σ = true stress in ksi

ϵ = true strain

The true plastic strain at fracture is 0.5. Determine the following: true fracture strength, total true strain at fracture, strength coefficient, strain hardening exponent, 0.2% offset yield strength, percent reduction of area, maximum elastic strain energy (work per unit volume that the metal can do acting as a spring), cyclic hardening or softening.

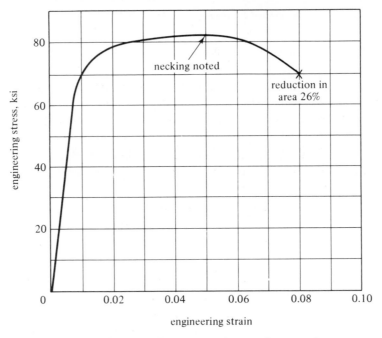

Fig. P4-31. Engineering stress-strain curve for a metal.

4-31. Determine the following properties from the engineering stress-strain curve shown in Fig. P4-31: Young's modulus, 0.1% offset yield strength, 0.2% offset yield strength, ultimate strength, true fracture strength, fracture ductility.

4-32. What could be the reason for the large difference in the curves in Fig. P4-32?

4-33 to 4-56. Figures P4-33 to 4-56 show monotonic and cyclic stress-strain curves for several metals. Some of these are for the same metal with different initial conditions caused by thermal or mechanical treatments. Determine the following: E, approximate range of monotonic strain hardening exponent, and 0.2% offset yield strengths for monotonic (σ_y) and cyclic (σ'_y) loading. Compare and discuss the results for different metals (including different initial conditions but identical compositions) whenever you work on more than one of these problems.

NOTES: 1. The cyclic stress-strain curves are not unique but depend on the cyclic loading history of each specimen.
2. The end points of the curves are not necessarily representing fracture.
3. There are other important mechanical properties for design purposes that cannot be obtained from these curves.
4. Each of the metals may be produced in a different condition from those shown here, and this may affect the mechanical behavior significantly.

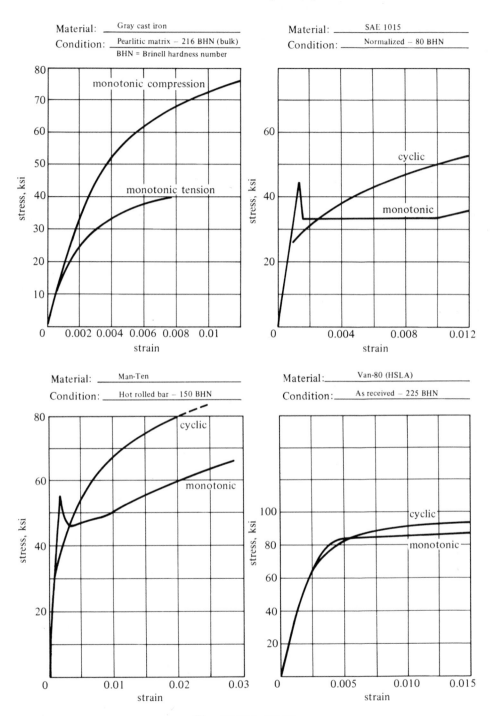

Material: _____Gray cast iron_____
Condition: ___Pearlitic matrix – 216 BHN (bulk)___
___BHN = Brinell hardness number___

monotonic compression

monotonic tension

stress, ksi

strain

Material: _____SAE 1015_____
Condition: ___Normalized – 80 BHN___

cyclic

monotonic

stress, ksi

strain

Material: _____Man-Ten_____
Condition: ___Hot rolled bar – 150 BHN___

cyclic

monotonic

stress, ksi

strain

Material: ___Van-80 (HSLA)___
Condition: ___As received – 225 BHN___

cyclic

monotonic

stress, ksi

strain

Figs. P4-32 to P4-35

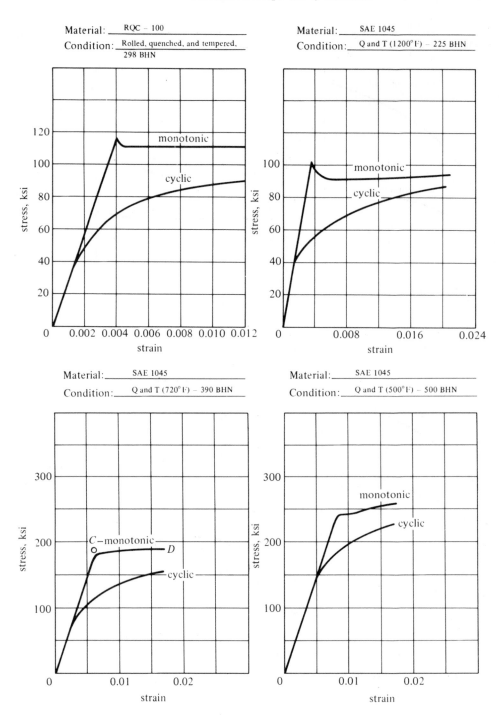

Material: RQC – 100
Condition: Rolled, quenched, and tempered, 298 BHN

Material: SAE 1045
Condition: Q and T (1200°F) – 225 BHN

Material: SAE 1045
Condition: Q and T (720°F) – 390 BHN

Material: SAE 1045
Condition: Q and T (500°F) – 500 BHN

Figs. P4-36 to P4-39

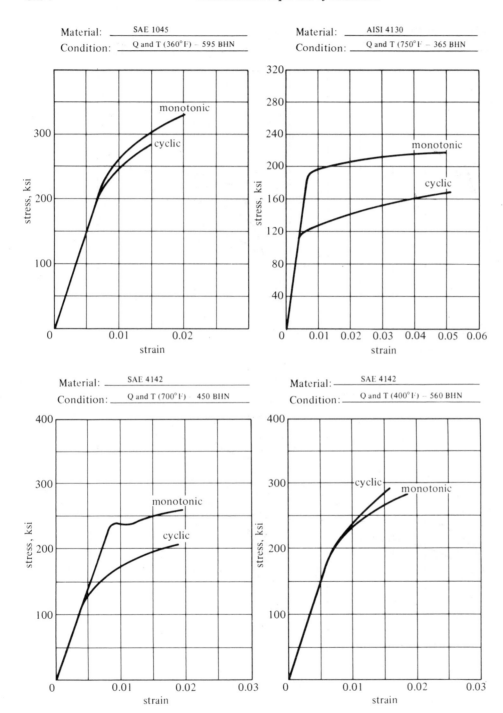

Figs. P4-40 to P4-43

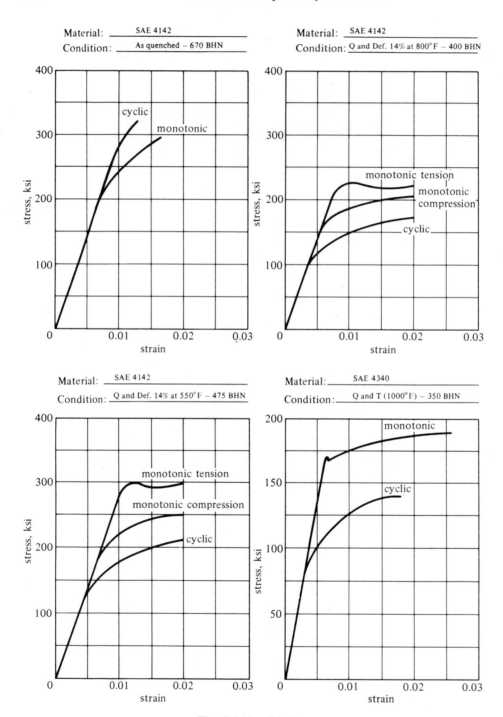

Figs. P4-44 to P4-47

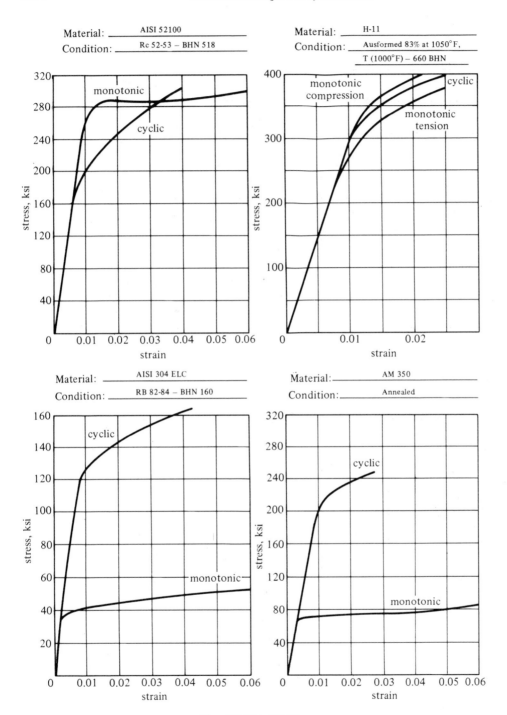

Figs. P4-48 to P4-51

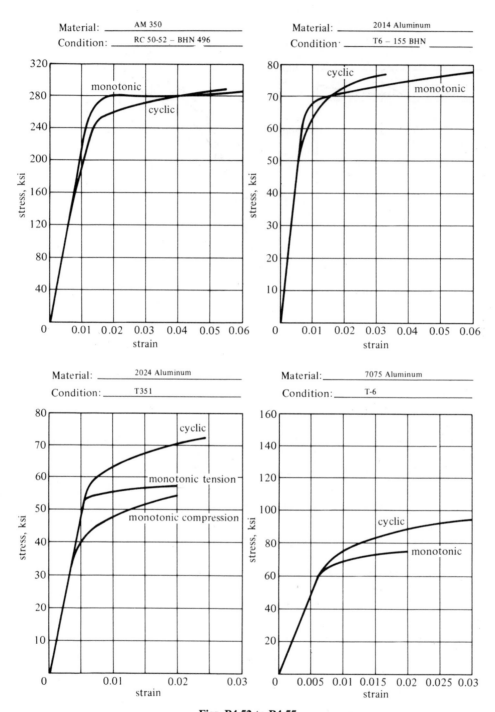

Figs. P4-52 to P4-55

Material: <u>LM-4M Cast Aluminum (British Std. 1490)</u>

Condition: <u>   As cast (specimens)   </u>

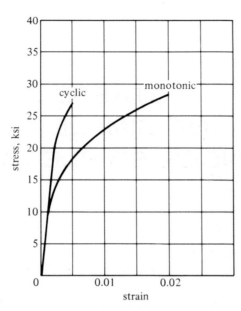

Fig. P4-56

5

STRESS ANALYSIS
OF SIMPLE MEMBERS
UNDER SIMPLE LOADING

The material presented so far can be used either directly or with minor extensions to solve many practical problems. This chapter shows that sometimes apparently unrelated problems can be solved using the same basic concepts. These also serve to introduce several important topics that are normally discussed in introductory courses but not so extensively as to warrant a full chapter for each.

The simplest practical problems in strength of materials are those that involve no bending or twisting of the member. Furthermore, the stresses must be linearly distributed in the member. These conditions imply certain requirements concerning the geometry of the member and the way the loads are applied to it. For example, a two-force member must be straight if bending is to be avoided. The block B in Fig. 5-1 is such a simple member: It is straight, and the compressive forces on it are collinear. On the other hand, the clamp C is not a simple member, even though the forces on it are the reactions from the block: The clamp bends as it is tightened. The stress in the block itself becomes difficult to analyze if there is a hole or other discontinuity in it. These more complicated problems will be discussed in some detail in later chapters.

Even in simple problems where the stresses and strains are related to each other in easily predictable ways it is helpful to distinguish between the independent and dependent variables. This is done by stating the control condition: load (or stress) control or deformation (or strain) control. The differences between these and some of their significances are shown in the following simple examples.

A wire holding a weight is in load control. The force applied to the wire does not depend on the wire in a static situation. The strain in the wire is the dependent variable (function of load, wire size, and material). Thus, various material properties of the wire become of interest depending on one's requirements. Only the

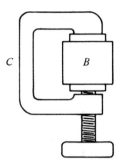

Fig. 5-1. Block in clamp.

fracture (or ultimate strength) properties are of interest if fracture is to be avoided. E, σ_y, and n become of interest if the extent of the deformation must be known. Another example of a member in load control is the rotating blade in a gas turbine.

The same wire (original length L_o) may be in a mechanism where it is stretched to length L_1 to make contact with another object. This is deformation or strain control. If it doesn't make any difference how much force is required in stretching the wire, the only property of interest is the total strain to fracture. If the wire must return to its original length after the deformation, and perhaps must do this repeatedly, the properties of interest are E, σ_y, and σ_y' (specifying an appropriately small offset).

Conditions of strain control are generally not so obvious as those of load control. Further simple examples are strain gages and electroplated metals. Both of these kinds of members are thin and weak relative to the member to which they are attached. Thus, a structural component A may be in load control (Fig. 5-2), but the gage B and the coating C are too weak to contribute to the load-carrying ability of member A. One could say that B and C are simply riding along A. The load P and the size and material of A determine the deformation of A, and this deformation is imposed on B and C, regardless of the stress generated in B and C. Overall, strain control (at least partially) is a very common control condition, but since this cannot be explored further in this text, see Ref. 25.

The ability to distinguish stress and strain control situations and to predict

Fig. 5-2. Different control conditions: A is in load or stress control; B and C are in deformation or strain control.

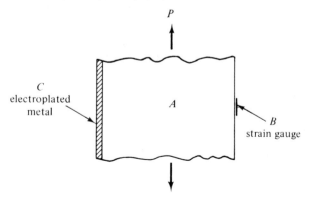

the material behavior in each case is valuable in modern design practices. The analyses and examples in this chapter show the basic patterns of setting up problems and obtaining solutions: The control condition is needed (but it is not always stated as a control condition) to determine what information is available at the beginning of the solution; the geometry and material properties of the member are used in the process of deriving further information.

5-1. TENSION OR COMPRESSION

The methods of solving problems that involve axial tension or compression are quite similar to one another. These will be demonstrated in the following examples.

EXAMPLE 5-1

A coaxial, multichannel transatlantic telephone cable has these general features of construction: The inner core is a bundle of nearly 50, tightly wound, high-strength steel wires. A continuous layer of copper is extruded to form a tight, well-adhering sheath around the steel wire. The copper is surrounded by a thick layer of polyethylene. This insulating layer is enclosed in another copper tube, which is covered by an outermost protective layer of polyethylene. This amazing cable is made in 20-mile long sections by automated equipment. The tensile strength of the

Coaxial, multichannel telephone cable for transatlantic service. *Photograph by B. Boyce, University of Wisconsin, Madison, Wis.*

cable is 20,000 lb, and it is almost entirely because of the steel core (copper and polyethylene have negligible strength in comparison). What is the control condition in the two layers of copper? What is the strain in each copper layer when a tensile load of 10,000 lb is applied to the cable? $E_{steel} = 30 \times 10^6$ psi, $E_{copper} = 17 \times 10^6$ psi. The effective area of the steel is 0.07 in.^{2}.

Solution

The control condition in the copper layers is strain control because they have negligible strength and must stretch along with the steel. The strain in the steel *and* in the copper is

$$\epsilon_{st} = \frac{\sigma_{\text{steel}}}{E_{\text{steel}}} = \frac{10,000}{(0.07)(30 \times 10^6)} = 4.8 \times 10^{-3}$$

EXAMPLE 5-2

A planned surgical implant is a solid cobalt-nickel-molybdenum-chromium alloy cylinder 2 cm in diameter inside a thigh bone 4 cm in diameter. Would a 100-kg person feel the difference in longitudinal stiffness (E) of this member during walking after recovery? To simplify matters, consider the deformation in a 10-cm long segment of the bone with and without the implant. In the latter case, the bone is hollow with $D_i = 2$ cm. $E_{\text{bone}} = 20$ GPa, $E_{\text{alloy}} = 160$ GPa.

Solution

The deformation is given by

$$e = \epsilon L = \frac{\sigma}{E} L = \frac{PL}{AE} \tag{5-1}$$

where $P = 1000$ N
$L = 10$ cm

X-ray picture of porous metal implant in a research animal; the long stem of the implant is embedded in the femur; similar prostheses in humans are important in treating arthritis. *Courtesy M. Dustoor, University of Wisconsin, Madison, Wis.*

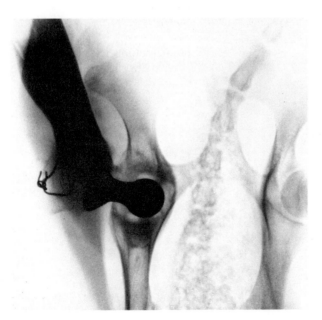
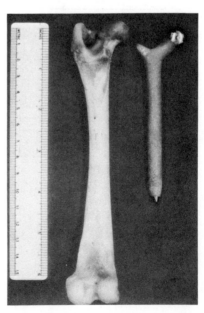

Without the implant,

$$A_1 = \pi(0.02^2 - 0.01^2) = 3\pi \times 10^{-4} \text{ m}^2,$$

$$E_1 = 2 \times 10^{10} \text{ Pa}, \qquad e_1 = 5.3 \times 10^{-6} \text{ m}$$

With the implant, $\qquad\qquad A_2 = 4\pi \times 10^{-4} \text{ m}^2$

and the effective modulus from the rule of mixtures is

$$E_2 = \frac{2(3) + 16(1)}{4} = 5.5 \times 10^{10} \text{ Pa}$$

so $\qquad\qquad\qquad\qquad e_2 = 1.45 \times 10^{-6} \text{ m}$

Both deformations are so small that a person would not feel the difference during walking.

EXAMPLE 5-3

     A straight bar has a rectangular cross section as shown in Fig. 5-3. The monotonic stress-strain curves for the metal are given in Fig. P4-32. The bar is to be subjected to a uniform axial tensile strain of 0.06%. This is the control condition. What is the stress in the bar, the force that must be applied, the location of the force, and the elongation of the bar?

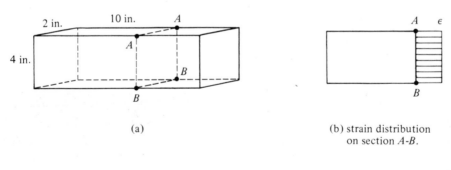

(a)

(b) strain distribution on section *A-B*.

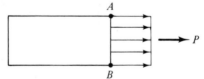

(c) stress distribution on section *A-B* and resultant force.

Fig. 5-3. Example 5-3.

Solution

The stress-strain curve reveals immediately whether the bar will fracture during the required straining. In this case fracture is not expected so it is reasonable to proceed with the solution. The stress can be read off the stress-strain curve, $\sigma = 10$ ksi. This is the axial stress in the bar at any point because the strain is uniform in the bar. The force P is the integral of the stresses over any cross section; in

this case $P = \sigma(\text{area}) = \sigma(4)(2) = 80{,}000 \text{ lb}$

The location of the line of action of the resultant force is the geometric center of the cross section by inspection. The elongation is

$$e = \epsilon L = 0.006 \text{ in.}$$

An alternative formula for the elongation is Eq. 5-1:

$$e = \frac{PL}{AE}$$

This formula is advantageous to use in certain problems but not here. It could not be used if the cross section was not constant or if the material was deforming plastically.

Note that the problem would have been equally meaningful if it had been stated as one of load control: A given force P is applied at the centroid of the cross section. What are the magnitudes and distributions of the stresses and strains, and what is the elongation? Of course, the answers to these are trivial now. Also, the same method is applicable if there is compression instead of tension.

EXAMPLE 5-4

Assume a variation of the preceding problem: The strains are measured on the bar (material RQC-100; see Fig. P4-36) and found to vary along the vertical dimension of the bar as shown in Fig. 5-4a. This is assumed as the control condition. What is the stress distribution in the bar, the magnitude and location of the resultant force, and the elongation?

Solution

The maximum and minimum stresses are found from the stress-strain curve, and the stress distribution is established readily as in Fig. 5-4b. It is obvious that the resultant force P is not at the centroid in this case,* but that it must be closer to A than to B. To find the location of P, its magnitude must be found first. This will be done in two ways.

---

*This is typical when the bar experiences some bending besides the axial deformation (one side of the bar deforms more than the other side). In general, noncentroidal axial loading causes a nonuniform stress distribution.

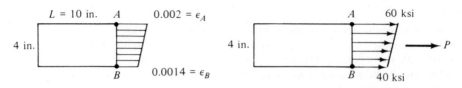

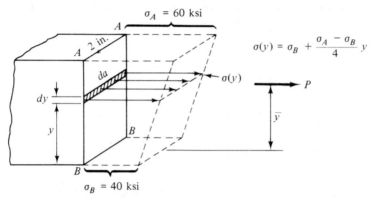

$$\sigma(y) = \sigma_B + \frac{\sigma_A - \sigma_B}{4} y$$

(c) stress distribution $\sigma(y)$ on element *da*
and resultant force for whole area.

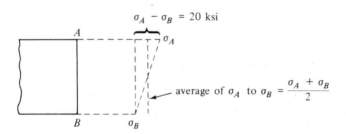

(d) average stress on section *A-B*.

Fig. 5-4. Example 5-4.

Direct integration is done using Fig. 5-4c. The force acting on the elemental

strip *da* is $dP = \sigma(y)da = \sigma(y)2dy$

where $\sigma(y)$ is the stress on the strip *da*; it is assumed a constant over *da*.

$$P = \int_B^A dP = 2 \int_0^4 \sigma(y)dy = 2 \int_0^4 (40 + 5y)dy = 400 \text{ kips}$$

The other method is based on the idea of averaging stresses over the cross section. The given stress distribution is considered as the sum of a uniform distribution σ_B and a triangular one with a peak $\sigma_A - \sigma_B$ as shown in Fig. 5-4d. Thus,

$$P = \sigma_B(4)(2) + \frac{\sigma_A - \sigma_B}{2}(4)(2) = 400 \text{ kips}$$

so the two methods yield the same result.

The location of the line of action of P is found using Fig. 5-4c again. The stress distribution is symmetric with respect to the vertical center line of the bar, so P must intersect this line somewhere. The problem is to find the distance $\bar{y}$ from the line B-B to the place where P is. This is done by writing moments in two different ways for the same cross section and equating them (this is valid since there can be only one moment physically). One way of writing the moment about the line B-B is

$$M_B = P\bar{y} = 400\bar{y}$$

The other way is to sum the infinitesimal moments caused by the stresses on elements such as da in Fig. 5-4c:

$$M_B = \int_B^A \sigma(y)(da)y$$

where the elemental force $\sigma(y)(da)$ is multiplied by the moment arm y. Equating the right sides of the last two equations, $\bar{y}$ is found:

$$\bar{y} = \frac{1}{400}\int_0^4 \sigma(y)day = \frac{1}{200}\int_0^4 (40 + 5y)ydy = 2.14 \text{ in.}$$

The calculation of the elongation presents a dilemma because the strains are not uniform over the cross section. A plausible idea is to work with the average

strain: $e = \epsilon_{\text{ave}}L = \dfrac{\epsilon_A + \epsilon_B}{2}L = 0.017 \text{ in.}$

Of course, the top layer of the whole bar experiences a larger elongation:

$$e_A = \epsilon_A L = 0.02 \text{ in.}$$

while the bottom layer has a smaller elongation:

$$e_B = \epsilon_B L = 0.014 \text{ in.}$$

Thus, there is a net elongation of the bar and a little curvature (one could say that this is caused by the noncentroidal loading). This seems to violate the condition of no bending in a simple problem that was stated in the beginning of this chapter. The problem is not serious since the loading is axial and the member was straight initially so its curvature under the loads is negligible.

EXAMPLE 5-5

A bar with the geometry shown in Fig. 5-5a is loaded in compression and the strain distribution is determined experimentally as that in Fig. 5-5b. The material is SAE 1045, 390 BHN. Assume that the monotonic stress-strain curve shown in Fig. P4-38 is also valid for compressive loading. What is the stress distribution in the bar, the magnitude and location of the resultant force, and the elastic stored energy?

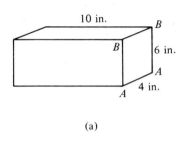

(a)

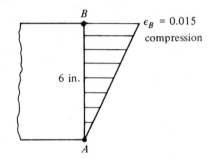

(b) strain distribution.

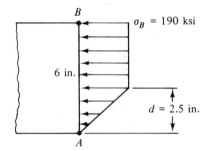

(c) stress distribution

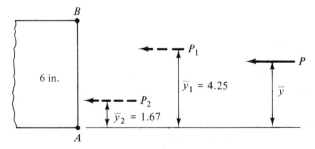

(d) lines of action of resultant forces.

Fig. 5-5. Example 5-5.

Solution

It is found from the stress-strain curve that $\sigma_A = 0$ and $\sigma_B = 190$ ksi. The strain varies linearly from A to B, but the stress is not linearly related to strain. To find the stress distribution, it is convenient to assume that the stress-strain curve has two linear parts: one from 0 to C, the other from C to D in Fig. P4-38. Thus, the stress increases linearly from zero to 190 ksi at the strain of 0.00625, and it remains 190 ksi for all strains larger than 0.00625 (Fig. 5-5c). The location (d) of the break in the stress distribution is found from the proportionality of strains:

$$\frac{0.015 \text{ in./in.}}{6 \text{ in.}} = \frac{0.00625 \text{ in./in.}}{d}, \qquad d = 2.5 \text{ in.}$$

The resultant force is the sum of those caused by the two segments of the stress distribution:

$$P = P_1 + P_2 = 190 \text{ ksi } (3.5 \text{ in.})(4 \text{ in.}) + \frac{190}{2} \text{ ksi } (2.5 \text{ in.})(4 \text{ in.})$$

$$= 2660 \text{ kips} + 950 \text{ kips} = 3.61 \times 10^6 \text{ lb}$$

The location ($\bar{y}$) of the line of action of P is found by writing moments in two different ways for the same cross section and equating them. This time it is convenient to use a simple method (whose validity could be proved by using integration): The resultant force acts at the centroid of a given stress distribution. From Fig. 5-5d,

$$P\bar{y} = P_1\bar{y}_1 + P_2\bar{y}_2$$

$$\bar{y} = \frac{2660 \text{ kips } (4.25 \text{ in.}) + 950 \text{ kips } (1.67 \text{ in.})}{3610 \text{ kips}} = 3.57 \text{ in.}$$

The elastic stored energy in a unit volume depends on the local stress and E:

$$\text{Energy per unit volume} = \frac{1}{2}\frac{\sigma^2}{E}$$

The total energy W in the bar is calculated in two steps. In the upper 3.5 in.,

$$\text{total energy, } W_1 = \frac{1}{2(30)(10^6) \text{ psi}}(190 \text{ ksi})^2(3.5)(4)(10 \text{ in}^3) = 7000 \text{ ft-lb}$$

For the lower 2.5 in. of the bar, use integration because the stress is not a constant throughout the volume. The stress is constant only in a thin slab of area 10 in. $\times$ 4 in. and thickness dy. The energy in this slab is a function of its location (y) above the lowest layer A:

$$dW_2 = \frac{1}{2}\frac{[\sigma(y)]^2}{E}(10 \text{ in.})(4 \text{ in.})(dy)$$

where $\sigma(y) = (190 \text{ ksi}/2.5)y$. The total energy is

$$W_2 = \int_0^{2.5} dW_2 = \frac{40 \text{ in}^2}{2(30)(10^6) \text{ psi}}\left(\frac{190{,}000 \text{ psi}}{2.5 \text{ in.}}\right)^2 \int_0^{2.5} y^2 dy = 1700 \text{ ft-lb}$$

The stored energy in the whole bar is

$$W = W_1 + W_2 = 8700 \text{ ft-lb}$$

This would be enough energy to propel a 150-lb person nearly 60 ft upward!

EXAMPLE 5-6

A 4-m long bar (cross section, 3 cm × 10 cm) of 2014-T6 aluminum alloy (see Fig. P4-53) is hinged at one end as shown in Fig. 5-6a. There are no horizontal forces on the bar in this position. The bar is slowly lowered into a vertical position. What are the maximum stress and strain and the elongation in the new position? What change in temperature would be necessary to have zero elongation in the vertical position? Assume that the hinge causes no change in the stress distribution in the bar.

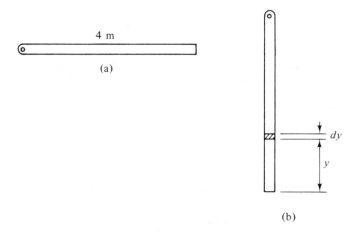

(a)

(b)

Fig. 5-6. Example 5-6.

Solution

The bar in the vertical position is in load control. The maximum stress is at the hinge where the stress is caused by the weight of the whole bar. It is found from a handbook that the aluminum weighs 0.1 lb/in³. Thus,

$$\sigma_{\max} = \frac{\text{total weight}}{\text{area}}$$

$$= 12{,}000 \text{ cm}^3 (0.1 \text{ lb/in}^3)(4.45 \text{ N/lb})\left(\frac{1}{16.4 \text{ cm}^3/\text{in}^3}\right)$$

$$\times \left(\frac{1}{30 \text{ cm}^2}\right)\left(\frac{10{,}000 \text{ cm}^2}{\text{m}^2}\right)$$

$$= 108.5 \frac{\text{kN}}{\text{m}^2} = 16 \text{ psi}$$

This stress is too small to enter in Fig. P4-53 for determining the maximum strain. E can be found, however, from the stress-strain curve:

$$E = \frac{50{,}000 \text{ psi}}{0.005} = 10^7 \text{ psi}$$

and

$$\epsilon_{max} = \frac{\sigma_{max}}{E} = \frac{16 \text{ psi}}{10^7 \text{ psi}} = 1.6 \ (10^{-6})$$

The total elongation is the sum of the strains integrated over the length of the bar. These strains vary from zero at the lower end to ϵ_{max} at the top of the bar. At any position (y) along the bar the strain in an infinitesimal gage length (dy) is

$$\epsilon(y) = \frac{\sigma(y)}{E} = \frac{\text{weight in length } y}{(\text{area})E} = \frac{wAy}{AE}$$

where w is the specific weight of the material. The total elongation is

$$e = \int_0^L \epsilon(y)\,dy = \int_0^L \frac{wAy}{AE}\,dy = \frac{wL^2}{2E}$$

$$e = \frac{0.1 \text{ lb/in}^3 (158 \text{ in.})^2}{2(10^7 \text{ lb/in}^2)} = 1.24(10^{-4}) \text{ in.} = 3.17(10^{-3}) \text{ mm}$$

The coefficient of thermal expansion is found from a handbook: $\alpha = 13 \times 10^{-6}$ per °F. From Eq. 3-4, a lowering of the temperature of the bar by

$$\Delta T = \frac{e}{\alpha L} = \frac{1.24 \times 10^{-4}}{13(10^{-6})(158)} = 0.06°\text{F}$$

is sufficient for countering the elongation caused by the weight in the vertical position. Thus, changes in length caused by normal fluctuations in temperature are much larger than that caused by the change in position.

EXAMPLE 5-7

What is the strength of a chain that is made of welded links of 5-mm diameter steel rods? The tensile strength of the metal is 800 MPa, and its shear strength is 500 MPa. Use the simplest concept of stress analysis. Should the answer be considered as higher or lower than the real strength of the chain?

Solution

There are two simple ways to solve this problem. Each link can be considered either as a two-branched tension member or as a rod in double shear (shearing on two parallel planes) at the point where the load is transmitted from a neighboring link. The total area that carries the load is twice the cross section of the 5-mm rod in both cases, so the shear strength is the limiting stress. On this basis, the maximum

load is $P = 500(10^6) \text{ N/m}^2 (2)\left(\frac{\pi}{4}\right)(0.005)^2 \text{ m}^2 = 19\,600 \text{ N}$

The real strength must be lower than this because the links tend to elongate under large loads, and there must be stresses in them other than those caused by simple tension and shear.

5-2.　FASTENERS IN SHEAR

Many kinds of fasteners such as bolts, rivets, and nails are used in engineering practice. The strength of a structure is limited by the strength of the connected joints in many cases. The analysis of a fastener's strength may be complicated if there are axial, shear, bending, and torsional loads on it simultaneously. The simple case of axial loading alone can be handled with the information presented already. The next step is to consider shear loading alone.

Figure 5-7a shows two thin plates held together by a single rivet. The rivet is being sheared at the interface of the plates, but the misalignment of the loads also causes some bending of the rivet. Assume here that the plates are thin so that the misalignment and bending are negligible.

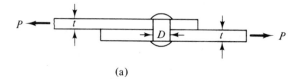

(a)

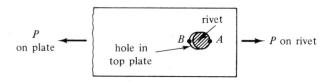

(b) cross section of top plate with rivet in it.

Fig. 5-7. Example 5-8.

There are two stresses in this basic problem that can be calculated easily. One is the *average* shearing *stress* τ *in the rivet* on the plane of the interface of the plates. This stress is acting on the cross-sectional area of the rivet, so

$$\tau = \frac{P}{(\pi D^2/4)}$$

The other stress is called a *bearing* or *contact* stress and it acts between the rivet and the plates. It is obvious from Fig. 5-7b that this stress can vary considerably over the contact area. It is always maximum at point A and minimum (sometimes zero) at point B. The distribution of this stress and the magnitude of the maximum value depend on the materials involved, the geometries of the parts, and the load. Only the average value of the bearing stress is easy to calculate because it is the force divided by the *projected* area of the rivet (or of the hole in the plate). Thus, the average bearing stress for the problem shown in Fig. 5-7 is

$$\sigma_B = \frac{P}{tD}$$

In many structures numerous fasteners are used to connect each pair of members. It is important in such cases to have the fasteners share evenly in the transmission of the load. This is difficult to achieve with a high level of confidence. The ability of the materials to deform plastically is helpful in distributing the load among the fasteners evenly. Uniform distribution of the load is often assumed in solving problems.

EXAMPLE 5-8

In Fig. 5-7 each plate is 1 mm thick and 20 cm wide. The load P is 17 kN. The available rivets are 5 mm in diameter and are believed to have a shear strength of 400 MPa. What is the minimum number of rivets necessary if each carries the same fraction of the load?

Solution

One rivet can support a load of

$$P_1 = 400 \text{ MN/m}^2 \left(\frac{\text{m}^2}{10^6 \text{ mm}^2}\right)\left(\frac{25\pi}{4} \text{ mm}^2\right) = 7.85 \text{ kN}$$

At least three rivets are necessary.

EXAMPLE 5-9

A flat bar A is connected to a U-shaped bracket as in Fig. 5-8 by a 1-in. diameter bolt that goes through the three plates. The latter are 0.5 in. thick. What are the bearing and shear stresses in the bar, bolt, and the bracket when a 1000-lb tensile load is applied to the bar?

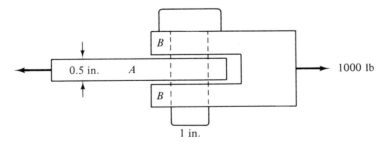

Fig. 5-8. Example 5-9.

Solution

Bar *A* causes a double shear (shearing on two planes) of the bolt, so the average shear stress in the bolt is

$$\tau_{ave} = \frac{500}{\pi(0.5)^2} = 637 \text{ psi}$$

This shear stress is acting in the bolt between *A* and *B*. The bearing stress in bar *A*, or on the bolt within *A*, is

$$\sigma_{ave_A} = \frac{1000}{(0.5)(1)} = 2000 \text{ psi}$$

The bearing stress in *B*, or on the bolt within *B*, is

$$\sigma_{ave_B} = \frac{500}{(0.5)(1)} = 1000 \text{ psi}$$

EXAMPLE 5-10

A flat plate specimen of a soft alloy is to be tested in tension by pulling on a round pin inserted through each end of the plate. What is the required diameter of the pins if the plate is 0.5 cm thick, the maximum tensile load is 6 kN, and the allowable average bearing stress is only 3 MPa? How does the answer affect the choice of material for the pins or the method of applying the tensile load to the plate?

Solution

Using $\sigma_{ave} = P/0.005D$, the diameter *D* of the pin is

$$D = \frac{6000}{(0.005)(3 \times 10^6)} = 0.4 \text{ m}$$

The shear load on the pin is 2×3000 N, so

$$\tau_{ave} = \frac{3000}{\pi(0.2)^2} = 24 \text{ kPa}$$

The shear stress in the pin is very low, so any cheap, common steel or aluminum would be adequate for the pin. The pin diameter is so large, however, that it is impractical. Another method should be found for applying the load to the specimen (for example, friction grip).

EXAMPLE 5-11

The uniform, circular door of a bank vault is designed to swing on two simple hinges (tongue and fork) that are arranged on a vertical center line tangent to the door. The door is 8 ft in diameter and weighs 50,000 lb. The hinges are 5 ft apart.

What is the minimum diameter of the hinge pins if the steel selected has a tensile and compressive strength of 120 ksi and a shear strength of 70 ksi? The bearing surface in each hinge is 4 in. long.

Solution

From statics,

$$2V_1 = 2V_2, \quad \text{or} \quad V_1 = V_2 = V \quad \text{(see Fig. 5-9)}$$

$$50,000(48) = 2V(60)$$

$$V = 20,000 \text{ lb}$$

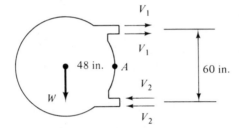

Fig. 5-9. Example 5-11.

The allowable average shear stress is

$$\tau_{ave} = \frac{V}{\pi R^2}, \quad \text{so} \quad R = \sqrt{\frac{20,000}{70,000\pi}} = 0.3 \text{ in.}$$

The bearing stress on a pin of this size is

$$\sigma_{ave} = \frac{2V}{LD} = \frac{40,000}{4(0.6)} = 16.7 \text{ ksi}$$

which is well below the allowable normal stress.

5-3. WELDED JOINTS

There are many techniques of welding used in engineering practice. With the exception of the rare cold welding in vacuum, these imply the local heating and fusing of members. The heat is from an electric current most often, but oxyacety-lene torches or friction (inertia welding) are also used. In practice, high strength of a weldment should seldom be taken for granted. The heating and cooling processes affect the microstructure of the weld region, dirt and gases may be entrapped, and the finished product is not free of internal stresses. In critical applications it is necessary to control the welding process carefully, to perform destructive sampling, and to inspect all weldments of the final product using nondestructive testing. For simplicity, it will be assumed here that the weldments are homogeneous, are free of internal stresses, and have uniform mechanical properties.

The principal kinds of welds are butt welds and fillet welds, and these are

(a)

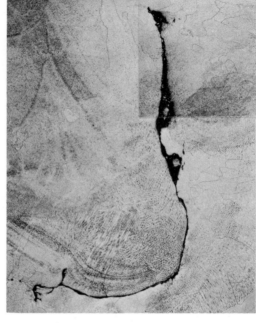

(b)

Incomplete fusion in a copper weld at different magnifications: (a) 2.5X, (b) 25X. *A. R. Pfluger and R. E. Lewis, editors,* Weld Imperfections (*Reading, Mass.: Addison-Wesley Publishing Company, 1968), p. 158.*

Radiograph of entrapped weld spatter flaws in stainless steel. *A. R. Pfluger and R. E. Lewis, editors,* Weld Imperfections (*Reading, Mass.: Addison-Wesley Publishing Company, 1968), p. 166.*

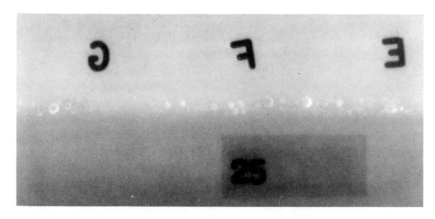

126

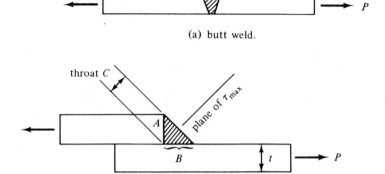

(a) butt weld.

throat C

plane of τ_{max}

A

B

t

P

(b) fillet weld (A and B are called *legs*).

Fig. 5-10. Common welds.

shown schematically in Fig. 5-10. The first of these is simple to analyze if the strength of the weld metal and the load-carrying cross section are given. The fillet weld is slightly more complicated. It may appear at first from Fig. 5-10b that area A is in tension and area B is in shear and that one of these is the critical region. The maximum shear stress is on a plane between A and B, however, where the weld metal is thinnest. Since shear strength is normally lower than tensile strength, this plane inclined to the plates is taken as the critical plane. For plates of identical thickness, t, the shearing area per unit length of weld is $t \sin 45°$. This is valid for welds parallel or transverse to the applied load.

EXAMPLE 5-12

Two plates of thickness t and width w are to be welded together to resist tensile loading. Compare the strength of a simple butt weld to that of a fillet weld as shown in Fig. 5-10b. Assume that the shear strength of the weld is 65% of the tensile strength, σ_u, in terms of stresses. Also assume that the strengths of the weld are the same as the respective strengths of the plates.

Solution

The tensile strength of the butt weld, assuming that the thickness at the weld

is t, is
$$P_{t_1} = tw\sigma_u$$

Its shear strength is slightly lower than this:

$$P_{s_1} = \frac{t}{0.707} w(0.65\sigma_u)$$

For the fillet weld,

$$P_{t_2} = tw\sigma_u, \qquad P_{s_2} = 0.707tw(0.65\sigma_u) < P_{s_1}$$

EXAMPLE 5-13

A simple ladder is to be made by welding rectangular rungs to two vertical bars that are attached to a tall container of chemicals. Each rung has a cross section 1.5 cm × 4 cm, and the wider side is vertical. What is the minimum length of weld on each end of a rung if the weld legs are 2 mm, the maximum load on each weld is 2 kN, and the shear strength of the weld is at least 80 MPa? Assume that in each weld the stresses are uniformly distributed. The arrangement is similar to that in Fig. 5-10b, with the opposing forces to be shown perpendicular to the plane of the paper.

Solution

The shear area per unit length of weld is $A = 0.2 \sin 45° = 0.1414 \text{ cm}^2$. The required length is

$$L = \frac{2000 \text{ N}}{0.1414(10^{-4}) \text{ m}^2/\text{cm} \, (80 \times 10^6) \text{ N/m}^2} = 1.77 \text{ cm}$$

EXAMPLE 5-14

A cylindrical tube of 2-in. outside diameter is inserted in another tube of 2-in. inside diameter, and a fillet weld is made circumferentially to join them. What is the minimum tensile strength of the joint (axial loading on the tubes) if the shear strength of the weld is 15 ksi? The legs of the weld are 0.2 in. A model for the problem is Fig. 5-10b.

Solution

The shear area is approximated by $A = 2\pi(0.2 \sin 45°) = 0.88 \text{ in}^2$:

$$P > 15 \text{ ksi } (0.88 \text{ in}^2) = 13.3 \text{ kips}$$

5-4. THIN-WALLED PRESSURE VESSELS

The stresses in certain pressure vessels can be analyzed using simple techniques. One of the conditions is to have relatively thin walls. A vessel is generally classified as thin-walled if its radius is at least 10 times larger than its wall thickness. Another limitation is concerning discontinuities. For example, at the ends of closed cylinders the end plates and the cylinder constrain each other. This is shown in Fig. 5-11. At the middle region of the cylinder things are relatively simple. A small element B feels a longitudinal stress, σ_l, and a circumferential stress σ_c. These stresses can be related to the internal pressure and the dimensions of the shell.

The stress σ_l is the easiest to find because it is caused by the internal forces acting on the end plates:

$$\sigma_l = \frac{pA}{D\pi t} = \frac{pD}{4t} \tag{5-2}$$

Construction of liquefied natural gas containers on a barge; note that for initial stress analysis the tanks are thin-walled; however, the walls are thick enough to cause brittle failures (see Chapter 13). *Engineering Case Library, Stanford University, and Chicago Stockyards Research Division. Courtesy Professor Henry O. Fuchs, Stanford University, and Dr. Sterling Beckwith, Menlo Park, Calif.*

Pressure vessel with 6-in. thick weld hydrotested to failure at 5000 psi; a split originated in a girth weld. *A. R. Pfluger and R. E. Lewis, editors,* Weld Imperfections (*Reading, Mass.: Addison-Wesley Publishing Company, 1968), p. 334.*

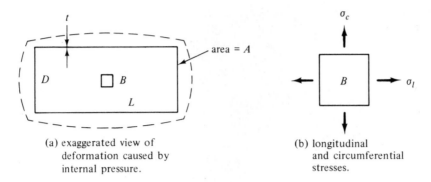

(a) exaggerated view of
    deformation caused by
    internal pressure.

(b) longitudinal
    and circumferential
    stresses.

Fig. 5-11. Thin-walled pressure vessel.

where pA = total force caused by internal pressure on an end plate

  $D\pi t$ = cross-sectional area of cylinder

Obviously, the radial deformation of the vessel must be small, and t must be small compared to D in Eq. 5-2.

The circumferential stress (also called hoop stress) can be determined by considering the equilibrium of a small semicircular element of the cylinder as shown in Fig. 5-12. The element is in equilibrium in the direction perpendicular to the paper because of σ_l acting in that direction. It can be proved, using elementary calculus, that the resultant forces caused by the internal pressure in Fig. 5-12a and

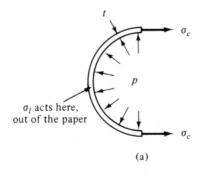

(a)

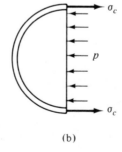

(b)

Fig. 5-12. Semicircular section from middle of cylinder shown in Fig. 5-11.

b are identical. Thus, the total force to the right is $2(\sigma_c t dl)$ and to the left it is $pDdl$ (Fig. 5-12b). These are equal for equilibrium in the horizontal direction, so

$$\sigma_c = \frac{pD}{2t} \tag{5-3}$$

The limitations are the same in using Eq. 5-2 and 5-3.

EXAMPLE 5-15

A new cable is designed for an advanced communications system. To prevent the cable from absorbing any moisture, it is planned to put the cable inside an air-tight jacket and pressurize this with dry nitrogen gas. Any flaws in the jacket would allow gas to escape (the drop in pressure can be used to signal the presence of a hole) but moisture could not get in if there is a sufficient supply of the dry gas. What is the required thickness of the jacket if its diameter is 4 cm, the tensile strength is 1 MPa, and the gage pressure is 2 kPa?

Solution

$$\sigma_c = \frac{pD}{2t}, \qquad 1 \text{ MPa} = \frac{0.002 \text{ MPa}(4 \text{ cm})}{2t}, \qquad t = 0.004 \text{ cm}$$

EXAMPLE 5-16

The research and development laboratory of a light bulb manufacturer is experimenting with bulbs that are filled with Krypton gas. Each of these bulbs is spherical with 2-in. diameter except for a small terminal base cemented to the glass. What is the minimum thickness of the glass if the pressure in the bulb is

(a)   5 psi below atmospheric pressure?
(b)   50 psi above atmospheric pressure?

Assume that the maximum safe stress in the glass is 5000 psi for both tension and compression.

Solution

A sphere has the same stresses, in all directions along the surface, as computed for σ_l in a cylindrical vessel. Hence,

$$\sigma = \frac{pD}{4t}$$

and

(a) $-5000 = \dfrac{-5(2)}{4t}, \qquad t = 0.0005 \text{ in.}$

(b) $5000 = \dfrac{50(2)}{4t}, \qquad t = 0.005 \text{ in.}$

EXAMPLE 5-17

A student experiments with simple models of structural components that could be used in an inflatable airplane. One such component is a 2-m long, 10-cm diameter, cylindrical tube, homemade from 0.001-cm thick Mylar. The tube is to be pressurized 70% above atmospheric pressure. What is the required shear strength of the adhesive for making the tube from the sheet material if the longitudinal seam of the tube has at least a 0.5-cm overlap? The Mylar edges glued together have a contact area 2 m long and 0.5 cm wide.

Solution

Consider the cross-sectional element in Fig. 5-13. For a section of length dl,

$$\tau(0.5)dl = \sigma t dl$$

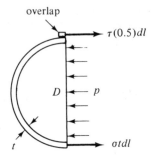

overlap

$\tau(0.5)dl$

D p

t $\sigma t dl$

Fig. 5-13. Example 5-17.

Summing forces,

$$pDdl = 2\tau(0.5)dl, \qquad \tau = 0.7 \text{ atm } (10^5 \text{ Pa/atm})(10 \text{ cm}) \text{ cm}^{-1} = 0.7 \text{ MPa}$$

EXAMPLE 5-18

Part of a proposed spacecraft is a windowlesss, closed cylinder. It is to be made of a thin shell of a filament-wound composite material. The high-strength fibers of

Fig. 5-14. Example 5-18.

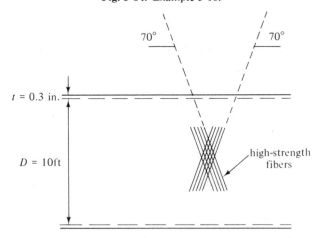

$70°$ $70°$

$t = 0.3$ in.

$D = 10$ft

high-strength fibers

the shell are in several layers, each fiber at 70° with the longitudinal axis of the cylinder (cross-plied fibers) as shown in Fig. 5-14. The internal pressure is 14 psi. Does the shell have adequate strength if the tensile strength in directions normal to the fibers is 6 ksi and the shear strength in directions parallel to the fibers is 4 ksi?

Solution

The principal stresses are

$$\sigma_l = \frac{pD}{4t} = \frac{14(10)(12)}{4(0.3)} = 1400 \text{ psi}, \qquad \sigma_c = \frac{pD}{2t} = 2800 \text{ psi}$$

Using a plot of Mohr's circle, one can prove that σ_c is the largest normal stress in the shell and that $\tau_{\max}$ is 700 psi. Both of these are well below the given strength values. Of course the stresses in the given directions of the fibers are even smaller than these.

Deep submergence rescue vehicle; the formulas for thin-walled pressure vessels must be used with caution for submarine vehicles; there are special problems with buckling of the hulls, thick walls, corrosion, and crack propagation even under repeated compressive loading. *Lockheed Missiles & Space Company, Sunnyvale, Calif.*

5-5. PROBLEMS

Sec. 5-1

5-1. A concrete column with a 10 in. × 10 in. cross section is standing in a nearly vertical position. The only external load is its weight. The slight leaning causes a strain distribution at the base of the column as shown in Fig. P5-1. What is the stress distribution and the magnitude and location of the resultant force? $E = 3 \times 10^6$ psi.

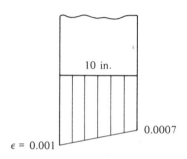

10 in.

$\epsilon = 0.001$

0.0007

Fig. P5-1

5-2. Calculate the length of the column in Fig. P5-1. What is the elongation if the column is put in a horizontal position on rollers?

5-3. Consider two long rods of different materials that are connected to form a single straight rod and hung vertically as shown in Fig. P5-3. Derive a formula for the total elongation of these rods under their own weights (A = area, L = length, ρ = density, E = Young's modulus).

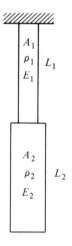

A_1
ρ_1 L_1
E_1

A_2
ρ_2 L_2
E_2

Fig. P5-3

5-4. Alter the formula derived in Prob. 5-3 to account for a concentrated weight at the free end of the rod.

*5-5. Attempt to derive a formula for the elongation of the circular, tapered column (Fig. P5-5) under its own weight.

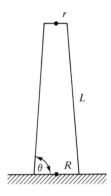

Fig. P5-5

*5-6. A concrete column is planned as in Fig. P5-5. $R = 30$ cm, $\theta = 88°$, density $= 2.4 \times 10^3$ kg/m³. What is the allowable height of the column if the maximum compressive stress is 20 kPa?

5-7. What is the allowable height of the column in Prob. 5-6 if there is a 100-kN load at the top of the column?

5-8. A rectangular bar of 2024-T351 aluminum alloy (see Fig. P4-54) has a cross section 4 cm $\times$ 10 cm. The strain distribution is shown in Fig. P5-8. Determine the stress distribution for static loading and the net force and its location on the cross-sectional area.

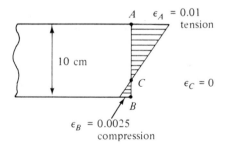

Fig. P5-8

5-9. Consider a bar made of AISI 4130 steel. The dimensions of the bar and the measured strains are the same as in Prob. 5-8. Assume that the monotonic and cyclic stress-strain curves (see Fig. P4-41) are applicable both in tension and compression. Determine the stress distribution and the magnitude of the resultant force on the cross section for (a) static loading; (b) cyclic loading.

5-10. Obtain an approximate solution for Prob. 5-9a if the material is LM-4M cast aluminum (see Fig. P4-56) instead of the steel.

5-11. A square bar of SAE 1015 steel (see Fig. P4-33) is 5 in. deep and it has the strain distribution shown in Fig. P5-11. Determine the approximate stress distribution for static loading and the magnitude and location of the resultant force.

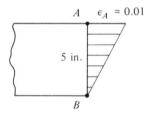

Fig. P5-11

*5-12. Consider Fig. P5-11 and recommend a different material for the bar to avoid yielding. Choose on the basis of the data given in Section 4-7 and assume that the cost of the material is proportional to its strength. Calculate the net force for the bar made from the material that you chose and compare it with the result of Prob. 5-11.

The following seven problems are related to a 1300-ft tall radio tower used for naval communications. The main structure has a triangular cross section as shown in Fig. P5-13. There are numerous guy wires arranged at several levels.

*5-13. The vertical legs of the tower are made from 30-ft long solid circular steel columns. The bottom columns are 6.5 in. in diameter, the top ones are 2 in. in diameter, and between these the diameter decreases stepwise (the diameter is constant for each 30-ft segment). What is the total deformation of the legs under their own weight? Assume that the connecting

Fig. P5-13

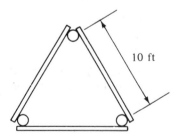

members and a full load of ice effectively double the load caused by the legs themselves but that the guy wires have zero tension in them (this is not true for a real tower).

*5-14. What is the total deformation of the tower caused by a 100°F decrease in temperature? Assume that there is no tension in the guy wires. For the steel, $\alpha = 6.5 \times 10^{-6}$ in./in. per °F.

*5-15. Consider only three of the guy wires of the radio tower, the ones attached 1000 ft aboveground. These are spaced 120° apart, and each makes a 45° angle with the tower. Each of these guys is a 2-in. diameter twisted steel cable (called a *bridge rope*). What is the total deformation in the tower caused by a 50,000-lb initial tension in each of the three guys?

*5-16. What is the change in tension in the guys of Prob. 5-15 if the temperature decreases by 100°F? The wires are steel with $\alpha = 6.5 \times 10^{-6}$ in./in. per °F.

*5-17. In high wind the maximum horizontal displacement of the tower at the 1000-ft level is 8 ft. What are the maximum elongation and change in tension in a guy caused by such a displacement?

*5-18. Each guy is made of several segments that are connected by porcelain insulator assemblies. Such an insulator is sketched in Fig. P5-18. Porcelain is relatively weak in tension, but the arrangement (U.S. Patent No. 2,625,581) assures that it is in compression even while the guy is in tension. The eyebolt A is 2.5 in. in diameter. What is the minimum thickness t of the head of the bolt if the shear strength is 50 ksi? What is the required diameter of the head if the compressive strength of the porcelain is 10 ksi? What is the elongation of the 35-in. long bolt under a load of 130,000 lb? The stress-strain curve is in Fig. P4-37.

Fig. P5-18

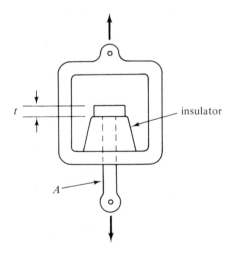

*5-19. Wind can induce rapid oscillations in a guy, and several million cycles may occur in a day. To avoid plastic deformations in the bolt (Fig. P4-37 and P5-18), what should be the maximum allowable fluctuating load in the guy?

The following seven problems are relevant to mountain climbing.

*5-20. What are the desirable mechanical properties of ropes to be used in mountain climbing? You do not have to be an expert climber to make up a reasonable list of properties.

*5-21. What does the elasticity of a rope depend on?

*5-22. Sketch a load versus deformation curve that you anticipate for nylon rope. What do you expect to happen if the load is constant on the rope for 1 hour, with the load being (a) 10% of the maximum load? (b) 50% of the maximum load?

*5-23. A 200-lb climber is using two 150-ft long nylon ropes in a free-rappel descent. Assume that his downward acceleration is 20 ft/sec² for 120 ft; then he brakes himself to stop relative to the rope in 10 ft (he is still suspended by the rope). Each rope has a breaking strength of 5500 lb, and the two ropes are in parallel so that the applied load is shared by them. Can the ropes sustain the loading? Can a similar descent be made again using the same ropes if they did not break in the first descent? This problem may be reconsidered after reading Section 12-5.

*5-24. In what ways, if any, could the length of a given rope be a factor in the strength of the rope?

*5-25. A desperate climber needs more rope, so he separates a rope into its three main strands and ties the three pieces end-to-end. According to a manual on mountaineering, the climber should be careful, because he could get a rope with much less than one-third of the strength of the original rope. Is this really possible?

*5-26. Devise a test that gives an indication of the loading history of a rope without damaging it.

Sec. 5-2

5-27. Two flat bars, each with a cross section 0.2 in. × 5 in., are to be riveted together with 0.25-in. diameter rivets. How many of these rivets are required as a minimum if the tensile load transmitted by the bars is 60,000 lb? The allowable shear stress is 20,000 psi, and the bearing stress is 40,000 psi for the rivets. Assume that there is no axial tensile load on the rivets.

5-28. Consider Prob. 5-27 again.
(a) How would you arrange the rivets to minimize the stresses in the bars?

What is the maximum average tensile stress in the bars in your arrangement?

(b)   How would you arrange the rivets to equalize the shear stresses in the rivets as much as possible? What is the maximum average tensile stress in the bars in this arrangement?

5-29.   Three plates are riveted together as shown in Fig. P5-29. Each plate is 2 mm thick and 30 cm wide. What is the minimum diameter of the rivets if their shear strength is 350 MPa? What is the largest average tensile stress in the plates?

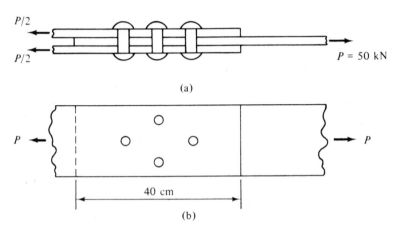

(a)

(b)

Fig. P5-29

5-30.   The plates shown in Fig. P5-29 are to be joined with a high-strength adhesive instead of rivets. What is the minimum shear strength of the adhesive that can be considered?

5-31.   A flat bar is attached to a structure by two 0.5-in. diameter rivets as shown in Fig. P5-31. What is the minimum acceptable shear strength of the rivets?

Fig. P5-31

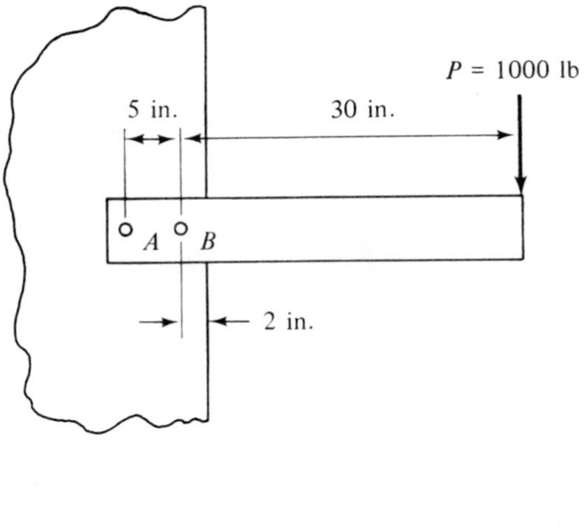

5-32. A clutch plate transmits a 5000-in.-lb torque by four bolts as shown in Fig. P5-32. What is the minimum diameter of the bolts if their shear strength is 25,000 psi?

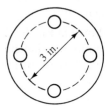

Fig. P5-32

5-33. A clutch plate such as that in Fig. P5-32 is designed to transmit a 2000-N·m torque. The bolts are arranged evenly on a 10-cm circle. What is the minimum number of 5-mm diameter bolts whose shear strength is 300 MPa?

5-34. Two tubes are joined with an adhesive as shown in Fig. P5-34. The shear strength of the adhesive is 20 MPa. What is the maximum axial load that the joint can transmit? What is the maximum allowable torque on the joint?

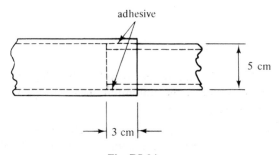

adhesive

5 cm

3 cm

Fig. P5-34

5-35. What is the minimum number of 4-mm diameter rivets with shear strengths of 80 MPa that could be used instead of the adhesive in Prob. 5-34?

Sec. 5-3

Consider a flat bar welded to a structure as shown in Fig. P5-36. In Prob. 5-36 to 5-43, τ is the shear strength of the weld metal and A is the leg of the weld ($A = B$ as in Fig. 5-10). Determine the unknown quantity.

5-36. $a = b = 4$ in., $\tau = 20$ ksi, $A = 0.25$ in., $P = ?$

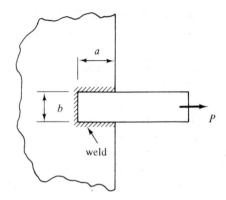

Fig. P5-36

5-37. $a = 10$ cm, $b = 5$ cm, $\tau = 100$ MPa, $A = 1$ cm, $P = ?$

5-38. $a = b = 5$ in., $\tau = 25$ ksi, $P = 10,000$ lb, $A = ?$

5-39. $a = 10$ cm, $b = 0$ (no weld on end), $\tau = 120$ MPa, $P = 20$ kN, $A = ?$

5-40. $a = 0$, $b = 5$ in., $\tau = 20$ ksi, $P = 3000$ lb, $A = ?$

5-41. $a = 0$, $b = 8$ cm, $\tau = 140$ MPa, $P = 10$ kN, $A = ?$

5-42. $a = 6$ in., $\tau = 25$ ksi, $P = 14,000$ lb, $A = 0.3$ in., $b = ?$

5-43. $b = 5$ cm, $\tau = 120$ MPa, $P = 50$ kN, $A = 0.5$ cm, $a = ?$

Sec. 5-4

5-44. Prove that the ways of representing the internal pressure in Fig. 5-12a and b are equivalent.

5-45. Can you determine the longitudinal stress, σ_l, from Fig. 5-12 or from another view of the same element?

5-46. How are the stresses σ_l and σ_c affected by location along the length of the cylinder (Fig. 5-11)?

5-47. Suppose the spacecraft discussed in Example 5-18 is assembled in space. How will it be affected, if it can be brought gently to the surface of the Earth, by (a) gravity? (b) the new environment?

5-48. A cylindrical pressure vessel is 10 ft long, is 4 ft in diameter, and has a wall thickness of 0.1 in. What is the allowable gage pressure if the maximum normal stress should not exceed 50 ksi?

*5-49. What are the axial elongation and the maximum enlargement of the diameter of the pressure vessel in Prob. 5-48 under the maximum pressure if the material is SAE 1045-390 BHN steel (Fig. P4-38)? $\mu = 0.3$.

5-50. A spherical pressure vessel has a 3-m diameter and a 4-mm wall thickness. What is the allowable gage pressure if the maximum normal stress should not exceed 250 MPa?

*5-51. What are the change in volume and the enlargement of the diameter of the vessel in Prob. 5-50 under the maximum pressure if the material is 2014-T6 aluminum (Fig. P4-53)? $\mu = 0.33$.

5-52. A cylindrical pressure vessel has a 6-ft diameter and a 0.2-in. wall thickness. The gage pressure is 280 psi. The vessel has to be assembled from two prefabricated halves. The plane of the seam could make an angle anywhere between 0 and 20° with respect to the center line of the vessel. What angle would you choose if the seam has a tensile strength (stress normal to the joint) of 80 ksi and a shear strength of 50 ksi?

*5-53. A thin-walled cylindrical pressure vessel (diameter D, wall thickness t) has four circular portholes (radius R) at its midsection, equally spaced on the circumference. What are the maximum normal and shear stresses at the edge of a porthole (between the metal shell and the glass window) when a gage pressure p is applied to the vessel?

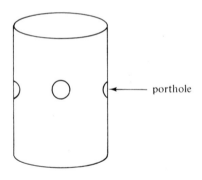

Fig. P5-53

*5-54. The ground antenna that was constructed in Maine for the Telstar satellite is a very large, movable structure. It is protected by a pressure-rigidized,

Fig. P5-54

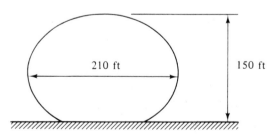

dielectric radome, shown in a cross-sectional view in Fig. P5-54. The radome material is 0.07 in. thick (for electrical reasons) and its total weight is 60,000 lb. What is the minimum gage pressure in the radome to maintain its nearly spherical shape? What is the minimum strength of the radome material on the basis of this pressure?

*5-55. A large sphere of strong material is designed as a balloon of essentially constant volume for high-altitude scientific research. What is the gage pressure that will cause a 5% change in volume of a 70-m diameter sphere (compared with having zero gage pressure)? The wall thickness is 3 mm and Young's modulus is 30 GPa; $\mu = 0.3$.

5-56. A cigar-shaped, nuclear-powered airship is designed to carry large loads. The material proposed for the fabric skin has a tensile strength of 500 MPa. The gage pressure is 1 kPa. What is the minimum wall thickness of a 30-m diameter airship?

5-57. An oil pipeline is 4 ft in diameter and has a 0.5-in. wall thickness. The pipe is mounted on rollers at regular distances to prevent large stresses caused by fluctuations in temperature. However, the viscosity and pressure of the oil depend on temperature. What is the largest change in stress in the pipe when the pressure increases to 1200 psi from the normal 900 psi?

6

SHEAR AND BENDING
MOMENT IN BEAMS

Imagine that you are an avid skier and that you would like to make a career of developing improved ski equipment. One of the reasonable projects is to reduce the weight and inertia of skis while making them sufficiently strong to avoid fractures or excessive deformations. The first step in working toward this goal is to establish the magnitudes of the forces and moments that may act at any point along a ski in a variety of situations. It is reasonable to assume at this time that the ski is a straight, flat bar, that it is never twisted about its longitudinal axis, and that all significant external forces are perpendicular to the plane of the bar. Such members are the simplest in the category of beams.

It is always important to know how a beam is supported. Figure 6-1 shows some of the models of the most common supports. These may occur in various combinations. The external loads may be moments and forces; the latter are concentrated or distributed. The applied loads and reactions must be such that all the equations for static equilibrium are satisfied.

6-1. SHEAR AND MOMENT FUNCTIONS

The internal force and moment at any transverse cross section of a beam can be determined with the aid of proper free-body diagrams. For example, consider the diagram for a ski in Fig. 6-2a. The load P and the distributed reaction $w(x)$ are assumed to satisfy the equilibrium requirements. If the whole member is in equilibrium, any part of it is also in equilibrium, assuming that the member is rigid. A partial free-body diagram of the right part of the ski is drawn in Fig. 6-2b. The external loads on this part are the force P and the reaction $w(x)$ acting over the distance $l_1 + l_2$. The shear force V_A and moment M_A are the internal reactions in

144

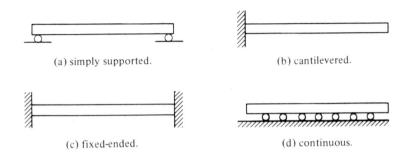

(a) simply supported.   (b) cantilevered.

(c) fixed-ended.   (d) continuous.

Fig. 6-1. Models for beams.

Fig. 6-2. Internal reactions in a beam.

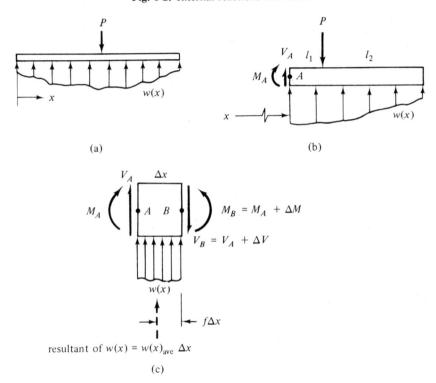

(a)   (b)

(c)

the beam caused by the external loads. The directions of these reactions are drawn by making assumptions about the magnitudes and locations of the external loads. Sign conventions for the internal force and moment are arbitrary (the important requirement is for consistency), but many people follow a custom. A moment that makes a beam concave upward (compression in top fibers, tension in bottom fibers) is taken as positive. A negative moment makes the beam concave downward. The positive sign is given to a shear force acting downward on the right side of a section or upward on the left side. Thus, both M_A and V_A are positive in Fig. 6-2b.

The internal shear force and the bending moment are functions of the loading on the beam. The relationships can be established by analyzing a short segment Δx such as that in Fig. 6-2c ($\Delta x < l_1$). For equilibrium in the vertical direction,

$$V_A - (V_A + \Delta V) + w(x)_{\text{ave}}\Delta x = 0$$

where $w(x)_{\text{ave}}$ applies to $w(x)$ between points A and B. Taking the limit as Δx approaches zero,

$$\frac{dV}{dx} = w(x) \qquad (6\text{-}1)$$

In words, the local variation in the shear force is equal to the loading $w(x)$ at any point in the beam where the load varies continuously.

The shear force itself is related to the change in bending moment along the beam. Taking moments about point B in Fig. 6-2c (with counterclockwise moments as positive),

$$(M_A + \Delta M) - M_A - V_A\Delta x - w(x)_{\text{ave}}\Delta x(f\Delta x) = 0$$

where $f\Delta x$ denotes the fraction of Δx where the resultant $w(x)_{\text{ave}}\Delta x$ of the elemental distributed force is acting. After dividing by Δx and rearranging,

$$V_A = \frac{\Delta M}{\Delta x} + w(x)_{\text{ave}}(f\Delta x)$$

Taking the limit as Δx approaches zero and omitting the subscript A (because the equation is valid anywhere for continuous loading),

$$V = \frac{dM}{dx} \qquad (6\text{-}2)$$

Another relationship worth noting is valid when the loading on a beam is continuous so that V can be differentiated:

$$\frac{dV}{dx} = \frac{d^2M}{dx^2} \qquad (6\text{-}3)$$

6-2. SHEAR AND MOMENT DIAGRAMS

A frequently occurring problem is to find the critical locations in a beam where the shear and bending moment are the largest. The signs of the largest moments are also of interest because these indicate what parts are in tension or compression. The sign of the shear force has practically no significance, as described for shear stresses. To find the critical locations along a beam, it is often necessary to plot shear and moment diagrams for the whole length of the beam. There are two practical ways of obtaining these diagrams, and they are about equivalent in the long run.

One method is to pick a few points of interest along the beam and work with the partial free-body diagrams from one end of the beam to each of these points. The values of shear and moment at each of these points are plotted on the appropriate diagram, and the points are connected by a reasonable curve in each diagram. For an example, consider a ski resting on two rocks as modeled in Fig. 6-3a. From equilibrium considerations,

$$P = R_1 + R_2, \qquad R_1 a = R_2 b, \qquad R_1 = \frac{P}{1 + \dfrac{a}{b}}$$

Fig. 6-3

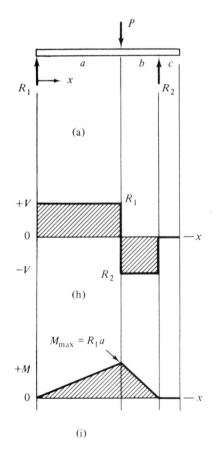

(a)

(h)

$+V$ R_1

0 $-x$

$-V$ R_2

$M_{\max} = R_1 a$

$+M$

0 $-x$

(i)

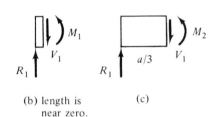

(b) length is near zero.

(c)

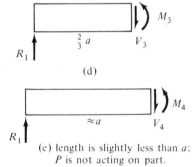

(d)

(e) length is slightly less than a; P is not acting on part.

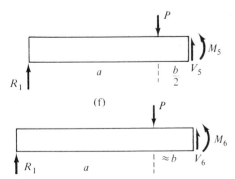

(f)

(g) length is slightly less than $a + b$; R_2 is not acting on part.

Several partial free-body diagrams are drawn, each including the reaction R_1 (this is arbitrary; one could work from the right end of the beam). Samples of these are shown in Fig. 6-3b through g, and the corresponding shears and moments are listed here:

$$V_1 = R_1, \qquad M_1 = 0 \qquad \text{(Fig. 6-3b)}$$

$$V_2 = R_1, \qquad M_2 = R_1\frac{a}{3} \qquad \text{(Fig. 6-3c)}$$

$$V_3 = R_1, \qquad M_3 = R_1\frac{2}{3}a \qquad \text{(Fig. 6-3d)}$$

$$V_4 = R_1, \qquad M_4 = R_1a \qquad \text{(Fig. 6-3e)}$$

$$V_5 = P - R_1, \qquad M_5 = R_1\left(a + \frac{b}{2}\right) - P\frac{b}{2} \qquad \text{(Fig. 6-3f)}$$

$$V_6 = P - R_1, \qquad M_6 = R_1(a + b) - Pb \qquad \text{(Fig. 6-3g)}$$

To the right of R_2,

$$V = 0, \qquad M = 0$$

Other elements can be evaluated similarly, and the data can be plotted on the shear and moment diagrams (they are best to draw below the free-body diagram of the whole beam). A small number of points may suffice for the plots if there is experience in sketching such diagrams. The cross-hatching helps in the visual evaluation of the plots. The sharp breaks in the plots are characteristic for a place where a discrete force is acting. Such regions are gray areas in knowledge because it is known that even concentrated forces act on finite areas, but the patterns of distribution for them are not known. Thus, the two diagrams are not precise where P, R_1, and R_2 are acting on the beam, but they are simple and perfectly adequate in practice.

The alternative procedure for obtaining the shear and bending moment diagrams is to write the functions of the shear and the moment in terms of the external loads and the distance x along the beam. These are based on a single free-body diagram for each part of the beam in which the loading varies smoothly. This means two regions for the beam in Fig. 6-3a, and a total of four functions for the whole beam:

$$V(x) = R_1 \qquad\qquad \text{in} \quad 0 < x < a$$
$$V(x) = -R_2 \qquad\qquad \text{in} \quad a < x < a + b$$
$$M(x) = R_1x \qquad\qquad \text{in} \quad 0 < x < a$$
$$M(x) = R_1x - P(x - a) \quad \text{in} \quad a < x < a + b$$

These can be evaluated at selected values of x and the lines drawn through the resulting points.

The shear and moment diagrams in Fig. 6-3 provide an opportunity to check a relationship presented in Section 6-1. From Eq. 6-2, $dM = V dx$, and

$$\int_{M_1}^{M_2} dM = \int_{x_1}^{x_2} V dx$$

from which $M_2 - M_1 =$ area under shear curve between x_1 and x_2. This means, for a simple beam, that the bending moment at a given point along the beam is equal to the area under the shear curve from the end of the beam (where $M = 0$) to the given point. The simplest example of this is the maximum moment in Fig. 6-3i, which is equal to the upper area $R_1 a$ of the shear diagram in Fig. 6-3h. Alternatively, $M_{max} = R_2 b$, which is the area of the lower part of the shear diagram in Fig. 6-3h.

Shear and moment diagrams can be considerably more complicated than these first two. The sketching of many different kinds should be practiced because the stress and deformation analysis of beams depends on these diagrams. The following examples show the wide range of diagrams that are possible.

EXAMPLE 6-1

Sketch the shear and moment diagrams for a ski simply supported at the ends, if the only load is its own weight, which is uniformly distributed. The ski is 2 m long and has a mass of 2 kg.

Solution

It is always best to plot those points first that can be determined by inspection. The free-body diagram in Fig. 6-4a shows the symmetry of the situation and the

Fig. 6-4. Example 6-1.

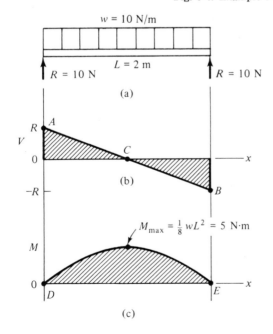

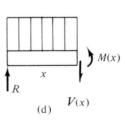

magnitudes of the reactions. With these, five points can be plotted at once. *A* and *B* are equal in magnitude to *R* (*A* for very small *x*, *B* for *x* nearly 2 m). They are opposite in sign according to the convention. *C* is zero at the center of the beam from a free-body diagram for half of the beam. *D* and *E* are both zero because there are no moments at the supports.

The shear diagram is linear from *A* to *B* because the magnitude of the distributed load on any part of the beam depends on its length *x* linearly. The moment diagram between *D* and *E* is more complicated. Consider the partial free-body diagram in Fig. 6-4d. This is valid everywhere between the supports because the loading is continuous. The internal moment is

$$M(x) = Rx - wx\left(\frac{x}{2}\right) = Rx - \frac{wx^2}{2}$$

where *wx* is the magnitude of the partial distributed load and *x*/2 is the moment arm from the resultant of this load to the section where *M*(*x*) is acting. The parabolic moment diagram is obtained by substituting a few values for *x* in the moment equation, plotting these, and connecting them with a smooth curve.

EXAMPLE 6-2

Redo Example 6-1 after adding a concentrated load of 250 N at midspan.

Solution

It is best to use superposition here. The dashed lines in Fig. 6-5b and c are for

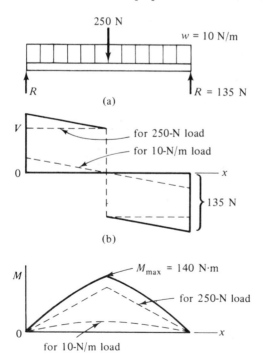

Fig. 6-5. Example 6-2.

the concentrated and distributed loads individually; they are not to scale because of large differences between them. In each diagram, the superposition is done by adding the two values at any given x. The resultant curves are drawn with solid lines. For this beam, the concentrated load is the most significant.

EXAMPLE 6-3

Redo Example 6-2 after moving the supports to the center of the beam where they are 20 cm apart.

Solution

Four points can be plotted at once because both the shear and the moment are zero at each end of the beam (Fig. 6-6). Other points of interest are mainly those sections where the concentrated forces are acting. The shears and moments can be evaluated one by one at these points. It is easy to verify that these values are extremes. Since smaller values are often unimportant, it is acceptable to draw the curves approximately (relying on experience) to include the calculated values. Note

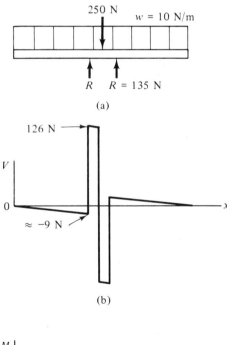

Fig. 6-6. Example 6-3.

that the diagrams in Figs. 6-5 and 6-6 are quite different even though the only difference between the beams is the positions of the supports.

EXAMPLE 6-4

Imagine that a 70-kg skier is evacuating a 40-kg person from a dangerous area.
Should the smaller person ride on the back of the larger so that all the load may be
concentrated at point *A* in Fig. 6-7a or try to stand on the ski at point *B* and just
hang on for balance? Assume that the skis are equally strong at all transverse cross
sections and that the worst condition occurs with simple supports at the ends.
Neglect dynamic loading but assume that all the load may be on one ski at the same
time.

Solution

The only bases for decision at this stage are the shear and moment diagrams
for the two configurations. The reactions are determined first.

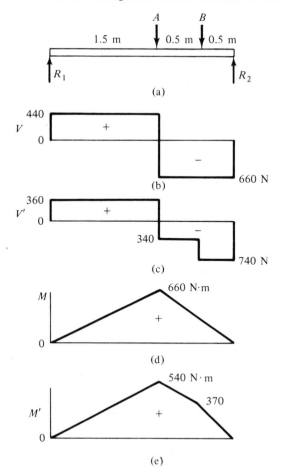

Fig. 6-7. Example 6-4.

$$R_1 + R_2 = 1100 \text{ N} \qquad \text{in both cases}$$

$$\left.\begin{aligned} R_1 &= \frac{1100(1)}{2.5} = 440 \text{ N} \\[6pt] R_2 &= 660 \text{ N} \end{aligned}\right\} \quad \text{for both riding at } A$$

$$\left.\begin{aligned} R_1' &= 360 \text{ N} \\ R_2' &= 750 \text{ N} \end{aligned}\right\} \quad \text{for riding separately}$$

It is seen from the diagrams in Fig. 6-7 that riding together is preferable if shear strength is critical, while riding separately is best if bending moments are to be minimized.

EXAMPLE 6-5

Consider a ski clamped on the back of a car as shown in Fig. 6-8a. The wind causes a distributed load that can be represented as a uniformly varying load on a cantilever beam (Fig. 6-8b). The axial component of the load on the beam is neglected. Draw the shear and moment diagrams.

Solution

The shear and the moment are zero at the right end of the beam (Fig. 6-8b). The shear at the clamp is the total force of the wind, 150 N. To calculate the

Fig. 6-8. Example 6-5.

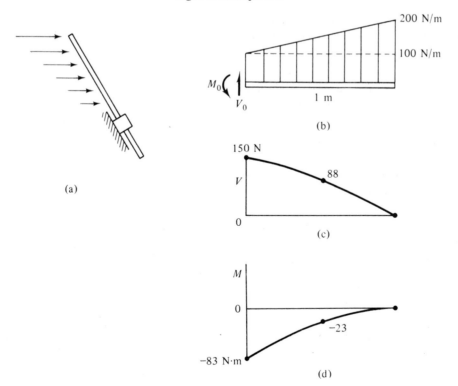

moment at the clamp, it is helpful to use the fact that the resultant of a distributed load acts at the geometric center of the distribution. For a right triangle, this is at one-third of the sides away from the right angle corner. With this idea,

$$M_0 = 100 \text{ N/m } (1 \text{ m})(0.5 \text{ m}) + 0.5(100 \text{ N/m})(1 \text{ m})(0.67 \text{ m})$$
$$= 83 \text{ N·m}$$

At the center of the span the total shear for the right half of the beam is

$$V_1 = 150 \text{ N/m } (0.5 \text{ m}) + 0.5(50 \text{ N/m})(0.5 \text{ m}) = 88 \text{ N}$$

The moment necessary for equilibrium of the right half of the beam is

$$M_1 = 150 \text{ N/m } (0.5 \text{ m})(0.25 \text{ m}) + 0.5(50 \text{ N/m})(0.5 \text{ m})(0.33 \text{ m})$$
$$= 23 \text{ N·m}$$

Since the loading varies smoothly, each diagram can be sketched as smooth curves going through the three respective points. The curves are also judged reasonable because the largest shear and moment are at the clamp. More points must be evaluated before drawing the curves if higher precision is required for the whole beam.

EXAMPLE 6-6

Consider a simply supported vertical panel under water, as modeled in Fig. 6-9a. Draw the shear and moment diagrams.

Solution

For convenience, the beam is imagined to be horizontal. The reactions are found from the equilibrium equations:

$$R_1 + R_2 = \frac{wL}{2} \qquad \text{(resultant of distributed load)}$$

The resultant of the distributed load acts at $x = \frac{2}{3}L$. Thus, taking moments about the left end of the beam,

$$R_2 L = \left(\frac{wL}{2}\right)\left(\frac{2}{3}L\right), \qquad R_2 = \frac{wL}{3}, \quad \text{and} \quad R_1 = \frac{wL}{6}$$

The shear and moment functions are obtained from Fig. 6-9b:

$$V(x) = R_1 - \frac{wx^2}{2L} = \frac{wL}{6} - \frac{wx^2}{2L}$$
$$M(x) = R_1 x - \left(\frac{wx^2}{2L}\right)\left(\frac{x}{3}\right) = \frac{wLx}{6} - \frac{wx^3}{6L}$$

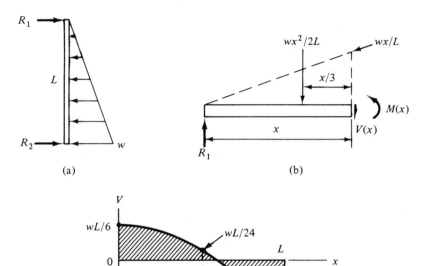

(a)   (b)

(c) shear diagram estimated
from three points; a
smooth curve is drawn
through the points be-
cause the load varies
smoothly over the
whole beam.

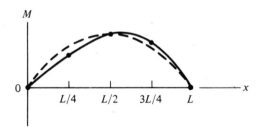

(d) moment diagrams
estimated from a
few points:

- - - $x = 0, L/2, L$

(symmetry is assumed
erroneously here)

—— $x = 0, L/4, L/2, 3L/4, L$

Fig. 6-9. Example 6-6.

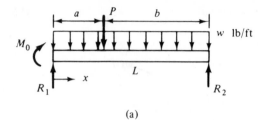

(a)

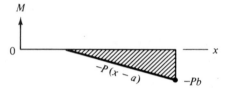

(b) moment diagram for R_1.

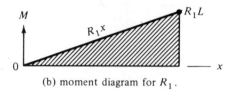

(c) moment diagram for P.

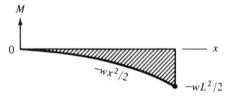

(d) moment diagram for w.

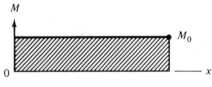

(e) moment diagram for M.

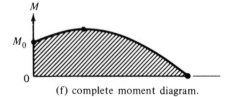

(f) complete moment diagram.

Fig. 6-10. Example 6-7.

For each point of the shear or moment diagram a value of x must be selected and used in these equations. Note how the accuracy improves when more points are added to a diagram that was erroneously assumed to be symmetric after plotting only three points (Fig. 6-9d).

EXAMPLE 6-7

Use superposition to find a quick but reasonable estimate for the complete moment diagram of the beam in Fig. 6-10a.

Solution

Separate moment diagrams are drawn for each external force or moment, and these are positioned with their origins on the same vertical line for convenience (Fig. 6-10b through e). The moments at the ends of the beam can be plotted on the complete moment diagram at once. These two points are accurate by inspection of the sketches in Fig. 6-10a through e. The next point that is most convenient to plot is the complete moment at $x = a$. With a little thought, the approximate complete moment curve can be drawn through the three points in Fig. 6-10f. Thus, the maximum moment is in the neighborhood of $x = a$.

Note that the sign could be chosen arbitrarily in the first individual moment diagram; however, all the individual diagrams have to conform to the same scheme for signs.

EXAMPLE 6-8

A large helicopter is 18 m long and has two rotors as shown in Fig. 6-11a, with the rotor shafts 14 m apart. The weight of the fuselage and payload is uniformly distributed along the longitudinal axis, and it is 5 kN/m. The maximum torque to each rotor is 8 kN·m. Draw the shear and bending moment diagrams for the helicopter. Assume that the power input to the two rotors is identical at any given time.

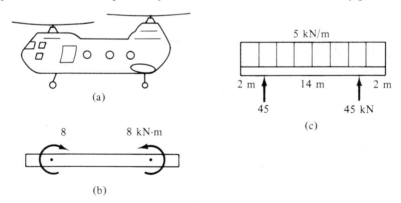

Fig. 6-11. Example 6-8.

Solution

       It is necessary to consider two views of the helicopter: the top view and the side view. The first of these is modeled in Fig. 6-11b. The rotations are opposite to eliminate a net moment and the resulting spinning of the fuselage. Since there are no shearing forces in this view, there is no shear diagram for it. The moment at any section between the rotor shafts is 8 kN·m, so the moment diagram is rectangular with a length of 14 m. The moment is zero from each rotor to the nearest end of the fuselage.

       The side view is modeled in Fig. 6-11c. This is essentially the same as the ski under its own weight resting on simple supports, so the reader can readily sketch the diagrams. For further guidance, it is given that the maximum shear and moment should not exceed 45 kN and 90 kN·m, respectively.

6-3. PROBLEMS

6-1.   How many separate functions of shear and moment must be used to draw the complete shear and moment diagrams for the beam shown in Fig. P6-1 ?

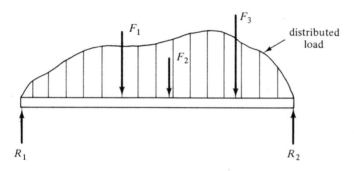

Fig. P6-1

6-2 to 6-31.   Write the shear and moment equations and sketch the shear and moment diagrams for each of the beams shown in Figs. P6-2 to P6-31.

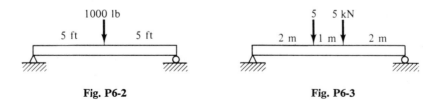

Fig. P6-2   **Fig. P6-3**

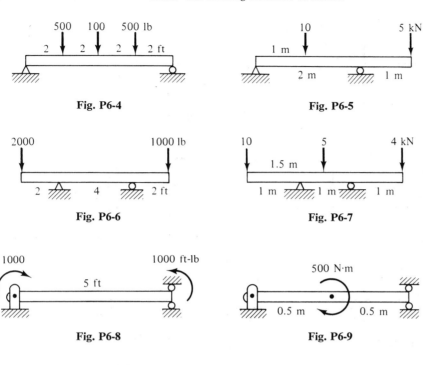

Fig. P6-4

Fig. P6-5

Fig. P6-6

Fig. P6-7

Fig. P6-8

Fig. P6-9

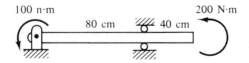

Fig. P6-10

Fig. P6-11

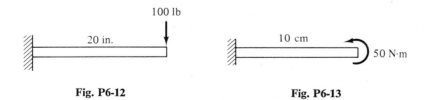

Fig. P6-12

Fig. P6-13

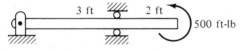

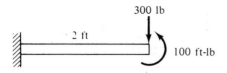

Fig. P6-14

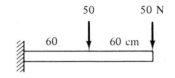

Fig. P6-15

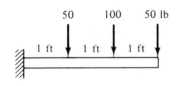

Fig. P6-16

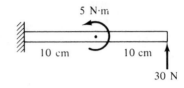

Fig. P6-17

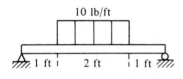

Fig. P6-18

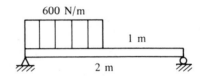

Fig. P6-19

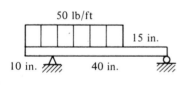

Fig. P6-20

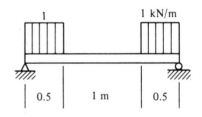

Fig. P6-21

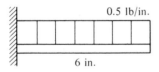

Fig. P6-22

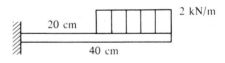

Fig. P6-23

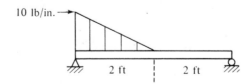

Fig. P6-24

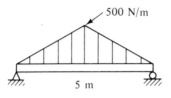

500 N/m

5 m

Fig. P6-25

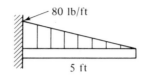

80 lb/ft

5 ft

Fig. P6-26

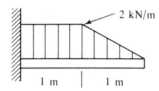

2 kN/m

1 m   1 m

Fig. P6-27

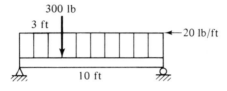

300 lb

3 ft

20 lb/ft

10 ft

Fig. P6-28

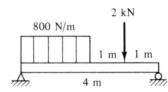

2 kN

800 N/m

1 m   1 m

4 m

Fig. P6-29

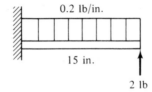

0.2 lb/in.

15 in.

2 lb

Fig. P6-30

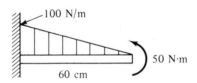

100 N/m

50 N·m

60 cm

Fig. P6-31

7

STRESSES IN BEAMS

SAGREDO: Simplicio, do you want to hear something exciting?

SIMPLICIO: Oh, no . . . not again!

SAGREDO: Oh, yes . . . and it is the greatest thing that ever crossed your narrow path.

SIMPLICIO: All right, all right. What is it? But be quick, I haven't got all day.

SAGREDO: Listen! I have found an error in some of the work of Il Maestro!

SIMPLICIO: An error in Il Professore's work? Impossible!

SAGREDO: Yes, an error. I was thinking about it very deeply for a few months. I am convinced now that I am correct and . . . and that he is wrong.

SIMPLICIO: What is it about?

SAGREDO: Remember when he was breaking all those cantilever beams? And his theory of internal forces in the beams?

SIMPLICIO: I remember, but not the details.

SAGREDO: Here. This is what he said. The internal force is largest near the wall, on top of the beam. At point A in this drawing.

SIMPLICIO: So? That's where the beams broke, didn't they? How could he be wrong?

SAGREDO: In the next step he made. He said that the internal force is largest at A and zero at point B, and in between these points the forces are larger than zero but less than maximum. Something like this.

SIMPLICIO: That makes sense. He has keen eyes, and saw that the fractures always started

NOTE: For more about these characters, see Galileo Galilei, *Dialogues Concerning Two New Sciences* (Evanston, Ill.: Northwestern University Press, 1968).

162

Fig. 7-1. Galileo's test facility for beams.

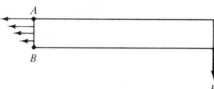

Fig. 7-2. Internal forces to act on the beam at the wall as suggested by Galileo.

at the top and never at the bottom of the beams. How can you argue against this? Also, I recall now that Il Professore had a theory to prove those internal forces between *A* and *B* are correct. They create a moment about point *B* that counteracts the moment of the weight *F* at the end of the beam. That's why the beam doesn't fall down until it fractures. This is great science, boy!

SAGREDO: Partially, it is great science. As it is, there are two things wrong with it . . .

SIMPLICIO: You don't dare . . .

SAGREDO: Calm yourself! I am not implying anything about you or Il Maestro. I did say these things were not easy to see, didn't I?

SIMPLICIO: All right. Continue.

SAGREDO: First of all, look at the last drawing again. The only forces acting horizontally on the beam are on the left end, correct? So, this beam should start moving to the left if those forces are really there as shown. Furthermore, considering the moments alone, the beam is not really in equilibrium that way, either. It seems to be at first sight, but it is not.

SIMPLICIO: Hmm . . . The first difficulty, that I can see. The second one, I am not with you yet. How do you think the internal forces act? There must be some.

SAGREDO: True. I think they are distributed this way. On the top half of the beam they are pulling on the beam to the left, on the bottom half they are pushing it to

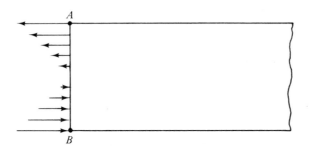

Fig. 7-3. Correct concept of distribution of internal forces at the wall.

the right. Thus, the beam doesn't want to go either right or left. And together they create a turning effect on the beam which opposes the turning effect of the load *F*. A neat theory, heh?

SIMPLICIO: Yes, indeed. But just a theory, nevertheless. How will you ever prove it?

SAGREDO: Of course, I have been worrying about that, too, for quite some time. And, finally, I have come up with something. This.

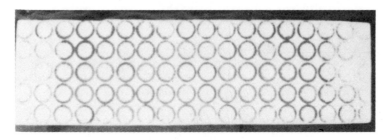

Fig. 7-4. Model of beam without external loads.

SIMPLICIO: What on earth is that?

SAGREDO: I bought one of the largest loaves of bread that Mamma Giorgione baked today, and cut this uniform beam from it. It's only the soft part. Then I painted these pretty circles on the beam. Before I go on, let me ask you a question. Do you think it is reasonable to assume that the beam has the same material and the same strength at every point? There is no crust or any foreign objects in it.

SIMPLICIO: Yes, it seems reasonable.

SAGREDO: Now, watch! I am going to bend this bread beam. See what happens? The top circles become stretched, the bottom ones are compressed, and the center ones are essentially unchanged. Go ahead, try it for yourself.

SIMPLICIO: Well, well. That is something. The magnitudes of the internal forces even seem to be the same at the top and at the bottom. Then, how come the fractures always start at the top?

SAGREDO: Stretching a material is not the same as pushing it together. Take two bricks. You can press them together with great force, even though they are separate, but it takes little effort to pull them apart.

Fig. 7-5. Model of beam deformed.

SIMPLICIO: The theory is becoming a little more interesting, but I would have to think about it some more. Have you told Il Professore already? No? When are you going to? I want to be there and see the sparks fly. But I just want to be an observer, understand?

SAGREDO: Yes. And I am sorry to disappoint you, but I am not planning such a show.

SIMPLICIO: Aha! You are not entirely convinced that your theory is correct.

SAGREDO: Yes, I am. But Il Maestro is getting rather old, and he has plenty of trouble at this time. I shall not cause any disruption in his lines of thought. Maybe in the future when his sky clears up a little. Then, we should argue again as in the old days. In the meantime, why don't you meet with the three young fellows and me? I have more to discuss about my ideas and what is definitely right in Il Maestro's work.

SIMPLICIO: All right. When?

SAGREDO: In the morning, two days hence.

7-1. THE FLEXURE FORMULA

The soft model of a beam with the circles painted on it is excellent to show qualitatively how the stresses are distributed inside a beam that is bent. The next step is to relate the external loading on the beam to the maximum stress in the beam. There is no external horizontal loading here.

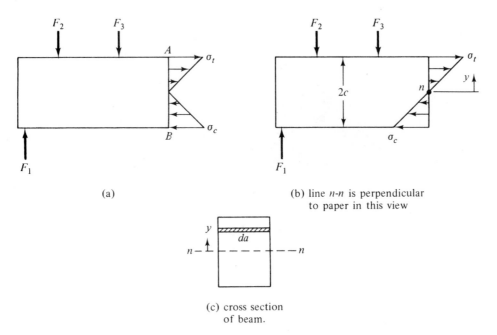

(a)

(b) line *n-n* is perpendicular to paper in this view

(c) cross section of beam.

Fig. 7-6. Internal normal stresses in beam caused by bending.

Consider part of a beam with cross section *A-B* as in Fig. 7-6a. The tensile and compressive stresses in this case are often shown as in Fig. 7-6b. Assume that the beam has not yielded anywhere, so the stress distribution is linear. The stress is zero at the so-called neutral axis. On a thin strip *da* the stress $\sigma(y)$ may be assumed constant (Fig. 7-6c).

The internal stresses have a clockwise moment M_I about the line *n-n* in Fig. 7-6b, which is determined as follows:

Normal force on area $da = \sigma(y)da$ where $\sigma(y) = \dfrac{\sigma_t}{c}y$

Moment of this force about *n-n* $= \sigma(y)yda$

Total internal moment caused by stresses $= M_I = \int \sigma(y)yda$

The integral can be simplified using these ideas: First multiply it by y/y:

$$M_I = \int \frac{\sigma(y)}{y}y^2da$$

The stresses vary linearly, and σ_t or σ_c can be treated as constants (because they are measurable), so

$$M_I = \int \frac{\sigma_t}{c}y^2da = \frac{\sigma_t}{c}\int y^2da$$

By definition, $\int y^2 da = I$, the moment of inertia of the whole area about its neutral axis (because y is measured from the neutral axis). On the basis of equilibrium, M_I must be equal in magnitude and opposite in direction to the net moment M caused by external loading at the same place in the beam. These lead to what is commonly called the *flexure formula*:

$$M = \frac{\sigma I}{c} \qquad (7\text{-}1)$$

where M = net moment at a given section in the beam

I = moment of inertia of the cross-sectional area about the neutral axis

σ = magnitude of stress at a distance c from the neutral axis

It is important to remember that Eq. 7-1 is valid only if there are no plastic deformations at the cross section considered.

Position of the Neutral Axis. Looking at the soft model with the painted circles on it in Fig. 7-5, it appears that the neutral axis is about in the middle of the beam (halfway between the top and the bottom). The questions are, how precise is this observation, and is it true in general? The key to the answers is that the horizontal forces on a cross-sectional face must be in equilibrium by themselves if there are no external horizontal forces on the beam (as in Fig. 7-6). This condition means that

$$\int \sigma(y) da = 0 \quad \text{or} \quad \int \frac{\sigma(y)}{y} y da = 0 \quad \text{or} \quad \frac{\sigma(y)}{y} \int y da = 0$$

since $\sigma(y)/y = \sigma_t/c = $ a constant, not equal to zero. Thus, $\int y da = 0$, where y is measured from the neutral axis. By definition, $\int y da$ is the first moment of the cross-sectional area with respect to the neutral axis, so

$$\int y da = A\bar{y} = 0$$

where A = area of the cross section

$\bar{y}$ = distance between the centroidal axis and the neutral axis.

It is clear that $\bar{y}$ must be zero. In other words, *the neutral axis is the centroidal axis*, and it can be found accurately if the geometry of the cross section is known.

Optimum Design of Beams. It is seen from the flexure formula how the geometry of a beam affects it as a structural member. A large moment of inertia of the cross section is desirable for resisting large external moments with low stresses in the beam. This has two implications in selecting and using beams.

If there is no choice between beams, the given beam should be oriented with respect to the applied loads in a way to maximize the moment of inertia. For example, a wooden board to be subjected to transverse forces has a higher moment of inertia (and a higher load-carrying ability for a given maximum stress) in the orientation shown in Fig. 7-7a than in 7-7b.

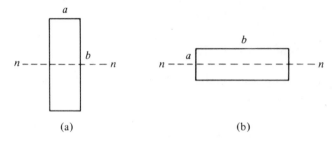

Fig. 7-7. Optimum orientation of cross section.

If there is a choice between beams or one can design the beam, it is most efficient to *use a beam that has little material near the neutral axis and has much material far from the neutral axis.* This increases the moment of inertia for a given cross-sectional area. Another way to put this is that material at and near the neutral axis is wasteful because it has low stresses (or none) in it, while material far from the neutral axis can be utilized fully with respect to its strength. I beams, box beams, channel-shaped beams, and tubular beams are common examples of efficient beams on the basis of the strength-to-weight ratio. The tubular shape of many bones is, as Galileo pointed out, a designer's marvel of nature.

Limitations in the Use of the Flexure Formula. The flexure formula is important in practice but it is frequently misused. The following limitations should be remembered.

1. The maximum stresses should be in the elastic range of the stress-strain curve of the material. If this is not true, the flexure formula can be used only qualitatively, such as showing that there should be some plastic strain in the beam or that the beam is grossly inadequate.
2. The modulus of elasticity should be the same in tension and in compression. This is reasonably true for many engineering materials but not for all.
3. The beam should not be curved sharply before the loads are applied.
4. There should be no holes or other sudden changes in the material at the cross section under consideration. The effects of these will be discussed under stress concentrations.
5. The beam should be loaded symmetrically with respect to the principal

Fig. 7-8. Examples of unsymmetrical loading.

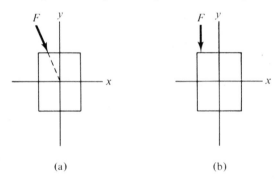

(a)   (b)

axes of the cross section. Figure 7-8 shows examples of unsymmetrical loading (*F* in Fig. 7-8a can be resolved into *x* and *y* components for which the flexure formula is valid individually; however, the neutral axis is seldom perpendicular to *F* in such cases).

6. The moment of inertia for a given cross-sectional area cannot be arbitrarily increased in practice. Parts of a beam that are too thin may collapse or wrinkle locally or cause the whole beam to twist instead of bend.

EXAMPLE 7-1

A number of long, slender beams of granite are stored at the construction site of a new building as shown in Fig. 7-9a. What is the maximum stress in a beam while it is stored and where does it occur in the beam?

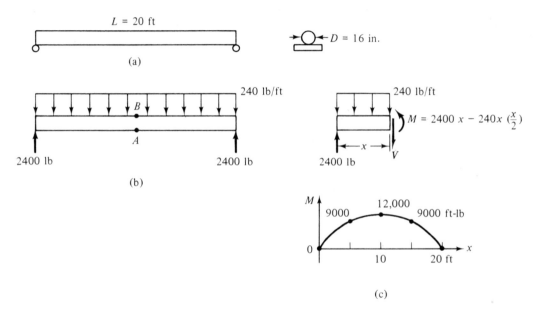

Fig. 7-9. Example 7-1.

Solution

There are two kinds of possibilities for maximum stress. One is the contact stress where the beam rests on a roller, the other is from bending under its own weight. The former is not possible to determine here because the size of the contact area and the distribution of the stresses on it are not known. The stresses from bending are calculated as follows:

Volume of beam = 48,000 in³
Weight of beam = 4800 lb
Each vertical reaction = 2400 lb

The maximum bending moment occurs at the center of the span: $M = 12,000$ ft-lb

The moment of inertia is $I = \dfrac{\pi r^4}{4} = 3220$ in⁴

The maximum stress is $\sigma = \dfrac{Mc}{I} = \dfrac{12,000(12)(8)}{3220} = 360$ psi

This is the magnitude of the maximum tensile stress which occurs at point A in Fig. 7-9b and of the maximum compressive stress which occurs at point B in the same figure. Both stresses are much lower than the respective strengths of the stone.

EXAMPLE 7-2

A cast-iron bar is designed to serve as a cantilever beam as shown in Fig. 7-10a. Is this beam designed properly? The metal's tensile strength is 50 ksi, its compressive strength is 100 ksi. Assume that the elastic modulus is the same in tension and compression (not in reality).

Solution

The largest bending moment is at the wall: $M = 50,000$ ft-lb. The location of the neutral axis must be found next. For this purpose, the total area A is broken

Fig. 7-10. Example 7-2.

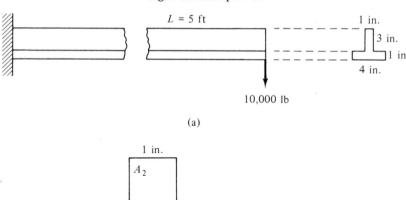

(a)

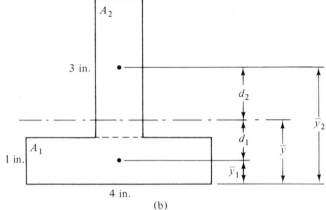

(b)

into two parts as in Fig. 7-10b. The centroid of A is $\bar{y}$, and it is found from the following equation.

$$A\bar{y} = A_1 \bar{y}_1 + A_2 \bar{y}_2, \qquad \bar{y} = \frac{4(0.5) + 3(2.5)}{7} = 1.36 \text{ in.}$$

Thus, the required moment of inertia is about a line that is 1.36 in. above the bottom edge of the beam. Moments of inertia are additive, so

$$I = I_1 + I_2$$
$$I_1 = I_{1_c} + A_1 d_1^2 = \tfrac{1}{12}(4)(1)^3 + 4(1.36 - 0.5)^2 = 3.29 \text{ in}^4$$
$$I_2 = I_{2_c} + A_2 d_2^2 = \tfrac{1}{12}(1)(3)^3 + 3(1.14)^2 = 6.15 \text{ in}^4$$
$$I = 9.44 \text{ in}^4$$

The maximum tensile stress is in the top layer of the beam, near the wall, and its magnitude is (assuming entirely elastic behavior)

$$\sigma_t = \frac{50,000(12)(2.64)}{9.44} = 167,500 \text{ psi}$$

The maximum compressive stress is in the bottom layer of the beam, and its magnitude is

$$\sigma_c = \frac{50,000(12)(1.36)}{9.44} = 86,400 \text{ psi}$$

Thus, the compressive stress is allowable, but the tensile stress is not. Either the load on the beam or the design of the beam must be changed. Note that cutting the load in half would not work but also turning the beam over would work.

EXAMPLE 7-3

Assume that the fuselage of the helicopter discussed in Example 6-8 is constructed with a stiffened thin skin where the skin provides most of the resistance to bending. Is it sufficient to have a 2-mm wall thickness if the allowable stress is 100 MPa? Assume a perfectly rectangular cross section 2 m in width and 2.5 m in height. Ignore any possible complications caused by windows, attachments, stiffeners, or buckling of the skin.

Solution

The stresses to check are those caused by the maximum moment of 90 kN·m. The moment of inertia to be used is for the area with the outer fiber distance c of 1.25 m. In the following, I_1 is for two rectangles for which the formula $\tfrac{1}{12}bh^3$ is valid. I_2 is for two thin rectangles that lie parallel to the neutral axis and so far from it that the definition of inertia (Ar^2) can be directly applied. Thus,

$$I = I_1 + I_2 = \tfrac{2}{12}(0.002)(2.5)^3 + 2(0.002)(2)(1.25)^2$$
$$= 1.77 \times 10^{-2} \text{ m}^4$$

The maximum stress is

$$\sigma = \frac{Mc}{I} = \frac{(90\ 000)(1.25)}{1.77 \times 10^{-2}} = 6.36\ \text{MPa}$$

The skin is sufficiently thick.

EXAMPLE 7-4

Members of an Arctic expedition team decide to use rectangular beams of ice for structural purposes. The cross section of the beams is chosen to be 10 in. wide × 15 in. deep. Ice has surprisingly low strength, so the maximum stress should be limited to 30 psi. What is the maximum length l of a simply supported beam if the expected peak load is 500 lb, concentrated anywhere along the beam?

Solution

The largest moment is caused by the load when it is at the center of the beam:

$$M_{max} = 125l, \qquad I = \tfrac{1}{12}(10)(15)^3 = 2812\ \text{in}^4$$

$$\sigma_{max} = 30 = \frac{125l(7.5)}{2812}, \qquad l = 90\ \text{in.}$$

EXAMPLE 7-5

A manufacturer of modern eyeglasses wishes to compare two designs for the bridge piece of a given eyeglass frame. One is a single beam with a square ($a \times a$) cross section. The other kind of bridge consists of two thin beams, each 1 mm × 1 mm, separated 1 cm center-to-center. What is the dimension a of the single beam that gives the same resistance to bending as the dual-beam bridge? Assume the same allowable stress for the two designs.

Solution

The same moment must be assumed for the two bridges. The problem is to compare the so-called section moduli, I/c. For the single bridge,

$$c_1 = \frac{a}{2}, \qquad I_1 = \frac{1}{12}a(a)^3 = \frac{a^4}{12}, \qquad \frac{c_1}{I_1} = \frac{6}{a^3}$$

For the dual bridge,

$$c_2 = 5.5\ \text{mm}, \qquad I_2 = 2\left[\frac{1^4}{12} + 1(5)^2\right], \qquad \frac{c_2}{I_2} = 0.1096\ \text{mm}^{-3}$$

The section moduli must be equal since

$$\sigma_{max} = M\frac{c_1}{I_1} = M\frac{c_2}{I_2}, \quad \text{so} \quad a = 3.8\ \text{mm}$$

The reader should analyze the bridges when bending is in a plane perpendicular to the plane assumed in this solution.

EXAMPLE 7-6

A wheelbarrow is designed for agricultural use where humans must do very heavy work. The distance from wheel to handle is 3 m (Fig. 7-11a). The two wood frame pieces are straight with constant circular cross sections. The maximum load may be concentrated or uniformly distributed, anywhere within the region 30 cm from the axle to 2 m from the axle. It is assumed that the force applied at each handle may be as much as 1.5 kN. What is the minimum diameter of the frame pieces if the stress should not exceed 30 MPa? Would there be a reason to try to make the frame members not constant in cross section?

Solution

It is necessary to determine the severest possible loading. This is shown in Fig. 7-11b. The maximum moment is

$$M_{max} = 1500(2.7) = 4050 \text{ N} \cdot \text{m}, \qquad \sigma_{allowable} = \frac{M_{max}r}{I_{min}}$$

$$3 \times 10^7 = \frac{4050r}{\pi r^4/4}, \qquad r = 5.56 \text{ cm}$$

Fig. 7-11. Example 7-6.

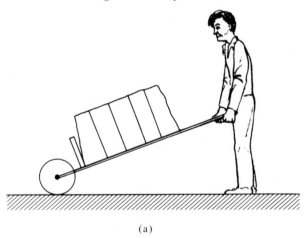

(a)

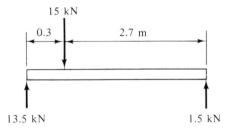

15 kN

0.3 | 2.7 m

13.5 kN   1.5 kN

(b)

This is too large for human hands, so the pieces should be reduced to about $r = 2$ cm at least at the grips or tapered from the grips to the cargo platform.

7-2. PLASTIC DEFORMATIONS IN BEAMS

Many important materials are sufficiently ductile to allow some plastic strains in bending without leading to fracture. Thus, an elementary understanding of yielding and resulting stresses in beams is desirable.

The discussion of this is based on the distribution of strains in bent beams. The vertical lines in Fig. 7-12a represent flat cross-sectional planes in a beam that is not loaded. When the beam is deformed in pure bending, these planes tend to remain flat but are not parallel to each other anymore. This means that the strains vary linearly through the depth of the beam at a given section. The stresses also vary linearly, according to the stress-strain curve, as long as there is no yielding.

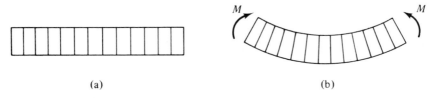

(a)   (b)

Fig. 7-12. Cross-sectional planes in pure bending.

If yielding is inevitable, it always occurs first in the outermost layers of homogeneous beams. The plastic deformation spreads inward as the external moment increases on the beam. Perhaps it is not difficult to accept that in this process the originally flat planes still tend to remain flat but the angles between them increase. Thus, the strains vary linearly through the depth of the beam even when part or all the beam has yielded. On the other hand, the stresses are linearly distributed only if there is no yielding. In a general case, the stress must be determined from the local strain in the beam (the geometry of the situation) and from the uniaxial stress-strain curve of the material. This means that two identical beams made from different materials have identical strain distributions if bent into the same shape, but the internal stresses (and the external loading required to accomplish the bending) are not the same in the two beams.

The strain and stress distributions in an idealized material are shown in Fig. 7-13. The material is assumed to be ideally elastic at small strains and ideally plastic at large strains (there is no real material exactly like this). It is best to think of the successive diagrams of this figure on the basis of imposing certain strains on the beam. A small bending moment is applied first, the beam is bent slightly, and the maximum strain ϵ_1 results from the change in geometry. The corresponding stress is σ_1; it was found from Fig. 7-13a. A larger moment is applied; the beam bends farther, and the peak strain is ϵ_2 and the peak stress is σ_2. Another increase in the bending moment causes a maximum strain ϵ_3 and a linear distribution of

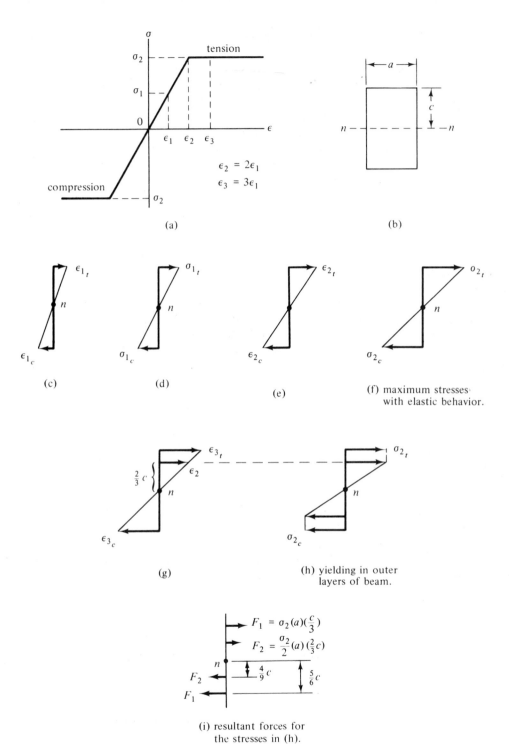

Fig. 7-13. Strain and stress distributions in a beam for increasing bending moments.

strains as in Fig. 7-13g. The material cannot have stresses, however, that are larger than σ_2 in Fig. 7-13a. Thus, the stress distribution becomes "clipped" as shown in Fig. 7-13h. It is possible to determine the distance from the neutral axis where the break in the stress distribution is. It is the point where the strain is ϵ_2. From similar triangles, this is at $\frac{2}{3}c$ in both directions from the neutral axis.

The total resisting moment generated by the stresses in Fig. 7-13h can be calculated, but it is slightly complicated. The method is to find the resultant force and its location for each segment of linear stress distribution. These resultants are shown in Fig. 7-13i. The total resisting moment is

$$M = 2F_1(\tfrac{5}{6}c) + 2F_2(\tfrac{4}{9}c)$$

This moment is slightly larger than the largest moment that can be applied to the beam without causing any yielding.

The preceding analysis is valid if the neutral axis does not shift because of yielding in the beam. In our case there is no reason to expect such a shift; the depth and width of the plastically deformed zone is the same at the top and the bottom of the beam, and the stress is the same at all points within these zones. In practice, it is difficult to make accurate analyses of plastic deformations in beams. Stress-strain curves are complex, and the depth of yielding is hard to predict.

7-3. NONHOMOGENEOUS BEAMS

There are many structural members that are not homogeneous; the flexure formula is not directly applicable to such beams. The special considerations necessary in the analysis are given here for three kinds of nonhomogeneous beams.

Symmetric Composite Beams. Consider a beam of material a with thin plates of a stronger, more rigid material b firmly joined to its top and bottom as shown in the cross-sectional view in Fig. 7-14a. As in Section 7-2, the concept of linearly varying strain is used. Consequently, the stress varies linearly in material a, but it increases abruptly in material b since $E_b > E_a$. The total moment at the cross section can be calculated using Fig. 7-14b (summation of two moments). For this, it can be

Fig. 7-14

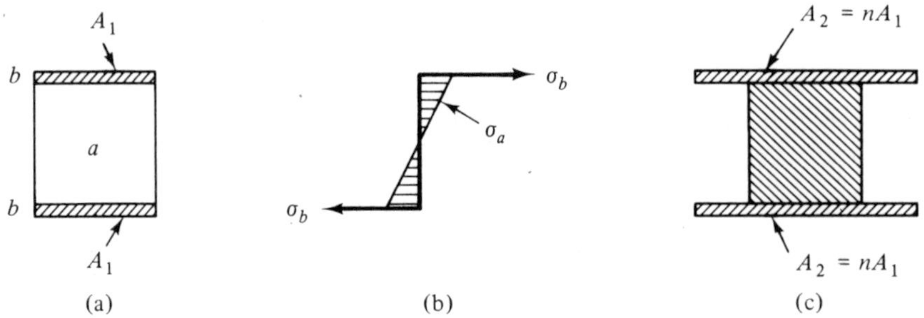

(a)   (b)   (c)

assumed that the strain in the thin plates is the same as the strain in the outer fibers of material a and that the stress is constant in the thin plates.

An equivalent procedure is to find the cross-sectional shape for a beam made of material a alone that behaves as the composite beam would. This is shown in Fig. 7-14c. The new area A_2 of material a must be larger than the area A_1 for material b because a given strain must result in the same force on them. Furthermore, areas A_1 and A_2 must be equivalent when moments are calculated, so they must be at the same distance from the neutral axis. For the equality of forces on A_1 and A_2,

$$E_b \epsilon A_1 = E_a \epsilon A_2 = E_a \epsilon n A_1$$

so

$$n = \frac{E_b}{E_a} \qquad (7\text{-}2)$$

where n is the ratio of the new area to the area replaced.

The flexure formula can be applied directly to the equivalent section shown in Fig. 7-14c. It must be remembered, however, that Fig. 7-14b is valid and fairly simple to use to calculate the total moment.

The reader is warned about using the flexure formula to calculate stresses. Calculating M using a "pretend" σ_a on a "pretend" area A_2 works. The "pretend" σ_a is *not* the actual stress σ_b, however, when the formula is inverted.

Nonsymmetric Composite Beams. It is slightly more complicated to analyze a nonsymmetric composite beam such as that in Fig. 7-14a without the bottom reinforcing plate. In this case the equivalent area A_2 is present only at the top in Fig. 7-14c. The moment can be calculated only after determining the position of the neutral axis. With this information, the flexure formula can be used for the homogeneous equivalent section.

A special case of nonsymmetric composite beams is a *reinforced concrete beam*. Concrete has practically no strength in tension, so reinforcing steel rods are often used only on the tension side of a beam as shown in Fig. 7-15a. Figure 7-15b shows the equivalent area* for the steel rods. The crosshatched areas are the effective areas for resisting a bending moment. The resultant compressive force C acts at the centroid of the assumed triangular stress distribution as shown in Fig. 7-15c. T must be equal and opposite to C.

*The equivalent area method for reinforced concrete is instructive in the use of mechanics principles. This conventional theory has been gradually replaced in design practice, however, during the last 2 decades by the *ultimate strength theory* for beams. This theory is based on the recognition that the compressive stress distribution at failure is not triangular as in Fig. 7-15c but approximately parabolic as in Fig. 7-15d. The magnitude and location of the resultant force C for such a distribution is determined in practice by referring to experimental results. This empirical approach has been used in building codes because it is most realistic in establishing structural safety. For further details about these theories and other aspects of reinforced concrete beams, see Phil M. Ferguson, *Reinforced Concrete Fundamentals* (New York: John Wiley & Sons, Inc., 1958); Chu-Kia Wang and Charles G. Salmon, *Reinforced Concrete Design*, 2nd ed. (New York: Intext Educational Publishers, 1973).

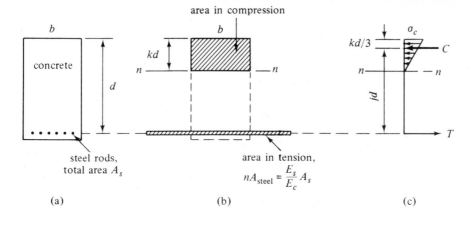

area in compression

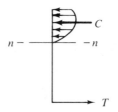

(d) real stress distribution
at failure.

Fig. 7-15

The position of the neutral axis is determined by equating the first moments of the crosshatched areas in Fig. 7-15b with respect to the neutral axis (which is assumed to be at the bottom of the upper area):

$$bkd\left(\frac{kd}{2}\right) = nA_s(d - kd) \tag{7-3}$$

This quadratic equation can be solved directly for kd.

The forces C and T form a couple with the moment arm jd, which can

be calculated now:

$$jd = d - \frac{kd}{3} \tag{7-4}$$

The next step is to relate the stresses in the concrete and in the steel to the external moment M applied at the given cross section. The forces C and T can be written in terms of stresses:*

$$C = \frac{bkd}{2}\sigma_c \tag{7-5}$$

$$T = A_s\sigma_s \tag{7-6}$$

*It is common practice to use the symbols f_c and f_s instead of σ_c and σ_s when working with reinforced concrete beams. The latter are used here to simplify matters for the student.

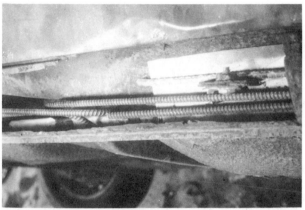

An unusual application of reinforced concrete: (a) The rocker panel of this car was so rusted that the door could not be closed. (b) Steel reinforcing rods in rocker panel opening. (c) The opening is filled with concrete (the repair was successful). *Courtesy J. Dreger, University of Wisconsin, Madison, Wis.*

where σ_c is the peak stress in the concrete (top layer in Fig. 7-15) and σ_s is the stress in the steel rods. The latter can be assumed constant over the cross section of any rod.

Without considering the relative strengths of the two materials,

$$M = C(jd) = T(jd) \tag{7-7}$$

There is no assurance in general, however, that the critical values of σ_c and σ_s are reached at the same time as the moment M is gradually increased (this would be ideal, and it can be achieved in design). When the strength of the concrete is critical (crumbling in compression before the rods fail in tension),

$$M_{\text{max}_c} = \frac{bkd}{2} jd\sigma_{cu} \tag{7-8}$$

When the tensile strength of the rods is critical,

$$M_{\text{max}_s} = A_s jd\sigma_{su} \tag{7-9}$$

where the subscript u denotes the ultimate strength or maximum allowable stress in a particular situation.

Another way to state the limitation for the applied moment is that it should not exceed the lower of the two maximum values computed using the last two equations. If there is a big difference between the computed maximum moments, this indicates that there is either too little or too much reinforcing steel in the beam.

Unequal Moduli in Single Material. Some beams are made of the same average material throughout; yet they are nonhomogeneous in bending if the elastic moduli in tension and compression are not equal. A typical example is cast iron. The flexure formula can be used for such a beam after the neutral surface is located and the equivalent section for a homogeneous beam is established.

EXAMPLE 7-7

A composite beam has a square cross section as shown in Fig. 7-16. The maximum allowable stresses are 3000 psi for the wood and 30,000 psi for the steel plates. What are the maximum bending moments for bending with respect to the x-axis and the y-axis? $E_w = 2 \times 10^6$ psi, $E_s = 30 \times 10^6$ psi.

Solution

For bending about the x-axis, the maximum strains allowable in the outer fibers are

$$\epsilon_{w_{\text{max}}} = \frac{3000}{2 \times 10^6} = 0.0015,$$

$$\epsilon_{s_{\text{max}}} = \frac{30,000}{30 \times 10^6} = 0.001$$

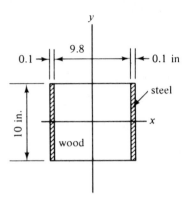

Fig. 7-16. Example 7-7.

Thus, the steel is limiting. Then $\epsilon(y) = 0.0002y$, so

$$\sigma_w(y) = 2 \times 10^6 \epsilon(y) = 400y, \qquad \sigma_s(y) = 30 \times 10^6 \epsilon(y) = 6000y$$

The total moment $M = M_w + M_s$. Using $M = \int \sigma(y)y\,da$,

$$M = \int_{-5}^{5} 400y^2(9.8)dy + 2\int_{-5}^{5} 6000y^2(0.1)dy = 4.27 \times 10^5 \text{ in.-lb}$$

For bending about the y-axis, again the steel is limiting, and $\epsilon(x) = 0.0002x$.

Then
$$\sigma_w(x) = 400x, \qquad \sigma_s(x) = 6000x$$

$$M = \int_{-4.9}^{4.9} 400x^2(10)dx + 2\int_{4.9}^{5.0} 6000x^2(10)dx = 6.08 \times 10^5 \text{ in.-lb}$$

EXAMPLE 7-8

A 0.5-mm thick layer of graphite-epoxy composite is joined to the bottom of a timber beam as shown in Fig. 7-17a. The applied moment of 1000 N·m is positive

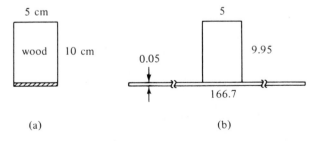

(a)   (b)

Fig. 7-17. Example 7-8.

so that the graphite filaments are in tension. What are the maximum stresses in the wood and in the graphite-epoxy? Assume in the analysis that the latter is not a composite itself but a homogeneous layer. $E_w = 15$ GPa, $E_{ge} = 500$ GPa.

Solution

The equivalent beam of wood is shown in Fig. 7-17b. The neutral axis is at $\bar{y} = 4.31$ cm from the bottom. I about this axis is 5.89×10^{-6} m⁴, so

$$\sigma_{W\,\text{tensile}} = \frac{1000(0.0426)}{5.89 \times 10^{-6}} = 7.23 \text{ MPa}$$

$$\sigma_{W\,\text{compressive}} = \frac{1000(0.0569)}{5.89 \times 10^{6}} = 9.66 \text{ MPa}$$

$$\sigma'_{ge\,\text{tensile}} = \frac{1000(0.0431)}{5.89 \times 10^{-6}} = 7.31 \text{ MPa}$$

The actual stress in the graphite-epoxy is

$$\sigma_{ge} = \frac{E_{ge}}{E_w}\sigma'_{ge} = 244 \text{ MPa}$$

Example 7-9

A reinforced concrete beam is 10 in. wide and 18 in. deep. The steel rods have a total cross-sectional area of 1.5 in² and they are in a single layer that is 2 in. from the bottom edge of the beam. Determine the maximum stresses in both materials when a bending moment of 40,000 ft-lb is applied. Assume $n = 12$.

Solution

The location of the neutral axis, kd, is determined first, using Eq. 7-3:

$$10kd\left(\frac{kd}{2}\right) = 18(16 - kd), \qquad (kd)^2 + 3.6kd - 57.6 = 0, \qquad kd = 6 \text{ in.}$$

The moment arm, jd, is

$$jd = 16 - \frac{kd}{3} = 14 \text{ in.}$$

The maximum compressive stress in the concrete is

$$\sigma_c = \frac{2M}{bkd\,jd} = \frac{80,000(12)}{10(6)(14)} = 1143 \text{ psi}$$

The stress in the steel rods is

$$\sigma_s = \frac{M}{A_s\,jd} = \frac{40,000(12)}{1.5(14)} = 22.86 \text{ ksi}$$

Example 7-10

A reinforced concrete beam is planned to have $b = 26$ cm, $d = 40$ cm, $A_s = 12$ cm², and $n = 15$. The maximum allowable stresses are $\sigma_{cu} = 6$ MPa and $\sigma_{su} =$

140 MPa. What is the maximum bending moment that can be applied? Is the amount of steel in the beam about right?

Solution

Starting the analysis as in Example 7-9,

$$26kd\left(\frac{kd}{2}\right) = 180(40 - kd), \qquad (kd)^2 + 13.8kd - 554 = 0$$

$$kd = 17.63 \text{ cm} = 0.1763 \text{ m}, \qquad jd = 40 - \frac{kd}{3} = 34.12 \text{ cm} = 0.3412 \text{ m}$$

The concrete is loaded to its limit by a moment of

$$M_{\text{max}_c} = \frac{bkdjd}{2}\sigma_{cu} = 0.13(0.176)(0.341)(6 \times 10^6)$$
$$= 46.8 \text{ kN·m}$$

The steel rods are loaded to their limit by

$$M_{\text{max}_s} = A_s jd\sigma_{su} = \frac{12}{10^4}(0.341)(140 \times 10^6) = 57.8 \text{ kN·m}$$

The smaller of these moments is the critical quantity. The amount of steel is a little too much.

EXAMPLE 7-11

Consider a uniform beam with a rectangular cross section as shown in Fig. 7-18a. The elastic moduli for the material are 20×10^6 psi in compression and 15×10^6 psi in tension. A positive moment of 10,000 ft-lb is applied, so the bottom layer is in tension. Determine the maximum tensile and compressive stresses.

Fig. 7-18. Example 7-11.

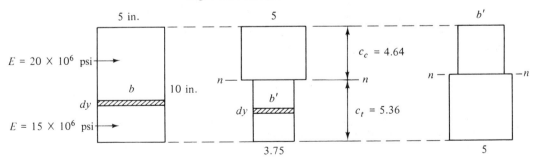

(a) (b) $E' = 20 \times 10^6$ psi everywhere. (c) $E' = 15 \times 10^6$ psi everywhere.

Solution

The neutral surface is located at the unknown distances c_c and c_t from the layers of maximum compression and tension, respectively (Fig. 7-18b). There are two equivalent areas that one can think of: the area with the compressive modulus throughout or the area with the tensile modulus throughout. The first is shown in Fig. 7-18b; the second, in Fig. 7-18c. The total depth is 10 in. in both cases, but the neutral axes have different locations. This solution will use the compressive modulus throughout.

It is required that the elemental force dF on any thin strip da parallel to the neutral axis be the same for the original area and the equivalent area (static equivalence for forces). This means that σda for the original area must equal $\sigma' ba'$ for the new area. The strain at a given distance from the neutral axis is the same in the two

cases, so
$$E\epsilon da = E'\epsilon da'$$

Since $da = bdy$ and $da' = b'dy$,

$$b' = \frac{E}{E'}b$$

The width of the equivalent area is proportional to the ratio of the old modulus E to the new modulus E', which is now the modulus for the whole cross section. On this basis, the width of the new area in Fig. 7-18b is 3.75 in.

The location of the neutral axis can be determined by finding the centroid:

$$5(c_c)\left(\frac{c_c}{2}\right) = 3.75(10 - c_c)\left(\frac{10 - c_c}{2}\right), \qquad c_c = 4.64 \text{ in.}$$

The moment of inertia of the new area is (using $\frac{1}{3}bh^3$ for each rectangle with respect to the neutral axis)

$$I = \tfrac{1}{3}(5)(4.64)^3 + \tfrac{1}{3}(3.75)(5.36)^3 = 359 \text{ in}^4$$

The stresses are calculated using the flexure formula. On the top of the beam,

$$\sigma_c = \frac{Mc_c}{I} = 1550 \text{ psi} \qquad \text{(compressive)}$$

and on the bottom of the beam

$$\sigma_t' = \frac{Mc_t}{I} = 1790 \text{ psi} \qquad \text{(tensile)}$$

It is important to remember that the stress σ_t' is for the equivalent area (thus, it is an imaginary stress). In the original beam the tensile stress is

$$\sigma_t = \frac{E}{E'}\sigma_t' = \frac{15}{20}(1790) = 1343 \text{ psi} \qquad \text{(tensile)}$$

Since some materials have relatively low shear strengths, it is necessary to investigate where the maximum shear stresses occur in beams subjected to bending. There are two possible causes of shear stresses in bending. One is that the normal stress calculated from the flexure formula has a shear stress associated with it. This stress is on planes that are at 45° with the cross-sectional planes, and its magnitude is $\tau = \sigma/2$.

The other causes of shear stresses are the internal shearing forces that exist in most members. Pure bending is relatively rare, even though the preceding discussions were based on that for convenience. In general, one should expect an internal moment M and a shear force V as shown in Fig. 7-19a. The shear stresses acting on the cross-sectional area are called transverse shear stresses. Recalling the nature of shear stresses, it is clear that there must be shear stresses on horizontal planes in the beam also. These are called *longitudinal shear* stresses, but they are equal in magnitude to the transverse stresses in the same region in the material. The longitudinal shear stresses are easy to visualize as similar to the frictional stresses between the slabs in Fig. 7-20 as they slide slightly against each other when the load is applied.

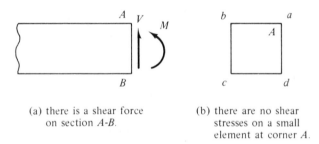

(a) there is a shear force
on section *A-B*.

(b) there are no shear
stresses on a small
element at corner *A*.

Fig. 7-19. Internal reactions in a beam.

The same concept that makes one realize that transverse and longitudinal shear stresses exist together leads to a difficult problem. For example, consider a small rectangular element at corner A in Fig. 7-19a. It must be free of shear stresses as shown in Fig. 7-19b. Face ab is the top free surface of the beam, and there are no stresses of any kind on it. Consequently, there can be no shear stress (but there may be a normal stress) on face ad. But what is the situation if it is known

Fig. 7-20. Model to show shear stresses in a beam.

(a)

(b)

that there is a shear force on face AB in Fig. 7-19a? The equilibrium conditions for shear stresses on a small element must be accepted as fundamental and they cannot be relaxed. Thus, there are no transverse or longitudinal shear stresses at points A and B in Fig. 7-19a.

The only way out of the dilemma, although not entirely satisfactory philosophically, is to say that the shear stresses increase from zero at the free surfaces to large values inside the beam. Figure 7-21 shows two schematic ways of showing this. Obviously, τ_{max} must be larger than

$$\tau_{ave} = \frac{V}{\text{cross-sectional area}}$$

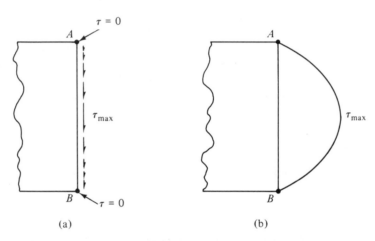

Fig. 7-21. Two ways of showing distribution of shear stresses on rectangular cross section of a beam; A-B is edge view of cross section.

Since the shear stresses are not constant over the cross section, it is necessary to seek a new way of determining their distribution. A fruitful approach is based on the concept that an internal shear force exists when the bending moment varies along the beam. The formal statement of this is

$$V = \frac{dM}{dl} \tag{7-10}$$

where l is measured along the beam.

The longitudinal shear stress τ_l can be determined as follows. A beam is loaded as shown in Fig. 7-22. Sections A and B are near each other, but the bending moment is slightly larger at B then at A: $M_B - M_A = dM$. Next, the stresses are drawn for a small element $abcd$ (Fig. 7-22c). The unknown shear stresses τ_A and τ_B are equal in magnitude. The unknown shear stress τ_l is assumed to be constant over the small distance dl. In Fig. 7-22d only the horizontal resultant forces on the element are shown. For equilibrium,

$$\tau_1 dl\, t = N_{bc} - N_{ad} \qquad (7\text{-}11)$$

$$N_{ad} = \int_{y_0}^{c} \sigma_A(y)\, da \quad \text{and} \quad N_{bc} = \int_{y_0}^{c} \sigma_B(y)\, da$$

where $\sigma_A(y)$ and $\sigma_B(y)$ are the normal stresses at a distance y from the neutral axis at sections A and B, respectively, and $da = t\, dy$. Recognizing that

$$\frac{\sigma_A(y)}{y} = \frac{\sigma_a}{c} = \text{a constant} \quad \text{and} \quad \frac{\sigma_B(y)}{y} = \frac{\sigma_b}{c} = \text{a constant}$$

$$N_{ad} = \frac{\sigma_a}{c} \int_{y_0}^{c} y\, da \quad \text{and} \quad N_{bc} = \frac{\sigma_b}{c} \int_{y_0}^{c} y\, da$$

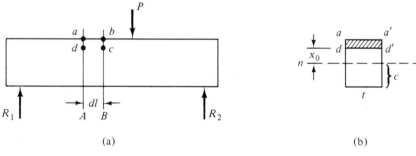

(a)   (b)

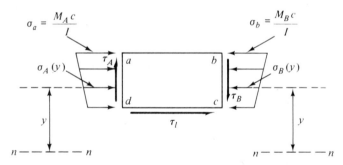

(c) enlarged free-body diagram of small
element from (a).

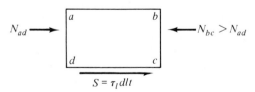

(d) resultant horizontal forces on
element in (c).

Fig. 7-22. Model for the determination of shear stresses in a beam.

The integral is the first moment of area $aa'dd'$ in Fig. 7-22b with respect to the neutral axis. It is commonly denoted by the letter Q. Thus, Eq. 7-11 can be written

as
$$\tau_l dlt = \frac{\sigma_b - \sigma_a}{c}(Q) \tag{7-12}$$

where
$$\frac{\sigma_b - \sigma_c}{c} = \frac{M_B - M_A}{I} = \frac{dM}{I}$$

With the introduction of Eq. 7-10, Eq. 7-12 can be rearranged to obtain a simple expression for the longitudinal shear stress:

$$\tau_l = \frac{VQ}{It} \tag{7-13}$$

where τ_l = the average shear stress on a horizontal plane (in these drawings) at an arbitrary distance y_0 from the neutral axis

V = total shear force on the cross-sectional area of the beam

Q = first moment of a segment of the cross-sectional area about the neutral axis; the segment is between y_0 and c (between plane on which τ_l is acting and outer face of beam)

I = moment of inertia of entire cross-sectional area about the neutral axis

t = width of plane on which τ_l acts

The shearing force per unit width or length is called q, the shear flow. From

Eq. 7-13,
$$q = \tau_l t = \frac{VQ}{I} \tag{7-14}$$

Shear Flow in Thin-walled Beams. A special case of shear stresses must be considered in beams that have thin flanges or walls. The reason for this is explained with the aid of Fig. 7-23. A channel section is bent so that its axis of symmetry is also the neutral axis. Consider the enlarged view of a small element in the top flange of this beam. If the bending moment varies along the beam, the normal forces on faces 1 and 3 are different. The shear force S on face 4 helps in maintaining equilibrium, but this gives rise to the shear force H on face 1. Of course, there must be shears also on faces 2 and 3, except when face 2 is the free edge of the flange. A similar element in the bottom flange has normal and shear forces in the opposite directions to those shown in the top flange.

The qualitative view of the shear flow distribution over the cross section of the channel is shown in Fig. 7-23b. The shear flow is zero at point A, gradually increases to a maximum at B, and gradually decreases to zero at point C. The distribution in the vertical part is similar to what it would be if there were no horizontal flanges, except that the shear stress does not drop to zero at the corners. The philosophical difficulty with the combined vertical and horizontal shears is also

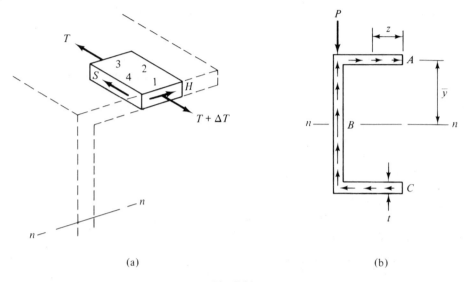

(a)   (b)

Fig. 7-23

similar to what has been discussed previously for the vertical shear stress distribution alone.

The quantitative determination of the shear flow at any given point P in the flange is based on the same principles as those used in the analysis of vertical shear. For point P at a distance z from the free edge of the flange (Fig. 7-23b),

$$q_H = \frac{VQ}{I} = \frac{Vtz\bar{y}}{I} \tag{7-15}$$

where $\bar{y}$, the distance from the neutral axis, is a constant for a given channel; this is in contrast to the vertical shear flow. Accordingly, the horizontal shear flow increases linearly from the free edge of the flange to the corner.

The ideas for the shear flow in the channel section can be extended to other shapes. The simplest of these is the I section, which should be visualized as two channel shapes placed back-to-back as in Fig. 7-24a. The box beam shape is two channels turned flange-to-flange (Fig. 7-24b). The cylindrical tube is similar to the box beam (Fig. 7-24c); the shear flow for this must be given in cylindrical coordi-

Fig. 7-24

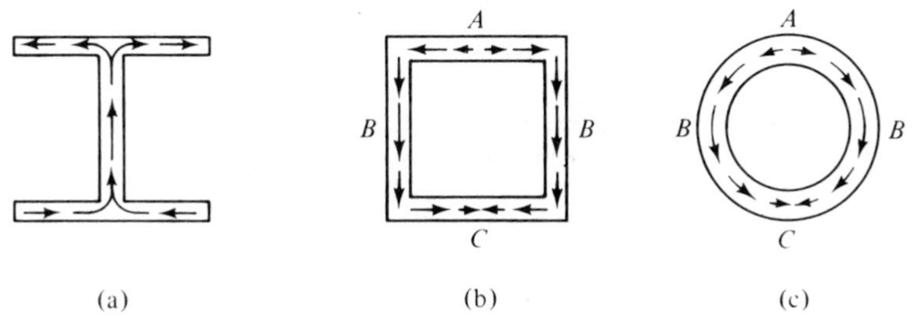

(a)   (b)   (c)

nates. For both tubular shapes, the shear flow is zero at point A, increases to the maximum at points B, and decreases to zero at point C on the other side.

Shear Center. It is seen from Fig. 7-23b that the beam is in equilibrium in the horizontal direction. The vertical shear flow is opposed by the external shear force. There is a net moment, however, caused by the total shear flow. This tends to twist the beam even in the absence of an external torque on it. The twisting can be prevented only if there is an external moment on the beam that opposes the moment caused by the shear flow. A simple way to achieve this is to apply the bending load eccentrically to cause not only bending but also twisting of the beam, the latter being of the right magnitude to cancel the twisting caused by the shear flow.

In most cases, the magnitude of the external load cannot be altered to influence the extent of twisting of the beam. Changing the location of its line of action is often possible. This is shown in Fig. 7-25. Twisting of the beam does not

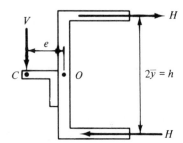

Fig. 7-25

occur when the resultant of the external loads passes through the shear center at point C. The required eccentricity, e, for this configuration is calculated by taking moments about point 0, eliminating the vertical shear flow from consideration:

$$e = \frac{hH}{V} \tag{7-16}$$

where H is the average shear flow in the flange times the length of the flange.

For members that have two axes of symmetry, the shear center is at the centroid.

EXAMPLE 7-12

Find the longitudinal shear stresses at the levels indicated on the cross section of a beam shown in Fig. 7-26a. The vertical shear is 10,000 lb.

Solution

At level 1, $\tau_l = 0$ because this is a free surface. Also, $Q = 0$ in Eq. 7-13. The moment of inertia is needed for finding the other stresses.

$$I = \tfrac{1}{12}(1)(8)^3 + \tfrac{1}{12}(2)(4)^3 = 53.3 \text{ in}^4$$

At level 2,

$$Q_2 = A_2 \bar{y}_2 = 1(3.5) = 3.5 \text{ in}^3 \quad \text{and} \quad \tau_{l_2} = \frac{VQ_2}{It_2} = \frac{10{,}000(3.5)}{53.3(1)} = 660 \text{ psi}$$

At level 3, $Q_3 = A_3 \bar{y}_3 = 2(3) = 6 \text{ in}^3$, but here t changes abruptly from 1 to 3 in. Let us evaluate τ_l just above the discontinuity and just below it:

$$\tau_{l_3} = \frac{10{,}000(6)}{53.3(1)} = 1130 \text{ psi} \quad \text{(above)}$$

$$\tau'_{l_3} = \frac{10{,}000(6)}{53.3(3)} \; 380 \text{ psi} \quad \text{(below)}$$

At level 4,

$$Q_4 = 3(2.5) + 2(1)(1.5) = 10.5 \text{ in}^3 \quad \text{and} \quad \tau_{l_4} = \frac{10{,}000(10.5)}{53.3(3)} = 660 \text{ psi}$$

At level 5,

$$Q_5 = 4(2) + 2(2)(1) = 12 \text{ in}^3 \quad \text{and} \quad \tau_{l_5} = \frac{10{,}000(12)}{53.3(3)} = 750 \text{ psi}$$

Fig. 7-26. Example 7-12.

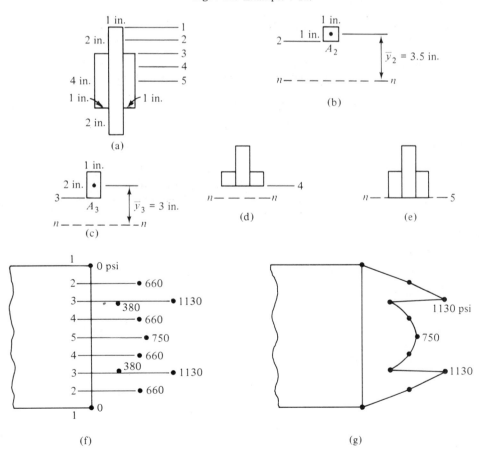

The relative magnitudes of the shear stresses are plotted with respect to location in the beam in Fig. 7-26f where the symmetry of the beam was also considered. In Fig. 7-26g a line is drawn through the 11 points plotted to show the approximate variation of shear stress with respect to depth in the beam.

If the beam is made of the same material throughout, one should be interested only in the maximum shear stress. Thus, with a little experience it is possible to reduce the number of calculations in this example. It is sufficient to find the shear stresses at level 3 just above the discontinuity and at level 5, because Q is always large at the neutral axis and t can be relatively small at a discontinuity.

EXAMPLE 7-13

What is the minimum acceptable shear strength of the bond between the wood and the layer of graphite-epoxy in Example 7-8 (Fig. 7-17)? Consider a transverse shear load of 2 kN.

Solution

I and $\bar{y}$ are the same as in Example 7-8. $V = 2$ kN, $t = 0.05$ m, and $Q = A\bar{y}$, where A and $\bar{y}$ refer to the original wood part. Q can be calculated in two ways: using the area above the joint in Fig. 7-17a,

$$Q = (0.05)(0.0995)(0.00715) = 3.56 \times 10^{-5} \text{ m}^3, \qquad \tau = \frac{VQ}{It} = 0.24 \text{ MPa}$$

Since the normal stress in the graphite-epoxy is 33.3 times as large as in the wood, so are the shear stresses. Hence the adhesive should be able to withstand

$$\tau = 33.3 \times 0.24 = 8 \text{ MPa}$$

EXAMPLE 7-14

Assume that a channel beam has 10-cm wide flanges, a web with $h = 25$ cm, and a thickness $t = 3$ mm. The shear force is 3 kN, acting perpendicular to the flanges. Calculate the maximum shear flow in the flanges, and determine the eccentricity of the loading necessary to prevent twisting of the beam.

Solution

$$I \cong \tfrac{1}{12}(0.003)(0.25)^3 + 2(0.003)(0.1)(0.125)^2$$
$$= 1.33 \times 10^{-5} \text{ m}^4$$
$$V = 3000 \text{ N}$$
$$Q = (0.003)(0.125)z = (3.75 \times 10^{-4})z \text{ m}^3 \qquad (z \text{ is according to Fig. 7-23b})$$
$$q = \frac{VQ}{I} = (8.46 \times 10^4)z \text{ N/m}$$
$$H = \int_0^{0.1} q\,dz = 423 \text{ N}$$

q_{max} occurs at $z = 0.1$ m

$$e = \frac{Hh}{V} = 0.035 \text{ m}$$

EXAMPLE 7-15

A channel beam has 4-in. wide flanges, a web with $h = 12$ in., and a thickness $t = 0.12$ in. A vertical shear force of 500 lb is applied in the plane of the web, perpendicular to the flanges. Determine the shear stress in the middle of the flanges and the net moment twisting the beam about its longitudinal axis.

Solution

$$I = \tfrac{1}{12}(0.12)(12)^3 + 2(0.12)(4)(6)^2 = 51.8 \text{ in}^4$$

$$Q = (0.12)(6)z = 0.72z$$

$$q = \frac{VQ}{I} = 6.95z \text{ lb/in.}$$

$$\tau \big|_{z=2} = \frac{q}{t} \bigg|_{z=2} = 116 \text{ psi}$$

$$H = \int_0^4 q\,dz = 55.6 \text{ lb}$$

$$T = Hh = 55.6 \text{ ft-lb}$$

7-5. CURVED BEAMS

The stress distribution is not linear in any beam that is sharply curved before the bending moment is applied even if the beam is homogeneous and behaves entirely elastically during loading. The reason is that the tensile and compressive stresses on fibers equidistant from the neutral axis are acting on fibers of different original length.

Consider the section of a curved beam in Fig. 7-27a where planes AB and CD are the radial planes before the beam is loaded. After loading, these planes rotate relative to each other as indicated by the angle $d\theta$. It is assumed that they remain plane during the loading (this is approximately true as can be proved with strain measurements). Accordingly, the deformations in fibers a and b are the same in magnitude, $e(a) = e(b)$, if a and b are equidistant from the neutral axis. Since the gage lengths of these fibers are different, the strains and the stresses in them are also different in magnitude.

From the definition of axial strain, $\epsilon(a) > \epsilon(b)$. Thus, $\sigma(a) > \sigma(b)$ if the elastic modulus in tension is the same as that in compression. This has the consequence that the neutral axis must shift from the centroid of the area toward the center of curvature O of the original beam (for equilibrium of forces). The expected stress distribution and position of the neutral axis are shown qualitatively, in contrast to the stresses calculated using the flexure formula, in Fig. 7-27b.

Handgrip exerciser; flaws in the straight section caused more severe stresses than the curvature of the member. *Photograph by R. Sandor, Madison, Wis.*

The quantitative determination of the stresses is based on the following analysis of Fig. 7-27a. The distance y to any fiber of interest is measured from the neutral axis; it is chosen positive toward the center of curvature.

The deformation $e(y)$ of a fiber at a distance y is given by

$$e(y) = -y\,d\theta$$

This is in accordance with no deformation at the neutral axis and compression above that axis.

To calculate the strain at any distance y from the neutral axis, it is necessary to know the fiber length $l(y)$ there. In terms of the radius of curvature and the

angle,
$$l(y) = (R - y)\theta$$

The longitudinal strain in $l(y)$ is

$$\epsilon(y) = \frac{e(y)}{l(y)} = \frac{-y\,d\theta}{(R - y)\theta} \tag{7-17}$$

Fig. 7-27

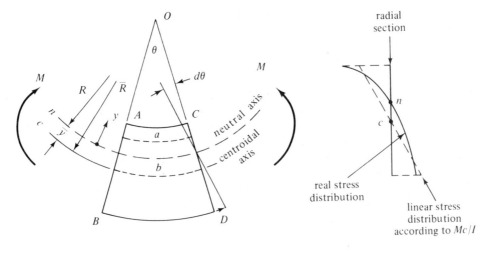

(a) initial curvature.

(b)

The next items of interest are the location of the neutral axis and the relation of the stress $\sigma(y)$ in any fiber $l(y)$ to the bending moment M. When there is no external axial force on the beam, the integral of stresses over the cross section is

zero: $$\int \sigma(y)da = E \int \epsilon(y)da = 0, \qquad -E \int \frac{yd\theta}{(R-y)\theta}da = 0$$

Since θ and $d\theta$ are constants with respect to this integration,

$$\int \frac{y}{R-y}da = 0 \tag{7-18}$$

The radius of curvature of any fiber $l(y)$ is $\rho = R - y$, and of the neutral axis it is R, a constant, so Eq. 7-18 can be written as

$$\int \frac{R-\rho}{\rho}da = R \int \frac{da}{\rho} - \int da = 0 \tag{7-19}$$

Solving for R, $$R = \frac{\int da}{\int da/\rho} \tag{7-20}$$

where both integrals are over the same cross-sectional area. R is the location of the neutral axis. Its evaluation can be quite complicated for all but some rectangular cross sections.

A similar procedure can be used to find an expression for the unknown stresses. The internal resisting moment is the summation of $\rho(y)yda$, for taking moments of elemental forces about the neutral axis. Thus,

$$M = \int \sigma(y)yda \tag{7-21}$$

In terms of the strain,

$$M = -\frac{Ed\theta}{\theta} \int \frac{y}{R-y}yda$$

With the integral written in terms of radii of curvature,

$$M = -\frac{Ed\theta}{\theta} \int \frac{(R-\rho)^2}{\rho}da$$

From Eq. 7-17, and using radii of curvature,

$$\sigma(y) = -\frac{E(R-\rho)d\theta}{\rho\theta}$$

In the last equation for the moment, $Ed\theta/\theta$ can be replaced by a term containing the radii and the unknown stress now written simply as σ:

$$M = \frac{\sigma\rho}{R-\rho}\int \frac{(R-\rho)^2}{\rho}\,da \qquad (7\text{-}22)$$

which can be expanded to give

$$M = \frac{\sigma\rho}{R-\rho}\left(R^2\int \frac{da}{\rho} - R\int da - R\int da + \int \rho da\right)$$

On the basis of Eq. 7-19,

$$R^2\int \frac{da}{\rho} - R\int da = 0$$

The third integral is equal to A, the whole cross-sectional area. The last integral is the first moment of the area with respect to the center of curvature, and it is equal to $\bar{R}A$ where $\bar{R}$ is the distance to the centroid of A. With these simplifications,

$$M = \frac{\sigma\rho}{R-\rho}(\bar{R}A - RA)$$

and the stress at a location denoted by the radius ρ,

$$\sigma = \frac{M(R-\rho)}{\rho A(\bar{R}-R)} \qquad (7\text{-}23)$$

In terms of the distance y from the neutral axis,

$$\sigma = \frac{My}{(R-y)Ay} \qquad (7\text{-}24)$$

The signs in this equation should be given as follows. A and R-y are always positive; $\bar{y}$ is always negative. With respect to Fig. 7-27a, $\bar{y} = -(\bar{R} - R)$. For positive M and y, the stress is negative (compressive) as can be expected from Fig. 7-27, and so forth.

The major limitation in using these equations is that the member should be deformed only elastically. The stresses caused by a net normal force acting on a cross section can be superimposed on those caused by the bending moment in most cases.

EXAMPLE 7-16

A curved beam has a rectangular cross section as shown in Fig. 7-28. The applied moment is 1000 ft-lb. Calculate the maximum stresses from the flexure formula as if the beam were straight. Calculate the stresses for centroidal radii of

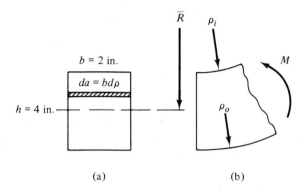

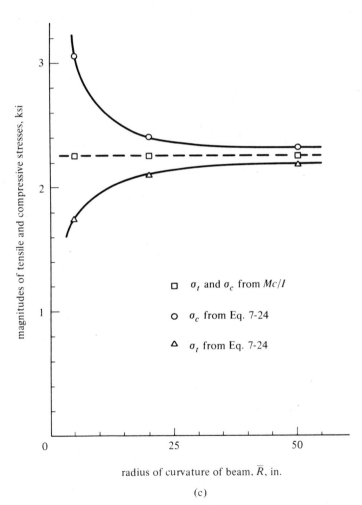

Fig. 7-28. Example 7-16.

$\bar{R} = 50$ in., 20 in., and 5 in. Plot all the stresses on the same diagram to show the effects of the different curvatures.

Solution

For the straight beam, the tensile and compressive stresses are the same in

magnitude: $\qquad \sigma_t = \sigma_c = \dfrac{Mc}{I} = \dfrac{1000(12)(2)}{\frac{32}{3}} = 2250$ psi

The next step is to find an expression for the radii of curvature of the neutral axis. With ρ_i and ρ_o as the limits of the integration,

$$R = \frac{\displaystyle\int_{\rho_i}^{\rho_o} da}{\displaystyle\int_{\rho_i}^{\rho_o} \frac{da}{\rho}} = \frac{hb}{b \displaystyle\int_{\rho_i}^{\rho_o} \frac{d\rho}{\rho}} = \frac{h}{\ln\left(\dfrac{\rho_o}{\rho_i}\right)}$$

For $\bar{R} = 50$,

$\qquad \rho_o = 52, \qquad \rho_i = 48, \quad$ and $\quad R = 49.9733$ in., $\qquad \bar{y} = -0.0267$ in.

For $\bar{R} = 20$,

$\qquad \rho_o = 22, \qquad \rho_i = 18, \quad$ and $\quad R = 19.933$ in., $\qquad \bar{y} = -0.067$ in.

For $\bar{R} = 5$,

$\qquad \rho_o = 7, \qquad \rho_i = 3, \quad$ and $\quad R = 4.721$ in., $\qquad \bar{y} = -0.279$ in.

Note that, in contrast to some calculations in strength of materials where rounding off answers is reasonable, it is necessary to perform very accurate calculations of the neutral axis because of the small changes in its position.

Using the appropriate value of y and $\bar{y}$, the stresses can be calculated. In each

case, $\qquad \sigma = \dfrac{My}{8(R-y)(\bar{y})} = \dfrac{12,000}{8} \dfrac{(\pm(h/2)+\bar{y})}{(\bar{R}\pm 2)(\bar{y})}$

since $R \pm y_{max} = \bar{R} \pm 2$ in.
For $\bar{R} = 50$,

$$\sigma_c = \frac{12,000}{8} \frac{(2-0.0267)}{(48)(-0.0267)} = -2310 \text{ psi} \qquad \text{(compressive)}$$

$$\sigma_t = \frac{12,000}{8} \frac{(-2.0267)}{(52)(-0.0267)} = 2190 \text{ psi} \qquad \text{(tensile)}$$

For $\bar{R} = 20$,

$$\sigma_c = \frac{12,000}{8} \frac{(2-0.067)}{(18)(-0.067)} = -2404 \text{ psi, C}$$

$$\sigma_t = \frac{12,000}{8} \frac{(-2.067)}{(22)(-0.067)} = 2103 \text{ psi, T}$$

For $\bar{R} = 5$,

$$\sigma_c = -3084 \text{ psi, C,} \qquad \sigma_t = 1750 \text{ psi, T}$$

Note that the calculated stresses can be significantly affected by small changes in $\bar{y}$. This is especially obvious when R is large. For example, unreasonable answers are obtained at $\bar{R} = 50$ using $\bar{y} = 0.025$ or 0.028.

The plots in Fig. 7-28c show that the results obtained using the flexure formula for a straight beam are increasingly in error as the radius of curvature decreases. The error is often tolerable for mildly curved beams.

7-6. PROBLEMS

Secs. 7-1 and 7-2

7-1. Considering moments alone, why is the beam shown in Fig. 7-2 not in equilibrium?

7-2. Determine the positions of the supports under the 20-ft long granite column to minimize the bending stresses in the column. How large a reduction in stresses could you obtain compared to the situation shown in Fig. 7-9a? The answer may be given as precise or approximate.

7-3. Calculate the ratio of maximum moments, M_a/M_b, for the two orientations of a given beam shown in Fig. 7-7.

7-4. Calculate the ratio M_p/M_e where M_p is the fully plastic bending moment (the whole cross section yields) and M_e is the maximum moment that does not cause any yielding. Assume that the stress-strain curve in Fig. 7-13a is valid and that the cross section is rectangular.

*In Prob. **7-5** to **7-44**, determine*
(a) *the maximum tensile and compressive stresses.*
(b) *the maximum normal stress at the cross section indicated by A if applicable.*
Assume that the beams are weightless and that horizontal forces are not acting on them.

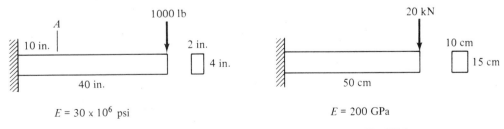

E = 30 x 10⁶ psi   E = 200 GPa

Fig. P7-5   **Fig. P7-6**

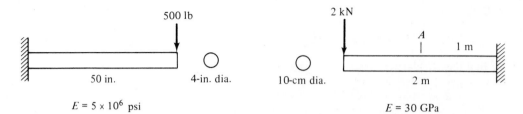

$E = 5 \times 10^6$ psi

Fig. P7-7

$E = 30$ GPa

Fig. P7-8

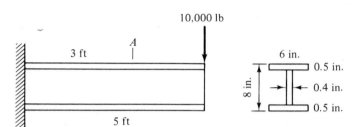

$E = 20 \times 10^6$ psi

Fig. P7-9

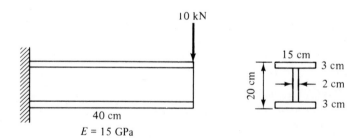

Fig. P7-10

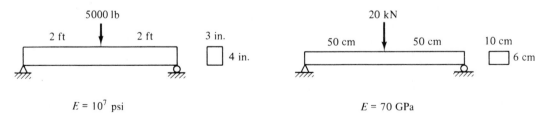

$E = 10^7$ psi

Fig. P7-11

$E = 70$ GPa

Fig. P7-12

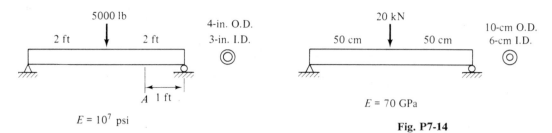

5000 lb

2 ft   2 ft

4-in. O.D.
3-in. I.D.

A 1 ft

$E = 10^7$ psi

Fig. P7-13

20 kN

50 cm   50 cm

10-cm O.D.
6-cm I.D.

$E = 70$ GPa

Fig. P7-14

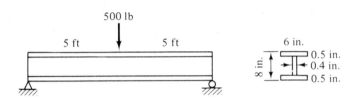

500 lb

5 ft   5 ft

6 in.

8 in.

0.5 in.
0.4 in.
0.5 in.

$E = 10^7$ psi

Fig. P7-15

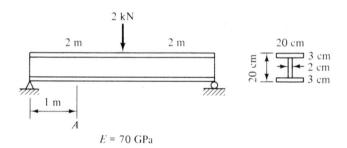

2 kN

2 m   2 m

1 m

A

20 cm

20 cm

3 cm
2 cm
3 cm

$E = 70$ GPa

Fig. P7-16

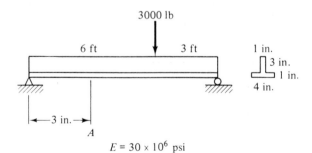

3000 lb

6 ft   3 ft

3 in.

A

1 in.
3 in.
1 in.
4 in.

$E = 30 \times 10^6$ psi

Fig. P7-17

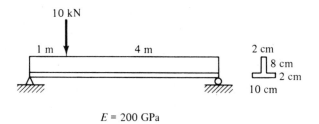

E = 200 GPa

Fig. P7-18

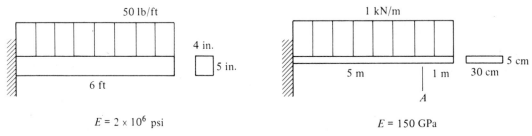

E = 2 × 10^6 psi

Fig. P7-19

E = 150 GPa

Fig. P7-20

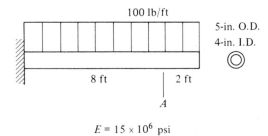

E = 15 × 10^6 psi

Fig. P7-21

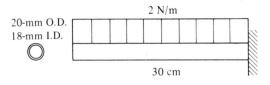

E = 120 GPa

Fig. P7-22

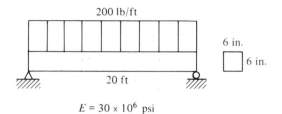

$E = 30 \times 10^6$ psi

Fig. P7-23

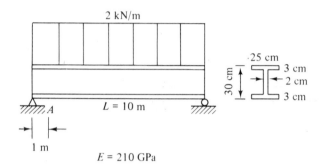

$E = 210$ GPa

Fig. P7-24

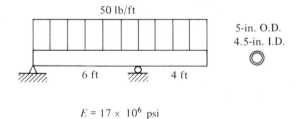

$E = 17 \times 10^6$ psi

Fig. P7-25

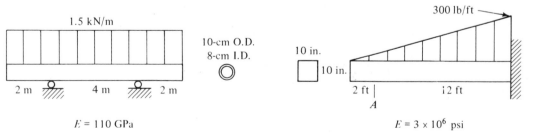

$E = 110$ GPa

Fig. P7-26

$E = 3 \times 10^6$ psi

Fig. P7-27

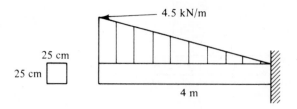

$E = 20$ GPa

Fig. P7-28

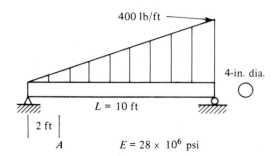

Fig. P7-29

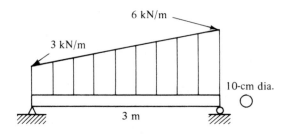

$E = 200$ GPa

Fig. P7-30

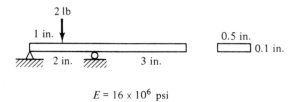

$E = 16 \times 10^6$ psi

Fig. P7-31

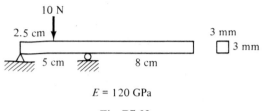

$E = 120$ GPa

Fig. P7-32

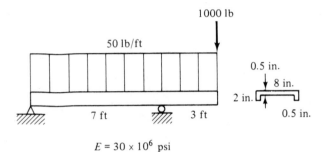

$E = 30 \times 10^6$ psi

Fig. P7-33

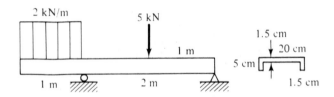

$E = 200$ GPa

Fig. P7-34

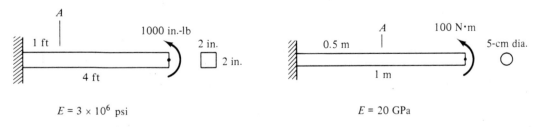

$E = 3 \times 10^6$ psi $E = 20$ GPa

Fig. P7-35   **Fig. P7-36**

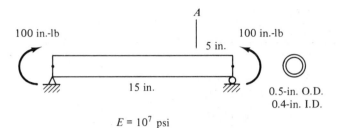

$E = 10^7$ psi

Fig. P7-37

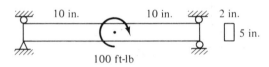

$E = 70$ GPa

Fig. P7-38

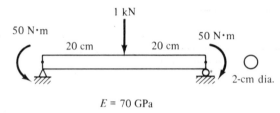

$E = 20 \times 10^6$ psi

Fig. P7-39

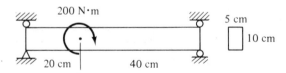

$E = 150$ GPa

Fig. P7-40

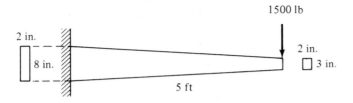

1500 lb

2 in.

8 in.

2 in.

3 in.

5 ft

$E = 20 \times 10^6$ psi

Fig. P7-41

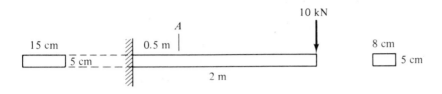

10 kN

A

15 cm

0.5 m

5 cm

2 m

8 cm

5 cm

$E = 150$ GPa

Fig. P7-42

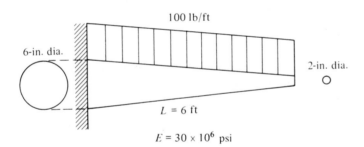

100 lb/ft

6-in. dia.

2-in. dia.

$L = 6$ ft

$E = 30 \times 10^6$ psi

Fig. P7-43

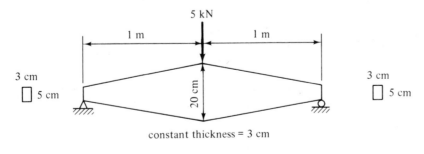

5 kN

1 m

1 m

3 cm

5 cm

20 cm

3 cm

5 cm

constant thickness = 3 cm

$E = 70$ GPa

Fig. P7-44

7-45. A wood beam 4 in. $\times$ 6 in. in cross section is reinforced with steel plates that are 0.05 in. thick and 6 in. wide. The resultant beam is similar to that in Fig. 7-16. Determine the maximum bending moment with respect to the x- and y-axes. $E_w = 2 \times 10^6$ psi, $E_s = 30 \times 10^6$ psi, $\sigma_{w\,max} = 2.5$ ksi, $\sigma_{s\,max} = 40$ ksi.

7-46. A wood beam 10 cm $\times$ 10 cm in cross section is reinforced all around with steel plates that are 1 mm thick. What is the increase in maximum bending moment after the reinforcement? $E_w = 15$ GPa, $E_s = 200$ GPa, $\sigma_{w\,max} = 16$ MPa, $\sigma_{s\,max} = 300$ MPa.

7-47. Redo Example 7-8 by adding a second layer of graphite-epoxy to the top of the beam.

7-48. Redo Example 7-8 assuming that the graphite-epoxy layer extends all around the beam.

7-49. A reinforced concrete beam has $b = 12$ in., $d = 20$ in., $A_s = 1.6$ in^2, $n = 14$. Determine the maximum stresses in both materials when $M = 70,000$ ft-lb.

7-50. A reinforced concrete beam has $b = 20$ cm, $d = 50$ cm, $A_s = 8$ cm^2, $n = 12$. Determine the maximum stresses in both materials when $M = 100$ kN$\cdot$m.

7-51. A reinforced concrete beam is planned with $b = 15$ in., $d = 36$ in., $A_s = 2.4$ in^2, $n = 15$, $\sigma_{cu} = 4000$ psi, $\sigma_{su} = 45$ ksi. Is the amount of steel reasonable in this beam?

7-52. A reinforced concrete beam is planned with $b = 30$ cm, $d = 55$ cm, $n = 12$, $\sigma_{cu} = 15$ MPa, $\sigma_{su} = 160$ MPa. Determine the optimum amount of steel in the beam.

7-53. A uniform beam has a rectangular cross section 12 cm $\times$ 20 cm. The elastic moduli are 110 GPa in compression and 80 GPa in tension. Determine the maximum tensile and compressive stresses when a moment of 10 kN$\cdot$m is applied. Assume the beam is oriented to minimize the stresses in it.

7-54. A uniform beam has a rectangular cross section 4 in. $\times$ 6 in. The elastic moduli are 18×10^6 psi in compression and 15×10^6 psi in tension. A moment of 10,000 ft-lb is applied so that the short sides have the maximum stresses. What is the largest possible error in each maximum stress if it is calculated directly from the flexure formula, assuming a uniform elastic modulus in tension and compression?

Sec. 7-4

7-55. Calculate the ratio of maximum and average transverse shear stresses τ_{max}/τ_{ave}, for a beam with a square cross section.

7-56. Determine the maximum longitudinal shear stress in selected beams from Probs. 7-5 to 7-44.

Determine the shear center for the following channel beams in which the shear flow is as in Fig. 7-23b.

7-57. Web $= 10$ in., flanges $= 4$ in., $t = 0.2$ in., $V = 2000$ lb.

7-58. Web $= 30$ cm, flanges $= 10$ cm, $t = 0.6$ cm, $V = 20$ kN.

7-59. Web $= 6$ in., flanges $= 1.5$ in., $t = 0.1$ in., $V = 600$ lb.

7-60. Web $= 3$ cm, flanges $= 3$ cm, $t = 0.12$ cm, $V_1 = 10$ N or $V_2 = 50$ N.

Sec. 7-5

7-61. A curved beam has a square cross section 3 cm $\times$ 3 cm. $M = 100$ N·m, and $\bar{R} = 12$ cm. Determine the maximum tensile and compressive stresses.

7-62. A curved beam has a square cross section 1 in. $\times$ 1 in. $M = 1000$ ft-lb, and $\bar{R} = 4$ in. Determine the maximum tensile and compressive stresses.

7-63. A curved beam has a rectangular cross section 2 cm $\times$ 5 cm. The short sides have the maximum stresses, which should not exceed 60 MPa. What is M_{max} if $\bar{R} = 10$ cm?

7-64. A curved beam has a rectangular cross section 1 in. $\times$ 6 in. The long sides have the maximum stresses, which should not exceed 3 ksi. What is M_{max} if $\bar{R} = 3$ in.?

8

DEFLECTIONS OF BEAMS

The deflections of beams are of interest for two reasons. They can be used quali-tatively to determine where tension and compression are in a beam. Exaggerated deflection curves can be used to visualize the locations of these normal stresses. For example, a beam supported at the ends as in Fig. 8-1a has tension at its bottom and compression at its top throughout the length. The same beam also supported at the center as in Fig. 8-1b has a more complex distribution of normal stresses along the

Fig. 8-1. Stresses in beams bent under their own weights.

length. The other interest in deflections is because their magnitudes must be known in some cases. The following discussion is of the quantitative aspects of the deflec-tions of beams that behave entirely elastically.

8-1. SLOPES OF DEFLECTION CURVES

The deflection curve is the curve of the neutral axis after the loads are applied to an originally straight beam. The slopes of lines tangent to the deflection curve may be of practical interest by themselves, but they also indicate the extent of the deflection. Consider two close cross-sectional planes in the unloaded beam in Fig.

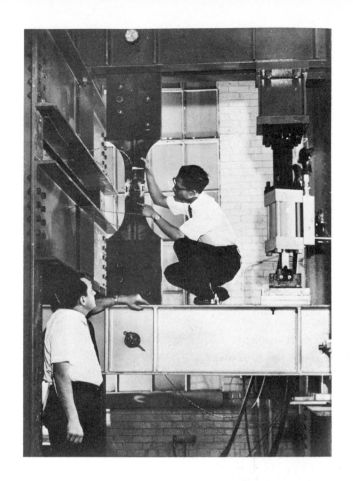

Large specimen in a test frame; the flexural rigidity of the frame must be high. *W. W. Sanders, Jr., "Structural Fatigue Testing at Iowa State University,"* Closed Loop, **2**, *No. 6 (1970), 12–14.*

Hydraulic actuator and load cell on moving frame for testing a wood joist floor. *M. D. Vanderbilt,* et al., *"Development and Verification of a Mathematical Model of Wood Joist Floors Using Computer Analysis and Closed-Loop Structural Testing,"* Closed Loop, **4**, *No. 1 (1973), 15–20.*

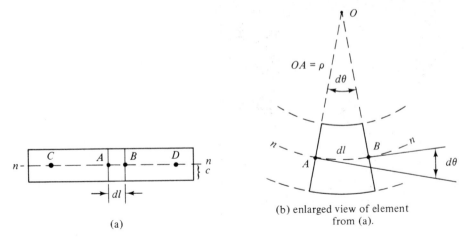

(b) enlarged view of element
from (a).

(a)

Fig. 8-2. Tangents to the deflection curve.

8-2a. The angle between these planes is $d\theta$ after the beam is deformed (the loads are not shown in Fig. 8-2b). The angle between the tangents to the neutral axis at points A and B is also $d\theta$. Consequently, the difference between the slopes at any distant points C and D is

$$\Delta\theta_{CD} = \int_{C}^{D} d\theta \qquad (8\text{-}1)$$

The changes in slopes can be related to the properties of the beam and the bending moment as follows. The element in Fig. 8-2b is enlarged in Fig. 8-3 where the original parallel position of the two planes is also shown. The total extension of a bottom fiber of original length dl is de. For small deformation of the beam (the sketches have to be exaggerated),

$$\frac{d\theta}{2} = \frac{de}{2}\frac{1}{c}$$

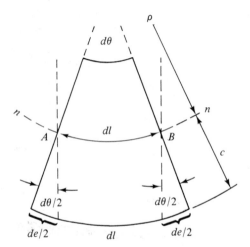

Fig. 8-3. Angular and linear deformations of the element in Fig. 8-2.

where c is the distance from the neutral axis to the outer layer. Since $de = \epsilon dl$, $\epsilon = \sigma/E$, and $\sigma = Mc/I$,

$$d\theta = \frac{M}{EI}dl \quad \text{and} \quad \Delta\theta_{CD} = \int_C^D \frac{M}{EI}dl \qquad \text{(8-2), (8-3)}$$

8-2. CURVATURES OF BEAMS

Equation 8-2 also means that the local curvature of the beam is directly proportional to the local bending moment and inversely proportional to Young's modulus and the moment of inertia of the cross section. A formal expression of this can be obtained after recognizing that $dl = \rho d\theta$ in Fig. 8-2b where ρ is the local radius of curvature. From Eq. 8-2,

$$d\theta = \frac{M}{EI}\rho d\theta$$

The result is

$$\frac{1}{\rho} = \frac{M}{EI} \qquad \text{(8-4)}$$

By definition, $1/\rho$ is the curvature of a line. The radius of curvature ρ is readily obtained from Eq. 8-4.

EXAMPLE 8-1

Thin fibers of smooth, pure silica are being evaluated for a device to be used at liquid helium temperature. The fibers are 10 μm in diameter and have a fracture strength of 2×10^6 psi at this temperature (these fibers are among the strongest materials known). What is the minimum radius of curvature of an originally straight fiber? What is the bending moment that can cause this curvature? $E = 10.5 \times 10^6$ psi.

Solution

$$c = 5 \ \mu\text{m} = 1.97 \times 10^{-4} \text{ in.}, \qquad \frac{1}{\rho} = \frac{M}{EI} = \frac{\sigma}{Ec}$$

$$\rho = \frac{Ec}{\sigma} = \frac{(10.5 \times 10^6)(1.97 \times 10^{-4})}{2 \times 10^6} = 1.03 \times 10^{-3} \text{ in.}$$

$$M = \frac{EI}{\rho} = \frac{(10.5 \times 10^6)\pi(1.97 \times 10^{-4})^4}{4(1.03 \times 10^{-3})} = 1.2 \times 10^{-5} \text{ in.-lb}$$

EXAMPLE 8-2

A band saw is made from 1-mm thick steel. What is the diameter of the smallest wheel that can be used to pass the band around it? The maximum strain in the band must be limited to 0.002. $E = 210$ GPa.

Solution

$$2c = 1 \text{ mm}, \qquad cd\theta = de = \epsilon dl = \epsilon p d\theta$$

so

$$\rho_{\min} = \frac{c}{\epsilon_{\max}} = \frac{0.0005}{0.002} = 0.25 \text{ m}$$

EXAMPLE 8-3

A cable-laying ship is designed to carry many large drums, each drum with 20 miles of 1.4-in. diameter coaxial telephone cable. The cable has a complex con-struction using steel, copper, and polyethylene. The maximum allowable strain at the surface of the cable is 0.03; at any point 0.2 in. below the surface it is 0.02. For the best utilization of space, a cable drum should have as small a core diameter as possible. What is the minimum core diameter that does not require overstraining the cable as it is wound on the drum? Assume that strain varies linearly in the cable even though it is a composite structure.

Solution

Note in Fig. 8-3 that the bottom fiber in the figure need not be the outer fiber of the structure. In general, since $cd\theta = \epsilon dl$ and $dl = \rho d\theta$, $\rho = c/\epsilon$ can be written

as

$$\rho = \frac{y}{\epsilon}$$

where ϵ is the strain at a distance y from the neutral axis. Then, for the subsurface

layer,

$$\rho_{\min} = \frac{0.5}{0.02} = 25 \text{ in.}$$

With the strain varying linearly, the strain is

$$\frac{0.7}{0.5}(0.02) = 0.028 \qquad \text{at } y = 0.7$$

when $\epsilon = 0.02$ at $y = 0.5$. Thus, the strain is within limits at the outer surface, and the minimum core diameter is 50 in.

8-3. DEFLECTION IN RECTANGULAR COORDINATES

The curvature of a line in rectangular coordinates is shown in calculus as

$$\frac{1}{\rho} = \frac{d^2y/dx^2}{[1 + (dy/dx)^2]^{3/2}} \tag{8-5}$$

The slope dy/dx is small compared to unity in small deflections, so

$$\frac{1}{\rho} \simeq \frac{d^2y}{dx^2}$$

and Eq. 8-4 becomes, with negligible error,

$$\frac{d^2y}{dx^2} = \frac{M}{EI} \tag{8-6}$$

Equation 8-6 is only a different form of Eq. 8-4 but by integrating it twice it is possible to find an expression in terms of linear dimensions. For this, assume that x is measured along the longitudinal axis of the beam and that the y coordinate is parallel to the external forces on the beam as in Fig. 8-4. This is a commonly used orientation for the coordinates, but the location of the origin and the directions of the axes are chosen arbitrarily for convenience in solving each problem.

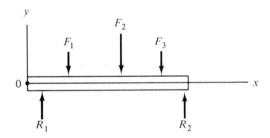

Fig. 8-4. Common orientation of coordinates.

For simplifying the integration, it is helpful to realize that E and I are constant along the length of the beam in most cases, but M is likely to be a function of x. With these assumptions,

$$EI\frac{dy}{dx} = EI\theta = \int M\,dx \tag{8-7}$$

$$EIy = \int \left(\int M\,dx \right) dx \tag{8-8}$$

Equation 8-7 gives the slope, and Eq. 8-8 gives the deflection of the beam at any point x in the beam. In most cases the integration is routine. One has to be careful to use a valid moment. The constants of integration are evaluated from boundary conditions that are known from inspection of the deflection curve. Figures 8-5 to 8-7 show examples of typical boundary conditions.

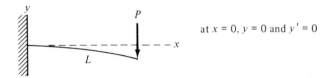

at $x = 0$, $y = 0$ and $y' = 0$

Fig. 8-5. Deflection of cantilever beam under a single force (beam is horizontal initially).

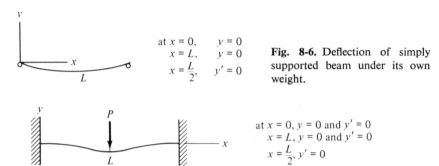

at $x = 0$, $y = 0$
$x = L$, $y = 0$
$x = \dfrac{L}{2}$, $y' = 0$

Fig. 8-6. Deflection of simply supported beam under its own weight.

at $x = 0$, $y = 0$ and $y' = 0$
$x = L$, $y = 0$ and $y' = 0$
$x = \dfrac{L}{2}$, $y' = 0$

Fig. 8-7. Deflection of beam with fixed ends.

In some cases there seem to be more constants to evaluate than there are independent boundary conditions. For an example of this, consider Fig. 8-8a. There is a moment M_a (function of x) that is valid in region a of the beam, and there is a different moment M_b (function of x) that is valid in region b. Equation 8-6 has to be applied to the two regions separately:

$$EIy_a'' = M_a, \qquad EIy_a' = \int M_a \, dx, \qquad EIy_a = \int \left(\int M_a \, dx \right) dx$$

and

$$EIy_b'' = M_b, \qquad EIy_b' = \int M_b \, dx, \qquad EIy_b = \int \left(\int M_b \, dx \right) dx$$

There are four constants to evaluate, but the only obvious boundary conditions are that, at $x = 0$, $y_a = 0$ and $y_a' = 0$.

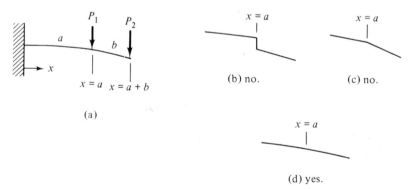

P_1 P_2

a

b

$x = a$ $x = a + b$

(a)

$x = a$

(b) no.

$x = a$

(c) no.

$x = a$

(d) yes.

Fig. 8-8. Deflection of cantilever beam under two discrete forces.

Other boundary conditions are based on the concept that the beam remains continuous and curving smoothly if it does not fracture, even if a single function cannot describe the moments throughout the beam. In other words, looking at the neighborhood of the beam at $x = a$, do not expect a sharp break in the curvature (Fig. 8-8b or c) but rather a smooth change from one radius of curvature (just left of $x = a$) to another (just right of $x = a$) as shown in Fig. 8-8d. This concept leads to the following new boundary conditions:

$$y_a = y_b \quad \text{and} \quad y_a' = y_b' \quad \text{at} \quad x = a$$

In words, approaching the point $x = a$ from the left or from the right, one finds the same deflection and the same slope in both approaches. The new boundary conditions are sufficient to evaluate the remaining constants.

Location of Maximum Deflection. In some cases the location of maximum deflection is obvious (cantilever beams with simple loading, symmetrically loaded beams). If it is not obvious, one can find the location fairly accurately simply by guessing. A reasonable sketch of the expectable deflection curve shows the region where the deflection is largest. For example, one should not expect the maximum deflection at the midpoint in the beam or under the load P in Fig. 8-9. Rather, it should be somewhere between these points as shown by the guessed deflection curve. The maximum error in deflection is only a few percent when determined on this basis.

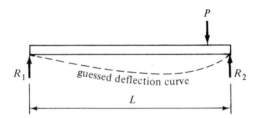

Fig. 8-9. Expected deflection curve of simply supported beam.

The accurate calculation of the location of maximum deflection in non-cantilever sections is based on the concept that $y' = 0$ where $y = $ maximum. Of course, in some beams there may be several places where $y' = 0$.

EXAMPLE 8-4

Find the maximum deflection of the granite beam that was discussed in Example 7-1.

Solution

The model for this problem is shown in Fig. 8-10a. The bending moment function that is valid everywhere in the beam is found from considering the equilibrium of part of the beam such as that in Fig. 8-10b. Thus,

$$EIy'' = 2400x - 120x^2$$
$$EIy' = 1200x^2 - 40x^3 + C_1$$
$$EIy = 400x^3 - 10x^4 + C_1 x + C_2$$

At $x = 0$, $y = 0$; at $x = L/2$, $y' = 0$. The first of these conditions introduced in the

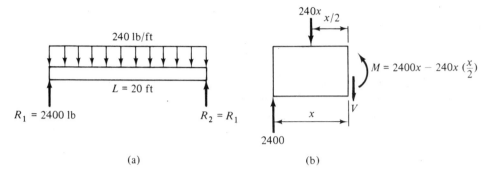

Fig. 8-10. Bending of granite column.

last equation shows that $C_2 = 0$. The other condition gives

$$0 = 1200\left(\frac{L}{2}\right)^2 - 40\left(\frac{L}{2}\right)^3 + C_1$$

With $L = 20$ ft, $\qquad\qquad C_1 = -80,000$ ft²lb

The maximum deflection is at $L/2$ by inspection:

$$y_{max} = \frac{1}{EI}\left[400\left(\frac{L}{2}\right)^3 - 10\left(\frac{L}{2}\right)^4 - 80,000\left(\frac{L}{2}\right)\right]$$

$$= \frac{1}{(6 \times 10^6)(3220)}(400,000 - 100,000 - 800,000)$$

$$\underset{\text{lb/in²}}{\uparrow} \quad \underset{\text{in⁴}}{\uparrow} \quad \underset{\text{ft³lb}}{\uparrow} \quad\quad \underset{\text{ft³lb}}{\uparrow} \quad\quad \underset{\text{ft³lb}}{\uparrow}$$

The units for each term are given at this stage in the calculation to show the potential problems if the units are not considered properly.

$$y_{max} = -\frac{5 \times 10^5 \text{ ft}^3}{1.93 \times 10^{10} \text{ in}^2}(1730 \text{ in}^3/\text{ft}^3) = -0.045 \text{ in.}$$

The meaning of the sign (up or down, right or left) in answers of this kind always should be evaluated with respect to what is expected in the physical situation.

EXAMPLE 8-5

A graphite-epoxy ski is 2 m long, its cross section is 1 cm × 8 cm. What is the approximate maximum deflection when it is simply supported and the concentrated load on it is 980 N? Also, calculate the deflections for thicknesses of 0.9 cm and 1.1 cm. $E = 200$ GPa. Assume there is no metal in the ski.

Solution

Assuming that the load acts at the center, each reaction is 490 N. The internal moment is $$M = 490x, \qquad 0 \leq x \leq \frac{L}{2}$$

The maximum deflection is at $L/2$ by inspection. Its magnitude is calculated from

$$EIy'' = M = 490x$$
$$EIy' = 245x^2 + C_1$$
$$EIy = 82x^3 + C_1 x + C_2$$

The boundary conditions are

$$y = 0 \qquad \text{at } x = 0$$
$$y' = 0 \qquad \text{at } x = \frac{L}{2} = 1 \text{ m}$$

So $$C_2 = 0, \qquad C_1 = -245$$

With $$I = \tfrac{1}{12}(0.08)(0.01)^3 = 6.67 \times 10^{-9} \text{ m}^4, \text{ and } x = 1 \text{ m},$$

$$(2 \times 10^{11})(6.67 \times 10^{-9})y = 82(1)^3 - 245(1)$$
$$y = -0.122 \text{ m}$$

If the thickness of the ski changes, everything but I remains the same.

For $$t = 0.9 \text{ cm}, \qquad I = 4.86 \times 10^{-9}, \quad \text{and} \quad y = 0.167 \text{ m}$$

For $$t = 1.1 \text{ cm}, \qquad I = 8.87 \times 10^{-9}, \quad \text{and} \quad y = 0.092 \text{ m}$$

Thus, in this particular case, increasing the thickness 10% decreases the deflection 25%; while decreasing the thickness 10% increases the deflection 37%.

EXAMPLE 8-6

Consider a wooden diving board with a person on its end (Fig. 8-11a). Find the slope of the board under the person. $E = 2 \times 10^6$ psi, $I = 5.1$ in^4.

Fig. 8-11. Bending of diving board.

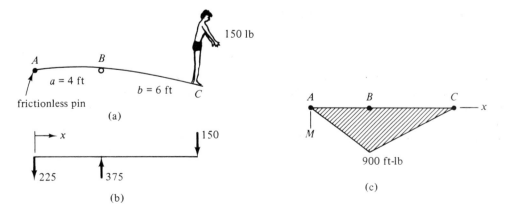

Solution

The free-body diagram and the moment diagram of the board are given in Fig. 8-11b and c. The equations valid for the two sections of the beam are

$$EIy_a'' = -225x$$

$$EIy_a' = -\frac{225x^2}{2} + C_1$$

$$EIy_a = -\frac{225x^3}{6} + C_1x + C_2$$

and

$$EIy_b'' = -225x + 375(x - 4)$$

$$EIy_b' = -\frac{225x^2}{2} + \frac{375x^2}{2} - 1500x + C_3$$

$$EIy_b = 25x^3 - 750x^2 + C_3x + C_4$$

The boundary conditions are

$$y_a = 0 \quad \text{at } x = 0$$

$$y_a = 0 \quad \text{and} \quad y_b = 0 \quad \text{at } x = 4$$

$$y_a' = y_b' \quad \text{at } x = 4$$

The main interest is in finding C_3, but this requires a few other steps. From the boundary conditions,

$$C_2 = 0, \qquad C_1 = \tfrac{225}{6}(4)^2 = 600 \text{ ft}^2\text{lb}$$

From $y_a' = y_b'$ at $x = 4$,

$$-\tfrac{225}{2}(4)^2 + 600 = -\tfrac{225}{2}(4)^2 + \tfrac{375}{2}(4)^2 - 1500(4) + C_3$$

$$C_3 = 3600 \text{ ft}^2\text{lb}$$

The required slope is at $x = 10$ ft,

$$y_b' = \frac{1}{EI}[75(10)^2 - 1500(10) + 3600] = \frac{-1}{(2 \times 10^6)(5.1)}(3900)\frac{\text{ft}^2\text{lb}}{(\text{lb/in}^2)\text{in}^4}(144 \text{ in}^2/\text{ft}^2)$$

$$y_b' = -0.055 \text{ rad}$$

EXAMPLE 8-7

A conveyor track is to be placed over the plating tanks in an automated facility for precision electroplating. A 15-cm deep steel I beam is proposed for the track. A two-wheeled carriage is to move on this with the maximum load of 3 kN. The wheel axles are 50 cm apart and share the load evenly. What is the maximum allowable spacing L of simple supports under the track if the maximum deflection of the track should not exceed 5 mm? $E = 200$ GPa, $I = 1000$ cm^4. The track is not continuous but is made up of sections of length L.

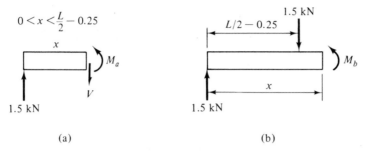

$0 < x < \dfrac{L}{2} - 0.25$

(a)

(b)

Fig. 8-12. Example 8-7.

Solution

The largest deflection occurs at $x = L/2$ when the center of the carriage is at $L/2$. From Fig. 8-12a,

$$EIy_a'' = M_a = 1500x$$

$$EIy_a' = 750x^2 + C_1$$

$$EIy_a = 250x^3 + C_1x + C_2$$

With the boundary condition $y = 0$ at $x = 0$, $C_2 = 0$. From Fig. 8-12b,

$$EIy_b'' = M_b = 1500x - 1500\left[x - \left(\frac{L}{2} - 0.25\right)\right]$$

$$EIy_b' = 1500\left(\frac{L}{2} - 0.25\right)x + C_3$$

$$EIy_b = 750\left(\frac{L}{2} - 0.25\right)x^2 + C_3x + C_4$$

The boundary conditions are

$$y' = 0 \qquad \text{at } x = \frac{L}{2}$$

from symmetry, and

$$y_a = y_b \qquad \text{at } x = \frac{L}{2} - 0.25$$

$$y_a' = y_b' \qquad \text{at } x = \frac{L}{2} - 0.25$$

Using the first of these,

$$C_3 = -1500\left(\frac{L}{2} - 0.25\right)\frac{L}{2}$$

After some algebra, the other two conditions give

$$C_1 = -750\left[\left(\frac{L}{2}\right)^2 - (0.25)^2\right], \qquad C_4 = 250\left[\frac{L}{2} - 0.25\right]^3$$

$$y_{\text{allowable}} = 0.005 \qquad \text{at } x = \frac{L}{2}$$

so

$$(2 \times 10^{11})(10^{-5})(-0.005) = 750\left(\frac{L}{2} - 0.25\right)\left(\frac{L}{2}\right)^2 - 1500\left(\frac{L}{2} - 0.25\right)\left(\frac{L}{2}\right)^2$$
$$+ 250\left(\frac{L}{2} - 0.25\right)^3$$

This cubic equation in $L/2$ can be transposed to

$$-500\left(\frac{L}{2}\right)^3 + 46.9\left(\frac{L}{2}\right) + 9996 = 0$$

The real solution is $L/2 = 2.725$ so $L = 5.45$ m.

8-4. MOMENT-AREA METHOD

In some cases the tangents to the deflection curve can be used advantageously to determine the deflections. The procedure is based on Eq. 8-2 and 8-3.

Assume that a simply supported straight beam is loaded by an arbitrary distributed load. The curvature diagram for this beam is determined from the moment diagram (curvature = M/EI) and plotted in Fig. 8-13a. The exaggerated elastic deflection curve is in Fig. 8-13b. Lines t_A, t_B, t_C, and t_D are tangents to this curve at the points indicated.

Fig. 8-13

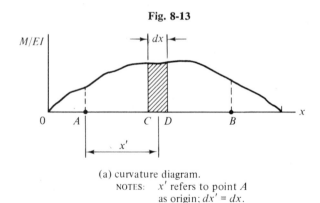

(a) curvature diagram.

NOTES: x' refers to point A
as origin; $dx' = dx$.

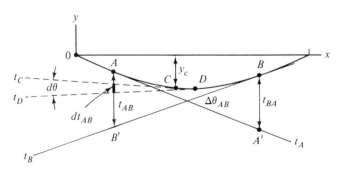

(b) elastic deflection curve.

According to Eq. 8-2, the angle $d\theta$ in Fig. 8-13b is equal to the shaded area under the M/EI curve. Similarly, the angle between the tangents at distant points A and B is the total area under the curvature diagram between A and B:

$$\Delta\theta_{AB} = \int_A^B \frac{M}{EI}dx$$

A useful concept for calculating deflections is the *tangential deviation*. For points A and B in Fig. 8-13, this is defined as the vertical distance t_{AB} between points A and B' where B' is on the tangent line t_B, directly below point A. The infinitesimal tangential deviation, dt_{AB}, is the dark segment of t_{AB} between the tangent lines t_C and t_D. It can be approximated as an arc length:

$$dt_{AB} = x'd\theta$$

Since $d\theta = (M/EI)dx$,

$$t_{AB} = \int_A^B \frac{M}{EI}x'dx \tag{8-9}$$

where x' always denotes the distance along the x-axis from the point of interest (here, point A), whose tangential deviation is to be found; x' can be replaced by x if this peculiarity is kept in mind. It should be noted from Fig. 8-13b that

$$t_{AB} \neq t_{BA}$$

except by chance.

The quantity $Mx'dx/EI$ is the moment of the infinitesimal shaded area about point A in Fig. 8-13a. The name *moment-area method* comes from this. In practice, it is not always necessary to integrate in Eq. 8-9 to calculate the tangential deviation. The equivalent method is to find the area and the centroid of the area under the curvature diagram between A and B first. The product of the area and the centroid's ordinate to point A equals the tangential deviation, t_{AB}.

The moment-area method is most useful in determining the actual deflection of a point on a beam, such as y_c in Fig. 8-13b. This can be done most conveniently when there is a point on the beam where the deflection and slope are known. The following examples illustrate how deflections can be calculated this way.

Sign Convention. In most cases, the direction of the deflection can be found by inspection. A consistent method is to call the deflections positive when the moments of positive areas are taken as positive, and vice versa.

EXAMPLE 8-8

A cantilever beam is straight and horizontal before the load is applied. What is the maximum deflection under the load P shown in Fig. 8-14? $E = 100$ GPa, $I = 200$ cm⁴.

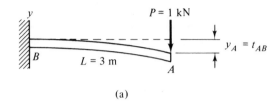

(a)

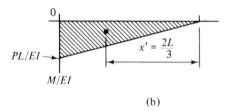

(b)

Fig. 8-14. Example 8-8.

Solution

The moment diagram and the curvature diagram are triangular as shown in Fig. 8-14b. The maximum curvature is PL/EI. The distance x' from the point of interest, A, to the centroid is $\frac{2}{3}L = 2$ m. Since the beam at point B is rigidly fixed, the tangent there is horizontal, and

$$y_A = t_{AB} = \frac{PL^2}{2EI}x' = \frac{(1000 \text{ N})(9 \text{ m}^2)}{2(10^{11} \text{ N/m}^2)(2 \times 10^{-6} \text{ m}^4)}(2 \text{ m})$$

$y_A = 4.5$ cm in the negative y-direction according to the convention.

EXAMPLE 8-9

A vertical pole is 30-ft tall and has $E = 10^7$ psi and $I = 12$ in⁴. What is the maximum deflection under a uniformly distributed wind load of 50 lb/ft?

Solution

The curvature diagram is parabolic as shown in Fig. 8-15b. After calculating the area and the location of the centroid, the deflection y_A is easily determined:

$$y_A = -\left(\frac{wL^3}{6EI}\right)\left(\frac{3}{4}L\right) = \frac{wL^4}{8EI} = -6.1 \text{ ft}$$

This is a large deflection and should be double-checked. Also, it must be realized that the answer may be somewhat in error since the theory was based on small deflections. Furthermore, the maximum stresses should be checked in such cases to make sure that the beam is at least behaving entirely elastically.

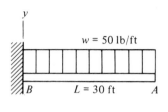

(a)

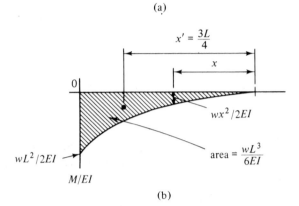

(b)

Fig. 8-15. Example 8-9.

EXAMPLE 8-10

A pure couple is to be applied at the free end of a cantilever member as shown in Fig. 8-16a. Derive an equation to calculate the moment from the deflection of point A and the properties of the beam.

Solution

The moment and curvature diagrams are rectangles. With y_A = deflection of

the free end,
$$y_A = \frac{ML}{EI}\left(\frac{L}{2}\right)$$

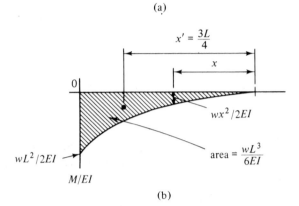

(a)

Fig. 8-16. Example 8-10.

downward deflection for the moment shown

and
$$M = \frac{2EIy_A}{L^2}$$

EXAMPLE 8-11

Compare the maximum deflections in two springs of identical material and geometry, under the same load. The only difference is that one is a cantilever with the load at the free end and that the other is simply supported with the load at the center.

Solution

On the basis of Example 8-8, the deflection of the cantilever is

$$y_c = \frac{PL^3}{3EI}$$

The simply supported beam will be analyzed in two slightly different ways. First, the complete curvature diagram is drawn in Fig. 8-17b. Because the slope at point *B* remains zero for all deflections,

$$y_s = (\text{area } ABC)\left(\frac{2}{3}\frac{L}{2}\right) = \frac{PL^2}{16EI}\frac{L}{3} = \frac{PL^3}{48EI}$$

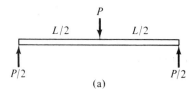

(a)

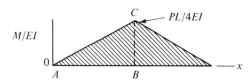

(b) curvature diagram.

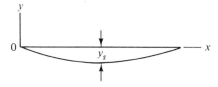

(c) deflection curve.   **Fig. 8-17**

Another method is based on the idea that this particular simply supported beam is equivalent to two identical cantilevers with a common rigid wall at point B and an upward load of $P/2$ at each free end. Then the equation for y_c above should give y_s if $P/2$ is substituted for P and $L/2$ for L.

$$y_s = \frac{P/2(L/2)^3}{3EI} = \frac{PL^3}{48EI}$$

The simply supported spring is much more rigid than the identical spring used as a cantilever.

8-5. DEFLECTION CAUSED BY SHEAR

The transverse shear acting on a beam causes a distortion that can be visualized as rectangles becoming rhomboidal (Fig. 8-18). The vertical displacement dv in a short element of length dl depends on the shear strain γ, which in turn is a function of the shear stress τ and the shear modulus G:

$$dv = \gamma dl = \frac{\tau}{G} dl$$

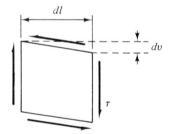

Fig. 8-18

The total vertical deflection in a beam is

$$v = \frac{1}{G} \int_{l_1}^{l_2} \tau dl \tag{8-10}$$

Since τ varies from zero to a maximum and back to zero over the cross section, there are philosophical problems with these concepts for all but the infinitesimal elements, as they were discussed for the shear stress variations before. When the average value of τ is used, the calculated shear deflection is an approximation that is adequate in most practical problems.

It is worthwhile to keep in mind that the total deflection caused by shear is normally much smaller than that caused by bending moments. In very short beams the moment is relatively small and the effect of the shear loads may have to be considered. This is especially true for wood, which has $E/G \approx 16$, while for steel or concrete it is only about 2.5.

EXAMPLE 8-12

A new elastomer is considered for vibration-isolating engine mounts. Each of these is a short, deep beam with a length of 1 in. and a cross section of 2 in. × 2 in. What is the deflection caused by a shear force of 200 lb? $G = 10^5$ psi.

Solution

Assuming that τ is the same everywhere in the short beam,

$$v = \frac{Vl}{AG} = \frac{200(1)}{4(100,000)} = 5 \times 10^{-4} \text{ in.}$$

This is a practically insignificant deflection, so there is no reason to seek a more accurate answer by using a more realistic function for τ in Eq. 8-10 than τ_{ave}.

8-6. SUPERPOSITIONS

The analysis of a beam that is subjected to many loads can be facilitated by using the method of superposition. The idea is to deal with the effects of each load separately and then to add the appropriate results algebraically. The method was illustrated by finding the moment diagram for a simply supported beam in Example 6-7.

In the calculation of stresses it is most convenient to work with the complete moment diagram. To determine deflections, it is sometimes advantageous (and instructional) to find the deflections for the individual loads and summing these for

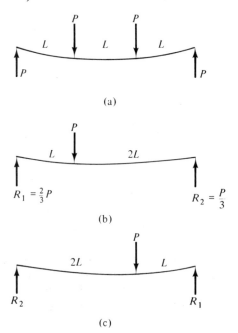

Fig. 8-19. Superposition of beam deflections.

the resulting deflections. The student is cautioned, however, that great care must be exercised in using boundary conditions for the individual deflection curves. It is best to acquire experience in selecting boundary conditions by solving simple problems in different ways (including superposition) and comparing the results before attempting to solve complex problems by superposition directly. For example, the situation shown in Fig. 8-19a can be conceived as the sum of those shown in (b) and (c). The deflection is zero at the ends of the beam in each of the three drawings, but the slopes are zero at different places in the three cases.

8-7. PROBLEMS

Sec. 8-1 and 8-2

Plot the approximate variation of curvature along the beam in Probs. 8-1 to 8-6.

8-1.	Figure P7-5	**8-2.**	Figure P7-12
8-3.	Figure P7-19	**8-4.**	Figure P7-26
8-5.	Figure P7-27	**8-6.**	Figure P7-36

Sec. 8-3

Determine the slope and deflection for the beams at the points indicated in Probs. 8-7 to 8-18.

8-7. Figure P7-6, at the center and at the end.

8-8.	Figure P7-11, at the center.	**8-9.**	Figure P7-13, at A.

8-10. Figure P7-18, under the load.

8-11. Figure P7-19, at the center and at the end.

8-12. Figure P7-24, at the center.

8-13.	Figure P7-27, at the end.	**8-14.**	Figure P7-30, at the center.
8-15.	Figure P7-36, at the end.	**8-16.**	Figure P7-39, at the center.
***8-17.**	Figure P7-42, at the end.	***8-18.**	Figure P7-43, at the end.

Sec. 8-4

Determine the deflection for the beams at the points indicated using the moment-area method in Probs. 8-19 to 8-26.

8-19.	Figure P7-5, at the end.	**8-20.**	Figure P7-8, at the end.
8-21.	Figure P7-13, at the center.	**8-22.**	Figure P7-20, at the end.
8-23.	Figure P7-23, at the center.	**8-24.**	Figure P7-27, at the end.
8-25.	Figure P7-36, at the end.	**8-26.**	Figure P7-37, at the center.

9

TORSION

Imagine that you are an orthopedic surgeon who had studied engineering before going into medicine. You have kept some of your original interest in engineering and regularly read about new developments in the physical sciences that may be relevant to your area. One such development is the use of porous metal implants to replace whole bones or parts of bones. A valuable property of these implants is that they can be made with a step change or a gradual change in density. One part of a member may be dense and strong while another part of it is weaker but allows the growth of new bone into the pores (Fig. 9-1). The dense powdered metal part can be up to 50% stronger than natural bone, and the interfacial bond (metal with bone in its pores) can be 30% stronger both in tension and in shear than bone alone. These are good reasons for exploring the possibilities of using metal implants.

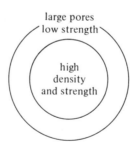

large pores
low strength

high
density
and strength

Fig. 9-1. Cross section of a porous material surgical implant.

Consider, for example, a torsional fracture in a leg bone. The main fracture surface is at about a 45° angle with the axis of the bone as in all brittle materials that fail in torsion (demonstrate this with a piece of chalk; see Fig. 9-2). If the surface is somewhat splintered, it is reasonable to think about an implant.

(a)

(b)

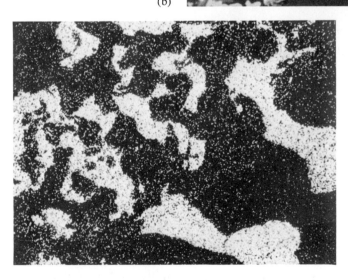

(c)

Porous metal surgical implant: (a) Micrograph shows variation of density from a maximum of 80% on the left to 30% at the outer layer on the right. (b) Secondary electron image shows the growth of new bone (gray) in the pores between the metal particles (white); (c) Ca-Kα X-ray image shows new growth is dense bone (white areas). *Courtesy M. Dustoor, University of Wisconsin, Madison, Wis.*

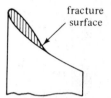

Fig. 9-2. Mode of fracture of brittle material in torsion.

What are the reasonable configurations for the implant? There are two. It could be shaped to conform to the fracture surface (Fig. 9-3) or made in a cylindrical shape to minimize the bond areas (Fig. 9-4). In either case the amount of metal used should be minimized and some bone would have to be removed (more in the second case) so that the length of the leg is not changed in the operation.

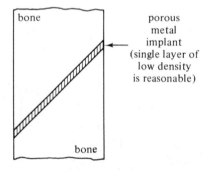

Fig. 9-3. Implant conforming to original fracture.

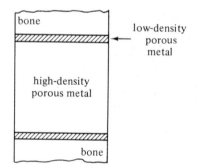

Fig. 9-4. Cylindrical implant.

Which configuration is the better one? The choice must be based primarily on the expectable loadings and the strength of the bonds after the patient resumes normal activities. The maximum strength attainable is known. There are three kinds of loading to be considered: axial, bending, and torsional. Since the fracture was in torsion, it is reasonable to consider this one first. Naturally, it appears that the cylindrical implant may be the easiest to analyze. This will be the main subject of discussion in the following sections.

A good way to start analyzing the internal stresses in a bar subjected to torsional loading is to imagine that the bar is in two separate pieces as shown in Fig. 9-5. The interface C between parts A and B is smooth and it is perpendicular to

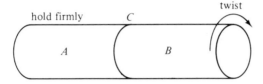

Fig. 9-5. Model for torsional deformations.

the common axis of A and B. If part A is held firmly in the left hand and part B is rotated as shown (clockwise, looking from B toward A) while pressing it against part A, frictional forces are acting at the interface C. Part A "feels" that it is being rubbed in the clockwise direction. The arrows in Fig. 9-6 show the direction of the frictional forces that are caused by part B on part A.

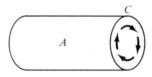

Fig. 9-6. Direction of frictional forces on rod A.

Next, consider the material of bar A near the face C. The small, nearly rectangular element in Fig. 9-7a is redrawn in Fig. 9-7b as viewed from the direction D. The arrow shows an elemental frictional force on the basis of Fig. 9-6. Obviously, Fig. 9-7b is not a proper free-body diagram. The correct diagram for static equilibrium is shown in Fig. 9-7c. This is an example of pure shear, the

Fig. 9-7. Equilibrium considerations of isolated corner element.

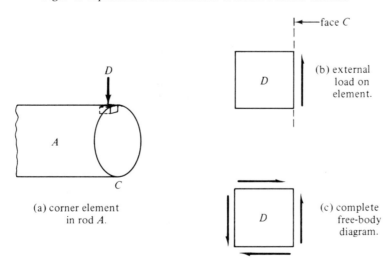

(a) corner element in rod A.

(b) external load on element.

(c) complete free-body diagram.

concept of which requires clarification. Figure 9-7c shows pure shear, but this element also "feels" internal tensile loading on planes other than the four sides shown (review Mohr's circle). This is the reason for occasional tensile failures in members subjected to purely torsional loading.

The frictional forces used in the preceding discussion are generally referred to as shear stresses (of course, the units of these must be distinguished). The concept of an internal shear stress caused by torsion is applicable to continuous rods as well as to the model used here.

Another aspect of the torsion problem that must receive attention is the possible variation in the magnitudes of the shear stresses at a given cross-sectional plane. Perhaps it is easy to accept that any gradients in the shear stress could depend only on the radial distance from the center of a cylindrical member. Is it realistic to expect any gradients in the shear stress, even as functions of the radial distance?

To try to visualize the causes of such gradients, if any, reconsider the rods in Fig. 9-5 and add two pegs to each rod near the face C. Each peg is mounted radially as shown in Fig. 9-8a. The angle α is the same for the two pegs in rod A and the two pegs in rod B. Next, assume that there are two rubber bands around the four pegs as shown in Fig. 9-8b. The rubber bands are shown in plan views in Fig. 9-8c. They form parallel rectangles before rods A and B are twisted relative to one another. Figure 9-8d shows the rubber bands after a clockwise torque is applied to rod B and slipping is allowed at the interface C. The top band deforms more severely. If the rectangles were of solid material (as the element in Fig. 9-7a), the shear stress would have to be largest at the greatest distance from the center line of rods A and B to cause the largest distortion. Conversely, the angular distortion and the shear stress would be zero at the center line.

Fig. 9-8. Model of torsional deformation.

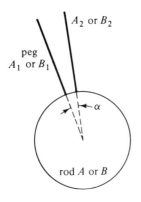

(a) pegs mounted on the rods.

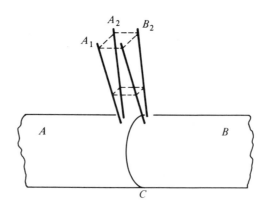

(b) rubber bands (dashed lines) stretched on the pegs.

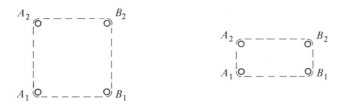

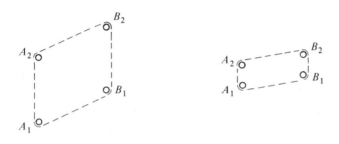

(c) plan views of rubber bands prior to
relative displacement of rods *A* and *B*.
Small rectangle is closest to center
line of rods *A* and *B*.

(d) plan views of rubber bands after
twisting rod *B* relative to rod *A*.

Fig. 9.8 (Continued)

A convenient way of showing the stress distribution at a given cross section in
a rod under torsional loading is shown in Fig. 9-9. At point 3 the shear stress is the
largest in the rod. For an ideally elastic rod the shear stress varies linearly from zero
at point 0 to the maximum at the outer fibers. The radial line 0-3 could be positioned
anywhere on the cross-sectional area; the stress distribution is the same for all
radial lines (see τ_3' in Fig. 9-9).

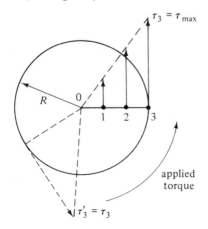

Fig. 9-9. Shear stress distribution
on a radial line.

9-2. PLASTIC FLOW IN TORSION

Since many important materials yield at high stresses, it is often necessary to consider inelastic behaviors as well. A useful model for this is the imaginary, ideally elastic-ideally plastic material. Assume the stress-strain curve shown in Fig. 9-10 is valid for such a material. To see how a rod made from this material might

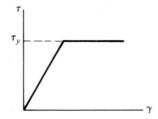

Fig. 9-10. Idealized shear stress-strain curve.

behave in torsion, assume that only a very small torque is applied at first. For example, the maximum shear stress is $\frac{1}{3}\tau_y$ (Fig. 9-11). There is no yielding, and the stresses at all points in the rod return to zero when the external torque is removed. A second loading to a peak of $\tau = \frac{2}{3}\tau_y$ produces similar results and so does any loading as long as the peak shear stress is less than τ_y.

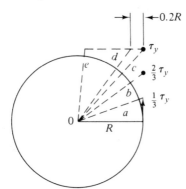

Fig. 9-11. Shear stress distributions for increasing torques: (a) peak stress $= \tau_y/3$. (c) peak stress $= \tau_y$. (d) yielding in outer part of rod. (e) nearly fully plastic condition.

When the peak stress reaches τ_y, the outer fibers of the bar yield. For another small increase in the torque, the outer layer that yields becomes deeper. Line d in Fig. 9-11 shows the stress distribution when 20% of the radial distance is the depth of the plastically deforming region. Further increases in the applied torque make the plastic zone penetrate deeper and deeper into the bar. Finally, a so-called fully plastic condition is reached when the stresses at all points of the cross section (except the center) are the same, namely τ_y.

Unloading the bar after any part of it has deformed plastically leads to residual stresses in the bar. These are internal stresses that exist even in the absence of external loads on the material. Such stresses may be harmful or beneficial. Generally, they are notorious because their magnitudes, signs, and precise locations are difficult to determine. They should be investigated whenever plastic deformations are expected near holes and other discontinuities or under any nonuniform stress distributions such as in bending and torsion. Fortunately, residual stresses can be ignored in the particular orthopedic problem that led to this discussion, because here yielding should not be allowed to occur.

9-3. THE TORSION FORMULA

There are enough concepts gathered now to analyze the torsion problem of the cylindrical implant in detail. The item of greatest concern is the relation between the externally applied torque and the resulting maximum shear and normal stresses that should not exceed the known strength values of the materials involved. The proposed implant has two densities but at a given cross section the density is constant. Therefore, the analysis should begin with the torsion of a homogeneous cylindrical rod.

Figure 9-12 shows the cross section of such a rod. First, it is necessary to consider the thin annular area. The shear stress τ_ρ on this ring is assumed to be constant everywhere. The total shearing force on the ring is

$$dF = \tau_\rho(\text{area of ring}) = \tau_\rho(2\pi\rho d\rho)$$

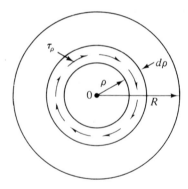

Fig. 9-12. Stress analysis for torsion member.

This force has a moment dM about the center 0,

$$dM = dF\rho = \tau_\rho(2\pi\rho^2)d\rho$$

The total moment M (about point 0) generated by the shear stresses over the whole cross-sectional area is the internal resisting torque, and it must be equal to the externally applied torque T for static equilibrium. Thus,

$$T = \int_{\rho=0}^{\rho=R} dM = \int_0^R \tau_\rho(2\pi\rho^2)d\rho \tag{9-1}$$

τ_ρ is a function of ρ, and it can be determined easily in the case of an ideally elastic member; $\rho/R = \tau_\rho/\tau_m$ where τ_m is the stress in the outermost fibers and is considered a constant. Making this substitution,

$$T = \int_0^R \frac{2\pi}{R} \rho^3 \tau_m d\rho = \frac{\pi R^3}{2} \tau_m$$

A rather useful variation of this equation is obtained if the right side is multiplied and divided by R,

$$T = \frac{\pi}{2} R^4 \frac{\tau_m}{R}$$

where $\pi R^4/2 = J$, the polar moment of the area. Finally,

$$\tau_m = \frac{TR}{J} \tag{9-2}$$

and, at any distance ρ from the center, $\tau_\rho = T\rho/J$.

The basic ideas used in this derivation are valid for a hollow cylindrical member, with the integral evaluated from R_{inside} to R_{outside}, or with J being valid for the actual area. See Appendix A for a review of moments of inertia.

Let us pause for a moment and marvel about the works of nature. A hollow bone is not only ideal to resist bending moments, as shown previously, it is also the best design for torsional loading. For a given cross-sectional area (which also means a given weight for the same length), J is larger for a hollow member than for a solid one. Therefore, the hollow member can resist a larger torque than the solid one of the same weight with the same maximum shear stress in each. In other words, there is a great advantage in removing some material from the center of a shaft where the stresses are low (Fig. 9-9) and putting this material on the outside where the strength of the material can be utilized.

9-4. MATERIAL PERFORMANCE

Consider the expected performance of a possible implant as shown in Fig. 9-4. When a torque is applied to the completely healed limb, the shear and normal stresses at any given distance ρ from the center line are the same, assuming the same geometry for all cross sections. Interestingly, this is true for the original bone, for the high-density part of the implant, and for the bone-metal composite at the interfaces. The reason for this is that the stresses depend only on the applied torque and the geometry of the member if the behavior is entirely elastic. Whether the member fails or not under the applied load depends on the strength of the material. In the case under discussion, if the bone doesn't fail, all parts of the implant should remain undamaged since the bone is the weakest component of the composite structure.

It is easy to see that considerations of any axial or bending loads lead to similar conclusions. The cylindrical implant (Fig. 9-4) is satisfactory as far as strength is concerned.

What could happen in the apparently more complex case of using a conforming implant (Fig. 9-3)? To gain understanding of the whole range of possibilities, consider two extreme cases by passing imaginary cutting planes through the member. These are planes A and B in Fig. 9-13. The respective cross-sectional views are shown in (b) and (c). In each, the dark area is the implant material and the rest of the area is bone. The shear stresses on both areas tend to have the distribution that is governed by the applied torque and the dimensions of the area. This is attained in reality if the stress-strain responses (in this case, the elastic moduli) of the bone and the implant are essentially the same.

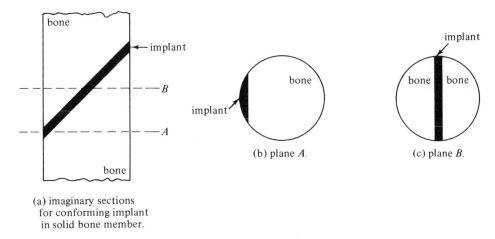

(a) imaginary sections
for conforming implant
in solid bone member.

(b) plane A.

(c) plane B.

Fig. 9-13. Views of sections.

If the material properties are different, it is best to seek an intuitive basis for finding the stress distributions. It is not expected that this will be easy. In a case like this, it is often possible to work out an analogy for a slightly different problem that can be understood relatively easily. For example, try to analyze a simple model for linear deformations as in Fig. 9-14. The five blocks are of rubber; they are glued together and to a common foundation. Block 2 has higher moduli of elasticity and rigidity than the other blocks. Shear forces of equal magnitude S are applied to the blocks as shown. The resulting shear deformation is γ, and it is about the same for each block because they constrain each other. This γ is smaller than the deformation would be for a single light block under the same unit load S because of the presence of the stronger dark block. On the other hand, the dark block deforms more than it would standing alone because its neighbors want to deform more and they have to move together.

The model can be altered in any number of ways (for example, one or more blocks without load S; a larger proportion of dark blocks), and the result will be the same qualitatively. There is a unique average deformation for a given arrangement of the blocks and the total loading on them, but the blocks tend to deform together. This is an example of deformation or strain control that is shown in a

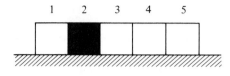

(a) blocks without load.

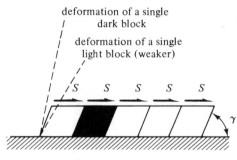

(b) blocks with shear loads.

Fig. 9-14. Model for shear deformation of composite material.

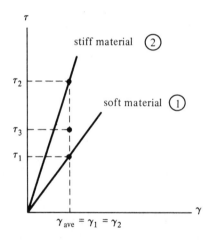

Fig. 9-15. Stresses in composite material under strain control.

different way in Fig. 9-15. If γ_{ave} is imposed on a pair of different materials that are deforming together, the stresses in the two materials (τ_1 and τ_2) will be different according to their stress-strain curves. This is similar to the axial strain control condition discussed earlier.

Return now to the original problem. Assume that the average shear strain on a thin concentric ring is γ_{ave}. Also assume that the shear stress on this ring would be τ_3 if the whole cross section was of bone. Next, imagine that the implant is present as in plane B in Fig. 9-13. The concentric ring now appears as in Fig. 9-16. This is similar to the linear model discussed earlier, so all parts of the ring should tend to deform the same amount when a torque is applied. The average shear strain in the ring is γ_{ave} in Fig. 9-15. The resulting local stresses in the bone (material ① in Fig.

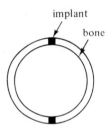

Fig. 9-16. Thin concentric ring from plane B in Fig. 9-13.

9-15) are lower than those in the implant (material ②). Neither of them is equal to τ_3, the average stress.

The implant is stronger than the bone, so the inequality in stresses is not necessarily harmful. Difficulty would arise if the material with the higher moduli did not also have an appropriately higher fracture strength than the softer material. If this is not clear, the reader should review the substantial differences between such important mechanical properties of materials as the modulus of elasticity and the yield and fracture strengths.

EXAMPLE 9-1

A tubular member should be made with an inside diameter of 2 cm and an outside diameter of 4 cm. The shear stress that should not be exceeded in the material is 80 MPa. What is the maximum torque that can be applied? The material does not yield at shear stresses less than 80 MPa.

Solution

$$T = \int_{R_i}^{R_o} \tau_\rho (2\pi \rho^2) d\rho$$

$$\frac{\rho}{R_o} = \frac{\tau_\rho}{80}, \qquad \tau_\rho = \frac{80}{0.02} \rho \text{ MPa}$$

$$T = 8(10^9)\pi \left(\frac{\rho^4}{4} \right)_{0.01}^{0.02} \text{ N·m}$$

$$T = 2\pi(10^9)(0.02^4 - 0.01^4) = 942 \text{ N·m}$$

EXAMPLE 9-2

A solid cylindrical member is 2 in. in diameter. What is the ratio of the torques for fully plastic deformation (the whole rod yields) and for maximum elastic straining? Assume the material has an ideally elastic-ideally plastic stress-strain curve with a shear yield strength of 40 ksi.

Solution

For the elastic behavior,

$$T_e = \frac{\tau_m \pi R^3}{2} = \frac{(40 \times 10^3)\pi(1)^3}{2} = 6.28 \times 10^4 \text{ in.-lb}$$

For fully plastic deformation, $\tau_\sigma = \tau_y$ everywhere, so

$$T_{fp} = \int_0^R \tau \rho^2 2\pi d\rho = \int_0^1 (40 \times 10^3) 2\pi \rho^2 d\rho = 8.37 \times 10^4 \text{ in.-lb}$$

The ratio is $$\frac{T_{fp}}{T_e} = \frac{4}{3}$$

which is also valid for other solid cylindrical members.

EXAMPLE 9-3

A screwdriver is to be designed for a spacecraft, so its weight must be minimized. The largest expected torque on this tool is 10 N·m. The allowable stress is 700 MPa for elastic behavior, and the maximum outside diameter $2r_o$ is 1 cm. How much weight can be saved (in percentage of a solid tool shaft) if a tubular shaft is made?

Solution

For a solid shaft, $T = \tau \pi R^3/2$, so

$$10 = \frac{(7 \times 10^8)\pi R^3}{2} \quad \text{and} \quad R = 2.09 \text{ mm}$$

For a hollow shaft, $T = \tau J/r_o$, so

$$10 = \frac{(7 \times 10^8)(\pi/2)(0.005^4 - r_i^4)}{0.005} \quad \text{and} \quad r_i = 4.853 \text{ mm}$$

The weights per length are proportional to the cross-sectional areas. With $A_{\text{solid}} = 0.0437\pi$ cm² and $A_{\text{hollow}} = 0.0145\pi$ cm², the weight saved by using the hollow shaft is about 67%.

EXAMPLE 9-4

A solid circular shaft has an 8-cm diameter. The maximum shear stress in the steel shaft is 300 MPa to assure entirely elastic behavior. What is the allowable torque on this shaft? How much more torque can be applied to a hollow shaft of the same steel if its inside diameter is 8 cm and its weight per unit length must be the same as for the solid shaft?

Solution

The allowable torque on the solid shaft is

$$T_s = \frac{\pi R^3}{2} \tau_m = \frac{\pi (0.04)^3 (300 \times 10^6)}{2} = 30.16 \text{ kN·m}$$

The hollow shaft will have the same cross-sectional area (and the same weight) as the solid one if its outside diameter is about 11.3 cm. The allowable torque on the hollow shaft is

$$T_h = \frac{\tau_m J}{R} = \frac{\tau_m[(\pi R_o^4/2) - (\pi R_i^4/2)]}{R_o} = \frac{300 \times 10^6 (\pi/2)(0.0565^4 - 0.04^4)}{0.0565}$$

$$= 63.64 \text{ kN·m}$$

Thus, a 111% increase in allowable torque is obtained by making the specified hollow shaft from the same amount of material.

EXAMPLE 9-5

A solid circular shaft taken from a broken machine has a 3-in. radius; its stress-strain curve can be approximated by Fig. 9-10 with $\tau_y = 60/\pi$ ksi. It is determined experimentally that the shaft has yielded to a depth of 1 in. What was the maximum torque applied during the service life of this shaft?

Solution

The variation of the shear stress in the shaft cannot be described by a single function, so the torque must be computed in two steps. In the inner, elastic part of the shaft the maximum torque was

$$T_E = \frac{\pi(2)^3}{2}\left(\frac{60,000}{\pi}\right) = 240,000 \text{ in.-lb}$$

The outer part of the shaft that yielded could have resisted a torque

$$T_Y = \int_{R_i}^{R_o} \tau_\rho(2\pi\rho^2)d\rho = 2\pi\left(\frac{60,000}{\pi}\right)\int_2^3 \rho^2 d\rho = 760,000 \text{ in.-lb}$$

The total maximum torque on the shaft was $T_{max} = T_E + T_Y = 10^6$ in.-lb.

EXAMPLE 9-6

Many railroad cars are equipped with large electric generators. Occasionally, a generator freezes up, and this can damage a number of components in its drive train (shafts, gears, universal joints). A crude but practical way to localize and limit the damage is to allow a shaft to break at a predetermined location before any of the more complex members are damaged. This may be called a *mechanical fuse*.

Assume that in the "fuse" section the diameter of the solid shaft is 1.7 in. Each universal joint in the system transmits the torque by two cylindrical shear pins that are sheared on a 4-in. diameter circle; the pins are on the same center line that is perpendicular to the axis of the shaft (Fig. 9-17). The shaft and the shear pins are made from the same steel; the stress-strain curve is assumed flat-topped at $\tau = 50$ ksi. What is the minimum diameter of the shear pins if they should survive a torque that breaks the shaft? (Naturally, the opposite requirement may also be made: Break the pins first.)

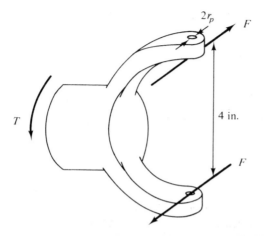

Fig. 9-17. Example 9-6.

Broken drive shaft of electric generator on a railroad car. *Photograph by R. Sandor, Madison, Wis.*

Solution

From Example 9-2, the torque in the fully plastic shaft is

$$T = \frac{4}{3} \frac{\tau_m \pi R^3}{2} = \frac{(50 \times 10^3)\pi(0.85)^3(2)}{3} = 6.43 \times 10^4 \text{ in.-lb}$$

With $4F = T,$ $F = 1.61 \times 10^4$ lb

Also, $\qquad F = \tau_y A = \tau_y \pi r_p^2, \quad \text{and} \quad r_p = \sqrt{\dfrac{1.61 \times 10^4}{\pi(50 \times 10^3)}} = 0.32 \text{ in.}$

9-5. ANGLES OF TWIST

In some cases the maximum stress resulting from torsion is acceptable, but the member twists too much. The angle of twist can be calculated easily for a homogeneous, cylindrical member if the maximum stress is in the range of elastic behavior. This is done by combining three simple relationships. Two of these are Eq. 3-3 and the generalized form of Eq. 9-2:

$$\tau_\rho = G\gamma_\rho \qquad (\text{from } \tau = G\gamma)$$

$$\tau_\rho = \frac{T\rho}{J}$$

The third is $\qquad\qquad\qquad \gamma_\rho = \dfrac{\theta\rho}{L} \qquad\qquad\qquad (9\text{-}3)$

where γ_ρ = shear strain at a distance ρ from the center line of the member
$\qquad \theta$ = relative angle of twist between parallel cross-sectional planes of the member that are a distance L apart

Equation 9-3 will not be proved in detail here. Its validity can be appreciated qualitatively by considering Fig. 9-8. The shear "strain" of the rubber bands depends directly on the distance from the center line and on the angle of slip between the two rods. It depends inversely on the distance between the pegs in one rod and those in the other.

Aiming to eliminate τ_ρ and γ_ρ, one obtains

$$G = \frac{T\rho/J}{\theta\rho/L} = \frac{TL}{\theta J}$$

and the angle of twist is

$$\theta = \frac{TL}{GJ} \qquad\qquad (9\text{-}4)$$

This is valid for a cylindrical member of uniform geometry and material under a given torque T. If the torque, the geometry, or the material are not constant over the length, appropriate functions of these should be used in integrating infinitesimal angular twists over the length. In the most general case,

$$\theta = \int_L d\theta = \int_L \frac{T}{GJ} dL \qquad\qquad (9\text{-}5)$$

EXAMPLE 9-7

Compare the twists in an intact leg bone (length L) and in one with a cylindrical implant. The implant is from $0.35L$ to $0.55L$. In this case the purpose of the analysis is not to reveal a design problem with the implant (because strength is more important than elastic deformations) but to find out if there might be any strange feeling in using the repaired leg in some situations.

Solution

The models used for the comparison are shown in Fig. 9-18. The same torque T is assumed in both cases. For the intact bone, the total angle of twist (rotation of face B with respect to face A) is

$$\theta_{AB} = \frac{TL}{G_b J}$$

where G_b is the shear modulus of bone.

The total angle of twist θ_{CF} in the repaired bone is obtained by summing the twists in the three discrete parts:

$$\theta_{CF} = \theta_{CD} + \theta_{DE} + \theta_{EF}$$

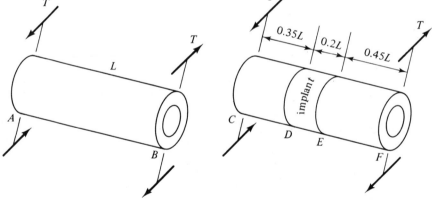

(a) model of bone
section in torsion

(b) bone section
with implant

Fig. 9-18. Example 9-7. (a) Model of bone section in torsion. (b) Bone section with implant.

Since the three parts have the same polar moments of inertia and the same torque T is acting on each,

$$\theta_{CF} = \frac{T(0.35L)}{G_b J} + \frac{T(0.2L)}{G_i J} + \frac{T(0.45L)}{G_b J} = \frac{T(0.8L)}{G_b J} + \frac{T(0.2L)}{G_i J}$$

where G_i is the shear modulus of the implant.

The implant is relatively short. Thus, if G_i is not too different from G_b, θ_{CF} will be indistinguishable from θ_{AB}.

EXAMPLE 9-8

A solid circular shaft made of a titanium alloy is 60 cm long and has a constant diameter of 4 cm. A change in the design requires a reduction in the diameter to 3 cm over half of the length of the shaft. What is the corresponding change in the angle under a torque of 1 kN·m? Assume elastic behavior and that $G = 45$ GPa. The torque is applied at the ends of the shaft.

The angle of twist for the uniform shaft is

$$\theta_u = \frac{TL}{GJ} = \frac{10^3(0.6)}{45 \times 10^9(0.02^4\pi/2)} = 0.053 \text{ rad}$$

The twist in the thicker half of the redesigned shaft is half of θ_u. The twist in the

thin half is $$\theta_t = \frac{TL/2}{GJ_t} = \frac{10^3(0.3)}{45 \times 10^9(0.015^4\pi/2)} = 0.084 \text{ rad}$$

The total twist over the length of the stepped shaft is

$$\frac{\theta_u}{2} + \theta_t = 0.11 \text{ rad}$$

EXAMPLE 9-9

A solid shaft is made from sintered metal and it is found that the density is not uniform in it. What is the twist in the 10-in. long, 1.6-in. diameter shaft if G varies linearly from 2×10^6 psi at one end to 2.6×10^6 psi at the other end? The applied torque is 300 ft-lb.

Solution

$$T = 300(12) = 3600 \text{ in.-lb}$$

With x the coordinate along the shaft,

$$G(x) = (6 \times 10^4)x + (2 \times 10^6)$$

$$\theta = \int_0^L \frac{T dx}{JG} = \int_0^{10} \frac{3600 dx}{(\pi/2)(0.8)^4[(6 \times 10^4)\,x + (2 \times 10^6)]} = 0.0244 \text{ rad}$$

Note that using the average G in TL/JG without integration gives $\theta = 0.0243$ rad.

EXAMPLE 9-10

What is the change in the angle of twist per unit length in a shaft if half the material is saved by drilling out the shaft? Assume the same torque in both cases.

Solution

Assume the outer radius is R and the inner radius is r. To save half the

material, $\pi R^2 - \pi r^2 = \dfrac{\pi R^2}{2}, \quad$ so $\quad r = 0.707R$

For the original solid shaft,

$$\frac{\theta}{L} = \frac{T}{JG} = \frac{T}{(\pi R^4/2)G} = \frac{2T}{\pi G}\frac{1}{R^4}$$

For the hollow shaft,

$$\frac{\theta'}{L} = \frac{T}{J'G} = \frac{T}{(\pi/2)(R^4 - r^4)G} = \frac{2T}{\pi G}\frac{4}{3R^4}$$

The hollow shaft twists 1.33 times as much as the original shaft.

EXAMPLE 9-11

A delicate torsional pendulum in the horizon scanner of a satellite tracking station has a 0.1-mm diameter sapphire filament for the torsion element. What is the maximum torque that can be applied if $L = 3$ cm, $G = 35 \times 10^6$ psi, $\theta_{max} = 10°$, and $\tau_{max} = 10^6$ psi?

Solution

If the stress is limiting,

$$T_{max} = \frac{\tau \pi R^3}{2} = \frac{(10^6)\pi(0.05/25.4)^3}{2} = 0.012 \text{ in.-lb}$$

If the angle of twist is limiting,

$$T_{max} = \frac{\theta JG}{L} = \frac{(10)(\pi/180)\pi(0.05/25.4)^4(35 \times 10^6)}{2(3/25.4)} = 1.22 \times 10^{-4} \text{ in.-lb}$$

The maximum torque that can be applied is 1.22×10^{-4} in.-lb.

9-6. POWER TRANSMISSION

Large numbers of torsion members are used for the transmission of mechanical power in rotary motion. The torque acting on such a member depends on the power, which is the work done per unit time. Since the work done by a torque T in a complete revolution is $W = 2\pi T$, the power P in N revolutions per unit time is

$$P = 2\pi TN \qquad\qquad (9\text{-}6)$$

There are several variations of this equation used in practice, depending on the

Simultaneous testing of four drive shafts in an electrohydraulic, torque-controlled machine. *Juengen Telschow, "Testing Drive Axles and Shafts with Flexible Couplings for Front Wheel Drive Vehicles,"* Closed Loop, **4**, *No. 4 (1974), 2–10.*

appropriate units for power and rate of rotation. For example, the common unit of power is the horsepower:

$$1 \text{ hp} = 550 \text{ ft-lb/sec} = 33,000 \text{ ft-lb/min}$$

With N given in revolutions per minute and P in horsepower,

$$T = \frac{63,000P}{N} \text{ in.-lb} \tag{9-7}$$

In SI units, 1 hp = 745.7 N·m/s = 745.7 W. One should be careful to distinguish the SI unit for force and the number of revolutions, both denoted by "N."

In practice, the horsepower formula is most often used to relate the shear stresses and dimensions of torsion members. More rarely, it is used to analyze the rigidity (angular twist) of members.

EXAMPLE 9-12

Design a solid shaft for the 5-W motor of an electric clock. The motor runs at 600 rpm, and the maximum shear stress for entirely elastic deformation is 150 MPa.

Solution

From Eq. 9-6,

$$5 = 2\pi T(600)(\tfrac{1}{60}), \quad \text{and} \quad T = 0.0796 \text{ N} \cdot \text{m}$$

$$T = \frac{\tau_m \pi R^3}{2} = \frac{(150 \times 10^6)\pi R^3}{2}, \quad \text{and} \quad R = 6.96 \times 10^{-4} \text{ m} = 0.7 \text{ mm}$$

The shaft's diameter must be at least 1.4 mm.

EXAMPLE 9-13

Assume that the maximum power from a human is 0.5 hp. What is the minimum diameter of the rear axle of a bicycle if the shear stress should not exceed 30 ksi? Assume that the axle is in pure torsion, the wheel is 26 in. in diameter, and the maximum power is transmitted at 1 mph.

Solution

$$1 \text{ mph} = \tfrac{1}{60} \text{ mi/min} = 88 \text{ ft/min}$$

The wheel has to turn at

$$N = \frac{88}{\pi(2.166)} = 12.93 \text{ rev/min} \text{ (rpm)}$$

$$T = \frac{63,000(0.5)}{12.93} = 2437 \text{ in.-lb}$$

For a solid shaft,

$$T = \frac{\tau \pi R^3}{2} = (3 \times 10^4)\frac{\pi}{2} R^3 = 2437 \quad \text{and} \quad R = 0.373 \text{ in.}$$

A 0.75-in. diameter axle would be sufficient.

EXAMPLE 9-14

A new pump designed for a remote area is to be powered by humans. The maximum power through the drive shaft is 0.1 hp, and it rotates at 100 rpm.

(a) Determine the diameter of the shaft if the maximum shear stress is 140 MPa.
(b) At a later time, electric power becomes available, so an electric motor is planned to drive the same pump. What should be the power rating of the motor if the shaft will be rotating at 1800 rpm after the modification?

Solution

(a) From Eq. 9-7,

$$T = \frac{63{,}000\,P}{N} = \frac{63{,}000(0.1)}{100} = 63 \text{ in.-lb}$$

$$\tau_{\text{allowed}} = 140 \text{ MPa} = 20 \text{ ksi}$$

$$T = \frac{\tau \pi R^3}{2} = 20{,}000\frac{\pi}{2}R^3, \quad \text{and} \quad R_{\min} = 0.126 \text{ in.}$$

(b) Since P is proportional to NT and $T = \tau J/\rho$ (τ, J, ρ are constant), increasing N by 18 allows increasing the power by 18: $P_{\max} = 1.8$ hp for the motor.

9-7. TORSION OF NONCIRCULAR MEMBERS

The mathematical analysis of *solid noncircular members* is complicated and will not be given here. For a qualitative understanding of the cause of the difficulty, scribe four lines on a rectangular eraser to outline a rectangular cross section in a plane perpendicular to the longitudinal axis. Twist the eraser without bending it, and note that the originally straight lines become distorted. This indicates warping of the cross-sectional plane. The reason for the warping is that the shear stress is zero at the corners, whereas it is large at other places on the surface (maximum at the middle of the wider sides). Torsion formulas for various noncircular members are available in several advanced tests.

On the other hand, the torsion of *thin-walled noncircular tubes* can be analyzed without too much difficulty. One basic idea is that there must be a continuous distribution of finite shear stresses everywhere in a thin-walled tube when a torque is applied to it. In other words, there must be a continuous shear flow on the closed path of a cross-sectional area as shown in Fig. 9-19. If the thickness of the tube is

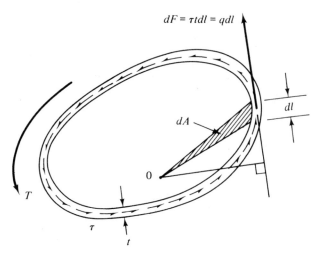

Fig. 9-19. Cross section of noncircular tube in torsion. Note: τ is not necessarily constant.

the same everywhere, the shear stress τ must be the same at all points; this can be seen by considering the free-body diagrams of any neighboring elements, each with the thickness of the wall. Continuing this line of thought, it is seen that the shear flow q is constant on the cross section and τ may vary if the wall thickness varies a

little:
$$q = \tau t = \text{constant}$$

This is supported by experimental evidence that, for entirely elastic behavior, the shear strain γ times t is constant around a given tube.

Consider next the shear force dF on the element dl and the moment dT caused by this force about an arbitrary reference point, 0, in Fig. 9-19. The integral of dT over the whole cross section equals the applied torque, and $pdl = 2dA$, so

$$T = \int pqdl = q \int pdl = 2q \int dA = 2qA$$

where A is the total area within the center line of the tube wall. Assuming that the shear stress does not vary over the small distance t,

$$\tau = \frac{T}{2At} \tag{9-8}$$

at any given point around the tube perimeter; t is the wall thickness at the given point, and A and T can be considered constants. Note that the equation is not useful for large deformations where sudden changes in A may occur as the tube collapses.

EXAMPLE 9-15

A thin-walled circular tube of radius R and wall thickness t is considered for a torsion member. Would it be a good idea to pull the tube through a die to give it a squarish cross section? Assume that the operation would not change the material properties or the circumferential distance of the tube. The maximum shear stress is τ, the same in both cases.

Solution

For the circular tube (c), the maximum torque is

$$T_c = \frac{\tau J}{R} = \frac{\tau 2\pi R^3 t}{R} = 2\pi \tau R^2 t$$

For the square tube (s),

$$T_s = \tau 2At$$

The perimeter is the same for the two tubes, so

$$A = \left(\frac{2\pi R}{4}\right)^2, \qquad T_s = 2\tau t \left(\frac{2\pi R}{4}\right)^2 = \frac{2\tau t R^2 \pi^2}{4}$$

which is only 79% of what the circular tube can resist.

EXAMPLE 9-16

Consider the possibility of torsion (this is not caused by rotation of the rotors) about the longitudinal axis of the helicopter discussed in Examples 6-8 and 7-3. If the maximum shear stress caused by torsion alone should not exceed 100 MPa, is it safe to have torques up to 2 kN·m? Ignore any complications caused by discontinuities in the shell or the possibility of wrinkling in the thin skin.

Solution

Using $\quad \tau = T/2At,$

$$T = 10^8 (2)(2)(2.5)(0.002) = 2 \text{ MN·m}$$

This is much higher than the expected torque of 2 kN·m.

9-8. PROBLEMS

Secs. 9-1, 9-2, 9-3

9-1 to 9-12. In each of these problems, determine the torque for the given distribution of shear stresses on the cross-sectional area of the shaft.

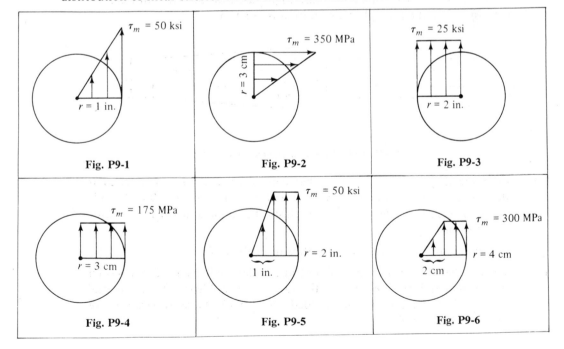

Fig. P9-1   Fig. P9-2   Fig. P9-3

Fig. P9-4   Fig. P9-5   Fig. P9-6

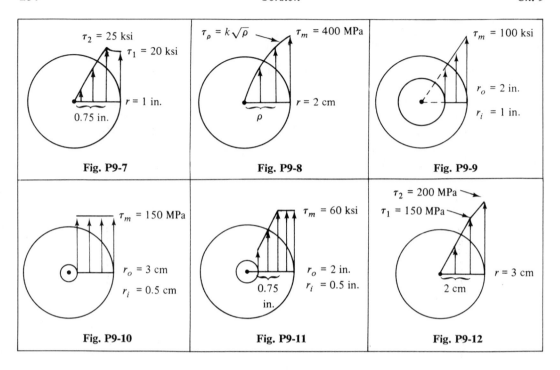

$\tau_2 = 25$ ksi $\tau_1 = 20$ ksi $r = 1$ in. 0.75 in. **Fig. P9-7**	$\tau_\rho = k\sqrt{\rho}$　$\tau_m = 400$ MPa $r = 2$ cm ρ **Fig. P9-8**	$\tau_m = 100$ ksi $r_o = 2$ in. $r_i = 1$ in. **Fig. P9-9**
$\tau_m = 150$ MPa $r_o = 3$ cm $r_i = 0.5$ cm **Fig. P9-10**	$\tau_m = 60$ ksi $r_o = 2$ in. $r_i = 0.5$ in. 0.75 in. **Fig. P9-11**	$\tau_2 = 200$ MPa $\tau_1 = 150$ MPa $r = 3$ cm 2 cm **Fig. P9-12**

9-13. In sketches of the members shown in Fig. P9-13, indicate plastic deformation at two stages of the torsional loading: (a) just after the initiation of yielding; (b) just before the fully plastic condition is reached.

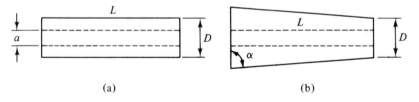

Fig. P9-13. Cylindrical and tapered tubes in torsion.

9-14. Compare the maximum torques that can be applied to a solid circular shaft (dia. $= D$) and to a hollow shaft with outside diameter D and inside diameter $D/2$. Consider (a) entirely elastic behavior and (b) fully plastic deformations.

9-15. Determine the strength-to-weight ratios in Prob. 9-14.

9-16. What is the stress distribution in a solid circular shaft made of mild steel if (a) the material outside 90% of the radius has yielded; (b) the material outside 10% of the radius has yielded. Work with the stress-strain curve shown in Fig. P9-16.

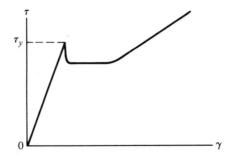

Fig. P9-16. True stress versus true strain for mild steel in shear.

*9-17. How would you position an electrical strain gage on a cylindrical torsion member to obtain the most useful data on the deformations? How would you relate the data obtained from the strain gage to the torque in the member?

Sec. 9-5

9-18 to 9-31. In each of these problems, calculate the relative angular displacement of the parallel planes denoted by the letters *A* and *B*.

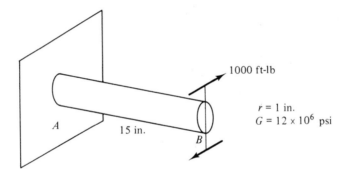

$r = 1$ in.
$G = 12 \times 10^6$ psi

Fig. P9-18

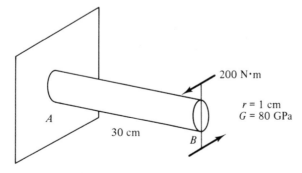

$r = 1$ cm
$G = 80$ GPa

Fig. P9-19

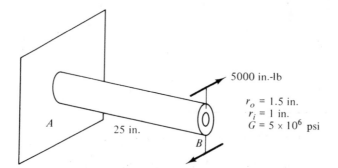

Fig. P9-20

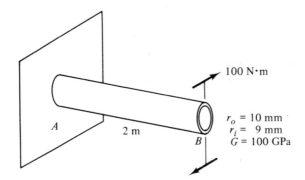

Fig. P9-21

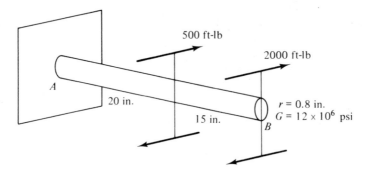

Fig. P9-22

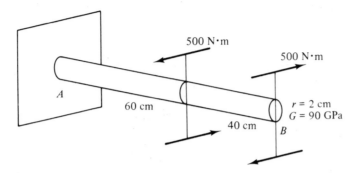

500 N·m

500 N·m

A

60 cm

40 cm

B

$r = 2$ cm
$G = 90$ GPa

Fig. P9-23

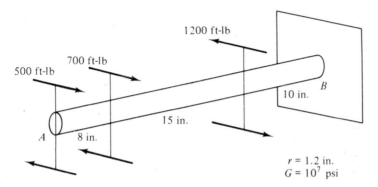

1200 ft-lb

700 ft-lb

500 ft-lb

B

10 in.

A

8 in.

15 in.

$r = 1.2$ in.
$G = 10^7$ psi

Fig. P9-24

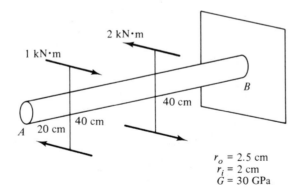

2 kN·m

1 kN·m

B

40 cm

A

20 cm

40 cm

$r_o = 2.5$ cm
$r_i = 2$ cm
$G = 30$ GPa

Fig. P9-25

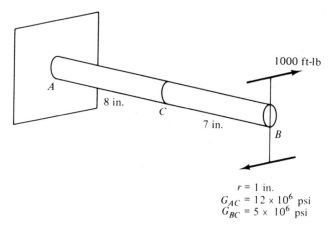

$$r = 1 \text{ in.}$$
$$G_{AC} = 12 \times 10^6 \text{ psi}$$
$$G_{BC} = 5 \times 10^6 \text{ psi}$$

Fig. P9-26

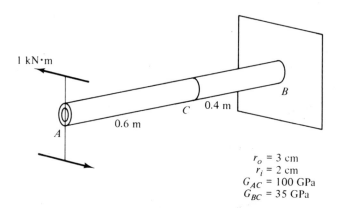

$$r_o = 3 \text{ cm}$$
$$r_i = 2 \text{ cm}$$
$$G_{AC} = 100 \text{ GPa}$$
$$G_{BC} = 35 \text{ GPa}$$

Fig. P9-27

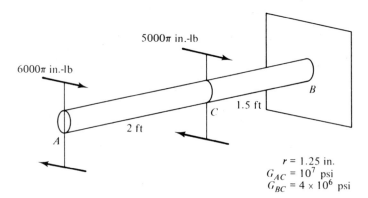

$$r = 1.25 \text{ in.}$$
$$G_{AC} = 10^7 \text{ psi}$$
$$G_{BC} = 4 \times 10^6 \text{ psi}$$

Fig. P9-28

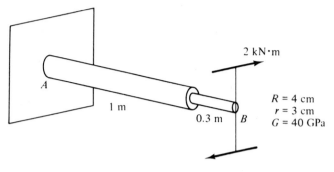

Fig. P9-29

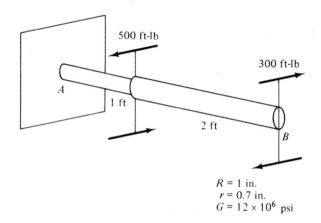

Fig. P9-30

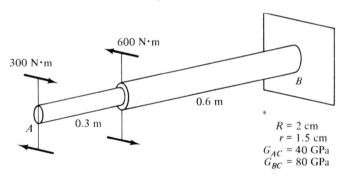

Fig. P9-31

9-32. What is the length of the bar in Prob. 9-23 that should replace the 40-cm piece to make $\theta_{AB} = 0$?

9-33. Change the 1-kN·m torque in Prob. 9-25 to make $\theta_{AB} = 0.1$ rad clockwise as viewed from end A.

Sec. 9-6

9-34. Design a solid shaft for a 20-hp electric motor that is to operate at 1200 rpm. The yield strength in shear is 50 ksi and should not be exceeded.

9-35. What is the power rating of a 2-cm diameter steel shaft rotating at 1000 rpm if the maximum elastic shear stress is 1.2 GPa?

9-36. What is the change in power rating of a 2-in. diameter shaft if N is changed from 1 to 100 rpm? $\tau_{max} = 60$ ksi. What is the general lesson from the result?

9-37. A solid shaft is 4 cm in diameter and has a yield strength in shear of 300 MPa. What is the required speed of rotation to transmit 100 kW?

9-38. A hollow shaft has $D_i = 1$ in., $D_o = 2$ in., and yield strength in shear of 40 ksi. What is its power rating at 1500 rpm?

9-39. A hollow shaft has $D_i = 2$ cm, $D_o = 4$ cm. It is to transmit 20 kW at 3000 rpm. What is the required strength of the shaft to avoid inelastic deformations?

Sec. 9-7

9-40. What is the maximum torque on a tube of semicircular cross section that was made from a circular tube as in Example 9-15?

9-41. Redo Prob. 9-40 for a tube with an equilateral triangular cross section.

9-42. Redo Prob. 9-40 for a tube with a rectangular cross section of sides $a \times 2a$.

9-43. Redo Prob. 9-40 for a tube with a rectangular cross section of sides $a \times 10a$.

10

STATICALLY INDETERMINATE
MEMBERS

Consider a circular table that is designed to stand on three vertical legs. If the locations of the legs and of the center of gravity are known, the forces in the legs can be calculated using the equilibrium equations of statics. The same table standing on four identical legs cannot be analyzed so easily, no matter how the legs are arranged. Of course, the four legs can be arranged so that the table has more stability than it could have with any arrangement of three vertical legs of the same length. This is a desirable feature of having an extra support. The price that one has to pay for such an advantage is the inability to determine the forces in any of the legs using the equilibrium equations. The floor is never perfectly flat, and the legs are not identical in length, so the legs cannot share the load in any predictable way. Obviously, this is not a serious problem for most tables and other furniture, but one is not always so fortunate in other situations.

Members that have more supports or constraints than the minimum required for static equilibrium are called *statically indeterminate*. There are many different practical situations where such conditions exist. The analysis of forces and deformations in statically indeterminate members is possible if a sufficient number of valid relationships can be found besides the equilibrium equations. The additional relationships are based on geometrical conditions that must be satisfied. The examples in this chapter illuminate some of the approaches that are useful in this area. New basic concepts are not introduced here.

10-1. STATICALLY INDETERMINATE AXIALLY LOADED MEMBERS

EXAMPLE 10-1

A rigid steel beam of $L = 10$ m and $W = 40$ kN is temporarily suspended in a horizontal position by three 5-m long, vertical wires. Two of the wires are 4 mm in diameter, and these are attached to the ends of the beam. The third wire is 5 mm in diameter, and it is attached at the center of the beam. The wires are of the same steel, with $E = 200$ GPa, $\sigma_y = 1.7$ GPa. How close are the wires to yielding in this situation?

Solution

From the free-body diagram in Fig. 10-1,

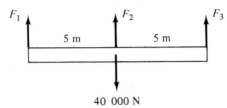

40 000 N   **Fig. 10-1.** Example 10-1.

$$F_1 = F_3 \quad \text{(because of symmetry)} \tag{a}$$
$$F_1 + F_2 + F_3 = 40\ 000 \tag{b}$$

A third equation is needed. For a rigid beam, $e_1 = e_2 = e_3$, which can be written as

$$\frac{F_1 L_1}{A_1 E_1} = \frac{F_2 L_2}{A_2 E_2} = \frac{F_3 L_3}{A_3 E_3}$$

Since $L_1 = L_2$ and $E_1 = E_2$,

$$\frac{F_1}{\pi r_1^2} = \frac{F_2}{\pi r_2^2}, \quad \text{and} \quad F_1 = F_2 \left(\frac{0.002}{0.0025}\right)^2 \tag{c}$$

Equations a, b, and c can be solved now.

$$F_1 = F_3 = 11\ 230, \qquad F_2 = 17\ 540$$
$$\sigma_1 = \sigma_3 = \frac{F_1}{\pi(0.002)^2} = 893\ \text{MPa}, \qquad \sigma_2 = \frac{F_2}{\pi(0.0025)^2} = 893\ \text{MPa}$$

The stresses are about half of the yield strength.

EXAMPLE 10-2

A concrete column with a cross-sectional area of 1 ft² is installed in a vertical position. It is attached firmly to rigid structures at its ends (Fig. 10-2a). The end

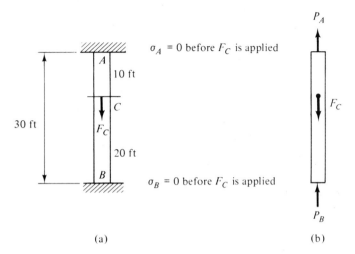

$\sigma_A = 0$ before F_C is applied

$\sigma_B = 0$ before F_C is applied

(a)   (b)

Fig. 10-2. Example 10-2.

structures are so rigid that the distance between them can be assumed a constant, regardless of any forces in the column. Also, assume that the stresses in the column at A and B are zero. Suppose that in the next step of the construction a downward force of 50,000 lb is applied symmetrically to the column at C (the resultant force is through the centroid of the cross section). What are the stresses σ_A and σ_B now?

Solution

The column is statically indeterminate because it has one more support than necessary for static equilibrium. It could hang from the upper support without touching anything at B or it could stand on the lower support without touching anything at A. But the column is attached firmly at both ends. There are two unknown forces and only one independent equilibrium equation from Fig. 10-2b:

$$P_A + P_B = F_C$$

The key to the solution is that the column remains continuous after the force F_C is applied. This means that if point C moves downward a distance e, the elongation of part AC is e and the deformation of part BC is also e. From these deformations the strains can be determined, and using these strains and the equilibrium equation the stresses can be calculated as follows:

$$\epsilon_{AC} = \frac{e}{L_{AC}} \quad \text{and} \quad \epsilon_{BC} = \frac{e}{L_{BC}}$$

$$e = \epsilon_{AC} L_{AC} = \epsilon_{BC} L_{BC}$$

$$\epsilon_{AC} = \frac{\sigma_{AC}}{E} \quad \text{and} \quad \epsilon_{BC} = \frac{\sigma_{BC}}{E}$$

since σ_{AC} is a constant along part AC and σ_{BC} is a constant along part BC. Furthermore,

$$\sigma_{AC} = \frac{P_A}{\text{area}}, \quad \sigma_{BC} = \frac{P_B}{\text{area}}, \quad \text{and} \quad \sigma_{AC} = \sigma_{BC}\frac{L_{BC}}{L_{AC}}$$

Thus,
$$\sigma_{AC}(\text{area}) + \sigma_{BC}(\text{area}) = 50{,}000,$$

$$\sigma_{BC}(\text{area})\left(\frac{L_{BC}}{L_{AC}} + 1\right) = 50{,}000$$

and
$$\sigma_B = \sigma_{BC} = \frac{50{,}000 \text{ lb}}{(1 \text{ ft}^2)(\frac{20}{10} + 1)(144 \text{ in}^2/\text{ft}^2)} = 116 \text{ psi}$$

$$\sigma_A = \sigma_{AC} = 2\sigma_{BC} = 232 \text{ psi}$$

σ_A is tensile and σ_B is compressive by inspection.

A formula that is frequently useful in analyzing axially loaded members is

Eq. 5-1:
$$e = \frac{PL}{AE}$$

where e = elongation of member (deformation should be entirely elastic)
P = axial force in member
L = length of member
A = cross-sectional area of member (should be a constant for the part under consideration)
E = modulus of elasticity

EXAMPLE 10-3

The concrete reservoir described in Example 3-8 is wrapped with an alloy steel wire until the steel is 0.5 in. thick along the cylinder's entire length. The tension in the steel causes a 650-psi compressive stress in the concrete at 70°F. What is the stress in the concrete at −297°F? $E_{steel} = 30 \times 10^6$ psi, $E_{concrete} = 4 \times 10^6$ psi, $\alpha_{steel} = 6 \times 10^{-6}$ per °F, $\alpha_{concrete} = 4.7 \times 10^{-6}$ per °F.

Solution

The problem is analogous to that of an axially loaded member (think of short, circumferential segments of the cylinder). Assume that the concrete and the steel are always at a common temperature and that they have zero strain at 70°F (after wrapping). Upon cooling, the circumference L of the steel would want to change by

$$\Delta L_{steel} = \pi D \alpha_{steel} \Delta T = \pi(38)(12)(6 \times 10^{-6})(-367) = -3.15 \text{ in.}$$

while that of the concrete would want to change by

$$\Delta L_{concrete} = \pi D \alpha_{concrete} \Delta T = -2.47 \text{ in.}$$

The steel would shrink more than the concrete if there were no constraint. The stresses induced by the misfit are determined from the deformations. The steel and the concrete go to the same final diameter upon cooling, so

Concrete liquid oxygen reservoir being prestressed with steel wires. Engineering Case Library, Stanford University. *Courtesy Professor Henry O. Fuchs, Stanford University, and H. C. Kornemann, Kenmore, New York.*

$$\Delta L_{\substack{\text{steel}\\ \text{therm.}}} + \Delta L_{\substack{\text{steel}\\ \text{mech.}}} = \Delta L_{\substack{\text{concrete}\\ \text{therm.}}} + \Delta L_{\substack{\text{concrete}\\ \text{mech.}}} = \Delta L_{\text{actual}}$$

This accounts for deformations caused by the temperature change and by the mechanical stress resulting from the misfit. In more detail,

$$\pi D\left(\alpha_{\text{steel}}\Delta T + \frac{\sigma_{\text{steel}}}{E_{\text{steel}}}\right) = \pi D\left(\alpha_{\text{concrete}}\Delta T + \frac{\sigma_{\text{concrete}}}{E_{\text{concrete}}}\right)$$

$$-3.15 + \sigma_{\text{steel}}(4.78 \times 10^{-5}) = -2.47 + \sigma_{\text{concrete}}(3.58 \times 10^{-4})$$

Another condition must be found to determine the two unknowns. For that, it is convenient to consider the equilibrium of a part of the cylinder, a semicircular section of any height h. The stresses σ_{concrete} and σ_{steel} are acting on this partial free body such that

$$2(h)(8)\sigma_{\text{concrete}} + 2(h)(0.5)\sigma_{\text{steel}} = 0$$

assuming that all the stresses are tensions (which was implicitly assumed in the equations involving the temperature change). Thus,

$$\sigma_{\text{steel}} = -16\sigma_{\text{concrete}}$$

Combining the two equations with the two unknowns results in

$$\sigma_{\text{concrete}}[(-16)(4.78 \times 10^{-5}) - (3.58 \times 10^{-4})] = 0.68, \ \sigma_{\text{concrete}} = -608 \text{ psi}$$

This should be added to the original prestress in the concrete at 70°F to find the total stress:

$$\sigma_{\text{concrete}} = -608 - 650 = -1258 \text{ psi} \quad \text{(compressive)}$$

10-2. STATICALLY INDETERMINATE BEAMS

EXAMPLE 10-4

A cantilever beam under uniformly distributed loading deflects as in Fig. 10-3a. This beam is statically determinate. Next, the free end of the beam is jacked up so that its final deflection (y_A) is zero (Fig. 10-3b). This may be desirable so there will be no deflection at point A or simply because this is equivalent to increasing the strength and stiffness of the beam. Determine the force F and the reactions at the wall.

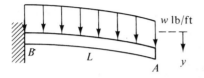

(a) determinate.

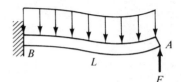

(b) indeterminate.

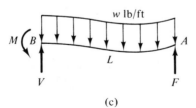

(c)

Fig. 10-3. Example 10-4.

Solution

The addition of the force F makes the beam statically indeterminate, but it also provides an additional boundary condition. The three unknown reactions in the free-body diagram of Fig. 10-3c can be determined as follows. First, the redundant force F is not considered and the deflection at the same place is calculated:

$$y_A = \frac{wL^4}{8EI} \quad \text{(under the loading of } w \text{ lb/ft alone)}$$

The force F must be large enough to cause the same magnitude of deflection to result in zero deflection at point A. The maximum deflection of a cantilever beam caused by a concentrated force F is

$$y_A = \frac{FL^3}{3EI} \quad \text{(under the loading of } F \text{ alone)}$$

The two deflections must be equal in magnitude, so

$$\frac{wL^4}{8EI} = \frac{FL^3}{3EI}, \quad \text{and} \quad F = \frac{3}{8}wL$$

The vertical shear at the wall can be calculated now by summing forces in the vertical direction:

$$\Sigma F_y = -\frac{3}{8}wL - V + wL = 0, \quad \text{so} \quad V = \frac{5}{8}wL$$

The moment at the wall is found equally simply:

$$\Sigma M_B = \frac{3}{8}wL(L) - wL\frac{L}{2} + M = 0, \quad \text{so} \quad M = \frac{wL^2}{8}$$

The stresses and deflection at any place in the beam can be calculated now using the conventional methods.

EXAMPLE 10-5

The tracks of a factory crane are horizontal I beams, each supported by two widely spaced columns. The distance is L between the supports of each beam. Determine whether there is a significant reduction of stresses in a beam if it is welded to each support instead of simply supported. Assume the welded connection to be ideally rigid, and make the comparison by using the beam's own weight w per unit length as the only load.

Solution

The comparison is made using Fig. 10-4. For the simply supported beam, $M_0 = 0$, and the maximum moment (and stress) is at the center of the beam:

$$M_{1\,\text{max}} = \frac{wL^2}{8}$$

Fig. 10-4. Example 10-5.

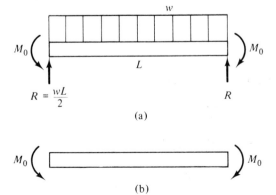

(a)

(b)

For the fixed-end beam, R is still $wL/2$, but it is not possible to solve for M_0 directly from equilibrium. An additional condition is that the beam has zero slope at the ends and that this is caused by the moments M_0. The moment is calculated by equating the slope at the end of a simply supported beam ($M_0 = 0$) with that under M_0 only (Fig. 10-4b) as follows. For the simple support,

$$y' = \frac{wL^3}{24EI}$$

For the pure bending,

$$y' = -\frac{M_0 L}{2EI}$$

The sum of the two loadings must give zero slope, so

$$0 = \frac{wL^3}{24EI} - \frac{M_0 L}{2EI}, \quad \text{and} \quad M_0 = \frac{wL^2}{12}$$

For the fixed-end beam, the maximum moment (and stress) is at the ends, and its magnitude is $M_{2\max} = M_0$.

Since stresses are proportional to the moments, the maximum stress is reduced by 33% if the track is welded to the supports.

EXAMPLE 10-6

The beam in Fig. 10-5 is horizontal and contacts the coil spring with zero force before the load is applied. Determine the reactions R at the simple supports after the uniform load is applied.

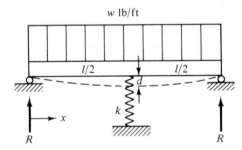

Fig. 10-5. Example 10-6.

Solution

$$M = Rx - \frac{wx^2}{2}, \quad 0 < x < \frac{l}{2}$$

$$EIy'' = Rx - \frac{wx^2}{2}$$

$$EIy' = \frac{Rx^2}{2} - \frac{wx^3}{6} + C_1$$

$$EIy = \frac{Rx^3}{6} - \frac{wx^4}{24} + C_1 x + C_2$$

At $x = 0$, $y = 0$, so $C_2 = 0$. At $x = l/2$, $y = -d$, $y' = 0$, so $C_1 = [w/6(l/2)^3] - [(R/2)(l/2)^2]$. From equilibrium in the y-direction,

$$2R + kd = wl, \qquad d = \frac{wl - 2R}{k}$$

The equation for deflection can be written with only one unknown, R:

$$-\frac{EI}{k}(wl - 2R) = -\frac{R}{3}\left(\frac{l}{2}\right)^3 + \frac{w}{8}\left(\frac{l}{2}\right)^4$$

$$R = \frac{w}{\frac{1}{3}(l/2)^3 - (2EI/k)}\left[\frac{1}{8}\left(\frac{l}{2}\right)^4 - \frac{EIl}{k}\right] = \frac{wl}{2}\left(\frac{3kl^3 + 384EI}{8kl^3 + 384EI}\right)$$

10-3. STATICALLY INDETERMINATE TORSION MEMBERS

EXAMPLE 10-7

Consider a shaft that is driven at constant speed by a motor with a known torque. The shaft drives two mechanisms as shown in Fig. 10-6a. The two mechanisms are governed to make the relative angular positions at points A and B always the same. Determine the maximum shear stresses in the two parts of the shaft, assuming ideally elastic behavior of the material.

Fig. 10-6. Example 10-7.

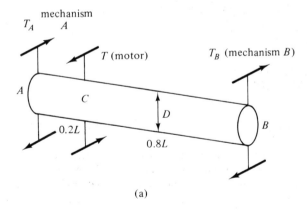

(a)

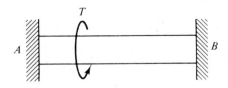

(b)

Solution

The problem can be treated as one involving static equilibrium. A useful model that simplifies the problem is a shaft attached at both ends A and B to ideally rigid walls (Fig. 10-6b). The equilibrium equation with two unknowns is

$$T_A + T_B = T$$

The additional fact that leads to the solution is that the shaft remains continuous after it is twisted. Thus, the angle at C with respect to A is the same as that with respect to B:

$$\theta_{AC} = \theta_{BC} \quad \text{or} \quad \frac{T_A(0.2L)}{GJ} = \frac{T_B(0.8L)}{GJ}$$

$$T_A = 4T_B, \quad \text{so} \quad T_B = \frac{T}{5} \quad \text{and} \quad T_A = \frac{4}{5}T$$

T_A is valid everywhere from A to C, and T_B is valid everywhere from B to C. The required shear stresses can be calculated routinely now.

EXAMPLE 10-8

The moving coils of modern electric meters have taut-band torsional suspensions (Fig. 10-7) that have less friction and are much more rugged than the finest devices with jewel bearings. For the given model, assume that the two torsional

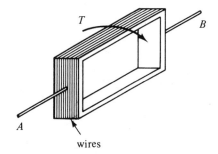

wires   **Fig. 10-7.** Example 10-8.

members are 0.1 mm in diameter and 5 mm long and are rigidly attached at A and B. The shear modulus is 100 GPa. The coil frame is 1 cm long, so the distance AB is 2 cm. What is the torque T (which has to do with the sensitivity of the meter) that can deflect the coil $\pm 60°$ from its equilibrium position? Assume that the axial tension in the filaments (which is necessary for the positioning of the coil) does not affect the torsional behavior.

Solution

Noting the symmetry, $T_A = T_B = T/2$. The same result is obtained using

$$\theta_A = \theta_B: \qquad \frac{T_A(0.005)}{(\pi/2)(5 \times 10^{-5})^4(10^{11})} = \frac{T_B(0.005)}{(\pi/2)(5 \times 10^{-5})^4(10^{11})}$$

For $\theta = 60°$,

$$60\left(\frac{\pi}{180}\right) = \frac{(T/2)(0.005)}{(\pi/2)(5 \times 10^{-5})^4(10^{11})}, \qquad T = 4.1 \times 10^{-4} \text{ N·m}$$

10-4. PROBLEMS

Sec. 10-1

10-1. A uniform, rigid bar is suspended by three steel wires of identical material and length, as shown in Fig. P10-1. What is the dimension a if the bar should remain horizontal after it is hung?

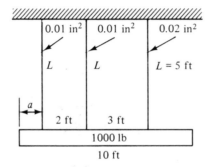

$E = 30 \times 10^6$ psi

Fig. P10-1

10-2. A uniform, rigid bar is suspended by three steel wires of identical material and length, as shown in Fig. P10-2. What are the stresses in the wires?

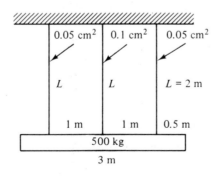

$E = 200$ GPa

Fig. P10-2

10-3 to 10-8. The uniform, rigid beam is pivoted at A. The wires are taut with negligible tension in them before a supporting structure is removed from under the beam. What are the stresses in the wires after the removal of the support?

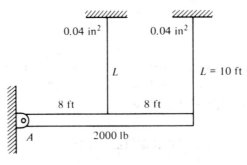

$E = 30 \times 10^6$ psi

Fig. P10-3

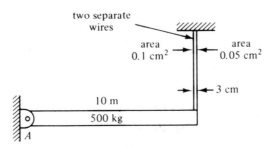

$E = 210$ GPa

Fig. P10-4

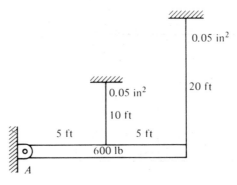

$E = 10^7$ psi

Fig. P10-5

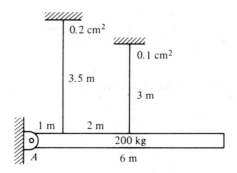

$E = 70$ GPa

Fig. P10-6

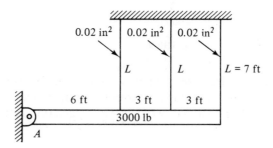

$E = 30 \times 10^6$ psi

Fig. P10-7

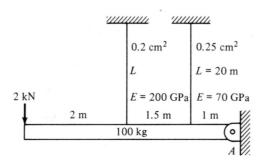

Fig. P10-8

*10-9. Three steel wires ($E = 30 \times 10^6$ psi) support a 2000-lb load. What are the stresses in the wires? Would it be advantageous to have the 10-ft wire with a 0.01-in² area and the other wires with 0.02-in² areas?

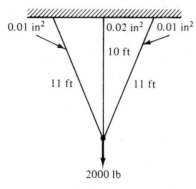

Fig. P10-9

*10-10. One steel and two aluminum wires support a 5-kN load. What are the stresses in the wires? Would it be advantageous to use an aluminum wire (0.1 cm²) as the vertical member and two steel wires (each 0.1 cm²) as the slanted members?

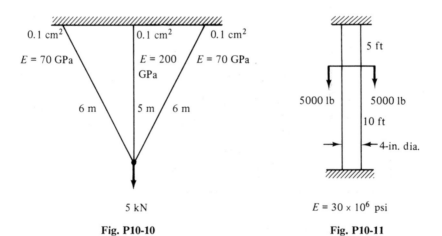

Fig. P10-10   **Fig. P10-11**

10-11. A uniform, round bar is attached to rigid supports at its ends. What are the normal stresses in the bar? The position where the loads are applied could be up to 3 ft lower than that shown in Fig. P10-11. Would such a change be reasonable?

10-12. A round bar of uniform material has a step change in diameter. What are the normal stresses in the bar? Evaluate the effect of inverting the bar (the loads are still applied at the place of change in section).

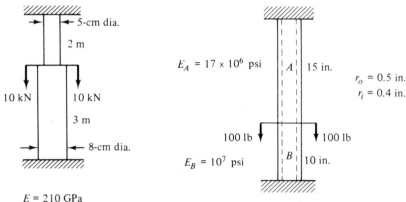

E = 210 GPa

Fig. P10-12

Fig. P10-13

10-13. Two cylindrical tubes of identical geometry but different materials are joined where the loads are applied. The ends are fixed rigidly. Determine the normal stresses in the tubes. Would it be advantageous to interchange the materials in the upper and lower parts if their tensile and compressive strengths (four strength values) were about equal?

*10-14. A round aluminum bar is to be loaded as shown, but its weight should be reduced as much as possible. This can be done by drilling out the bar. What are the desirable inside dimensions of the bar if the allowable stresses are 350 MPa in tension and 210 MPa in compression?

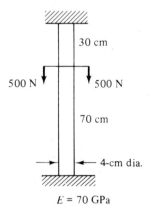

E = 70 GPa

Fig. P10-14

*10-15. Two round bars are joined and attached to rigid supports as shown. The stresses in the bars are essentially zero at 25°C. What are the stresses at −25°C? Assume that the coefficients of thermal expansion are 6.5×10^{-6} per °F for the steel and 13×10^{-6} per °F for the aluminum.

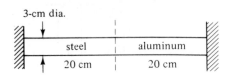

Fig. P10-15

Sec. 10-2

*10-16 to 10-29.   Determine all the external reactions for each of the elastic beams. Sketch the deflection curves. Assume that each beam has a rectangular cross section 5 cm × 7.5 cm (2 in. wide and 3 in. deep) and that $E = 10^7$ psi = 70 GPa in each case. Where there is a coil spring, the force is zero in the spring and the beam is horizontal before the load is applied.

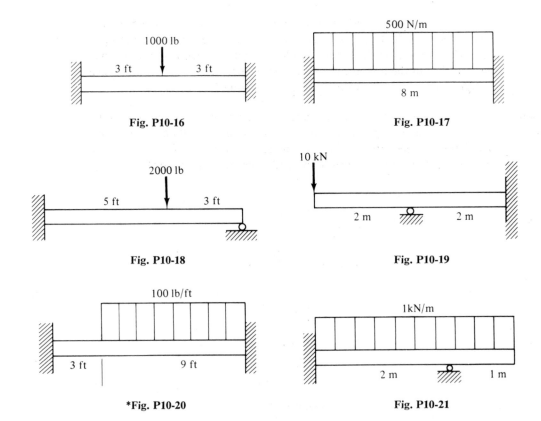

Fig. P10-16

Fig. P10-17

Fig. P10-18

Fig. P10-19

*Fig. P10-20

Fig. P10-21

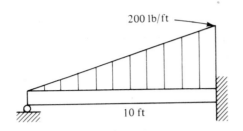

Fig. P10-22

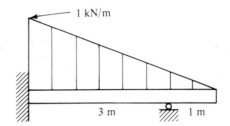

Fig. P10-23

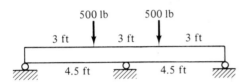

Fig. P10-24

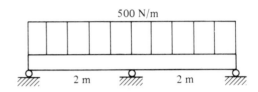

Fig. P10-25

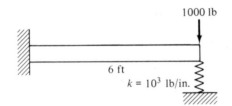

Fig. P10-26

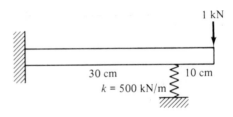

Fig. P10-27

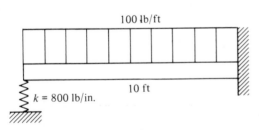

Fig. P10-28

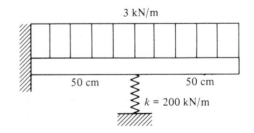

Fig. P10-29

Sec. 10-3

10-30 to *10-39.   Calculate the maximum shear stress in each of these problems. Assume elastic behavior and that the ends *A* and *B* are fixed. *T* is the torque applied at *C*. Where there are two torques, they are both clockwise when viewed from the end *B*.

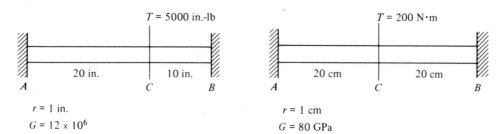

T = 5000 in.-lb

20 in.   10 in.

A   C   B

r = 1 in.

$G = 12 \times 10^6$

Fig. P10-30

T = 200 N·m

20 cm   20 cm

A   C   B

r = 1 cm

G = 80 GPa

Fig. P10-31

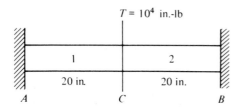

$T = 10^4$ in.-lb

1   2

20 in.   20 in.

A   C   B

r = 1 in.

$G_1 = 12 \times 10^6$ psi

$G_2 = 6 \times 10^6$ psi

Fig. P10-32

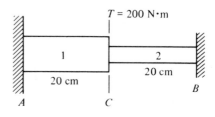

T = 200 N·m

1   2

20 cm   20 cm

A   C   B

$r_1 = 2$ cm

$r_2 = 1$ cm

G = 80 GPa

Fig. P10-33

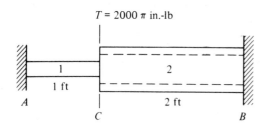

T = 2000 π in.-lb

1   2

1 ft   2 ft

A   C   B

$r_1 = 1$ in. (solid)

$r_{2_o} = 2$ in.

$r_{2_i} = 1.5$ in.

$G = 3 \times 10^6$ psi

Fig. P10-34

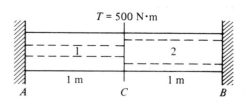

T = 500 N·m

1   2

1 m   1 m

A   C   B

r = 2 cm

$r_{1_i} = 1$ cm

$r_{2_i} = 1.5$ cm

$G_1 = 80$ GPa

$G_2 = 50$ GPa

Fig. P10-35

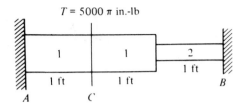

$T = 5000\,\pi$ in.-lb

1 ft   1 ft   1 ft

A C B

$r_1 = 0.8$ in.
$r_2 = 0.5$ in.

$G = 12 \times 10^6$ psi

Fig. P10-36

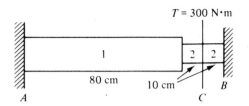

$T = 300$ N·m

80 cm   10 cm

A C B

$r_1 = 2$ cm
$r_2 = 1.5$ cm

$G = 40$ GPa

Fig. P10-37

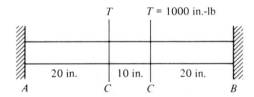

T $T = 1000$ in.-lb

20 in.   10 in.   20 in.

A C C B

$r = 0.5$ in.
$G = 4 \times 10^6$ psi

Fig. P10-38

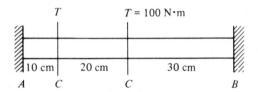

T $T = 100$ N·m

10 cm   20 cm   30 cm

A C C B

$r = 1$ cm
$G = 50$ GPa

Fig. P10-39

11

BUCKLING

Imagine that a friend hopes to be a champion pole-vaulter. He asks you to help him design a new pole, which has to be lighter than those available commercially. It has to be superior to others as a spring and it should be strong so that the probability of fracture is minimal. Your friend is particularly concerned about the fracture of poles. He shows you one that he broke himself. The broken pole looks somewhat frightening, but he isn't too concerned about the danger of being impaled on the splintered pole, only about the hard landing on the head or back in an uncontrolled fall. He has heard about the sudden, catastrophic failures when long, slender columns buckle and doesn't want this to happen to the newly designed pole.

At first it is not certain that the failures of vaulting poles can be classified as buckling similar to the failures of some structural members. It is necessary to learn the basic concepts of buckling to be able to decide what the problem is and what it is not. Also, understanding the problem would be helpful in selecting an appropriate laboratory test for the new pole.

11-1. ELASTIC BUCKLING

Buckling results from a state of instability. To understand the particular kind of instability involved, consider the model of a pole or column shown in Fig. 11-1. Figure 11-1a shows a straight, weightless pole that is firmly placed in a rigid foundation. The main section of the pole is vertical when there are no forces acting on it. Figure 11-1b is a free-body diagram of the pole when a small weight W is applied to the arm at point B. The force and moment reactions at point A should be self-explanatory. If the weight is doubled at B, the vertical reaction F_A is also doubled, but the moment reaction is more than doubled. This is because the pole

Fracture of fiber glass pole re-enacted.

leans farther away from the vertical as the weight is increased. Thus, the deformation of the pole itself causes an extra increase in the moment acting on it. Any increase in the external load has a similar effect, and the moment on the pole increases out of proportion with the applied load.

There is a critical load for the pole. A very small increment in the load above this will cause a runaway increase in the deflection, and the pole bends to the ground or breaks. These events always occur rapidly.

There are three main things that are noteworthy about the buckling of the pole. First of all, the maximum stress in the pole when the critical load is reached

Fig. 11-1. Model for instability of a column.

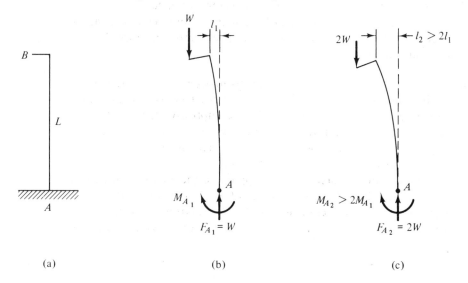

(a)   (b)   (c)

can be still below the yield or fracture strength of the material. Naturally, the pole could deform plastically or fracture as it collapses to the ground, but these are not necessary to start the unstable, large-scale deformation. This is called *elastic buckling*.

Another item of importance is that the external load does not have to be applied in an obviously eccentric way as in Fig. 11-1 to cause buckling. If the load is applied axially to a perfectly straight column at the centroid of the cross-sectional area, the stresses on the cross section are uniform, and there is no moment to cause buckling. Buckling may occur, however, even when these conditions are apparently satisfied. There are two common causes of *effectively eccentric loading*. Perfectly straight columns, especially long ones, cannot be expected to be produced by human hands. Thus, even a centroidally applied load causes a bending moment in the column, and this can result in unstable deformation. Even if the column could be made perfectly straight, an inhomogeneity in the material could make the loading effectively eccentric. This is illustrated in Fig. 11-2. In practice one can expect that a column will be neither perfectly straight nor homogeneous.

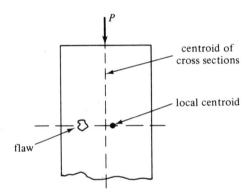

Fig. 11-2. Eccentricity of a column caused by an internal flaw.

The third item of interest about elastic buckling is that the effectively eccentric loading (which is always present to some extent) is critical in compression members that have certain shapes. Qualitatively, long and slender columns are more likely to buckle than short and stocky ones.

Euler's Equation. The concept of the critical load is formalized here with the aid of Fig. 11-3. A straight column is pinned at the ends, which are constrained to move along a given vertical line, the line of action of the compressive load, P. For an imaginary, perfect column, the load versus lateral deflection plot is the vertical coordinate in Fig. 11-3c. At any arbitrary load P_A there is only compressive deformation; y_{max} is always zero.

Imagine that a small lateral force F is also applied to the compressed perfect column as shown in Fig. 11-3a. This causes an internal moment $(-Py)$ at an arbitrary section whose lateral deflection is $-y$. If the situation is unstable, the column collapses. If it is stable, something interesting can be visualized. The

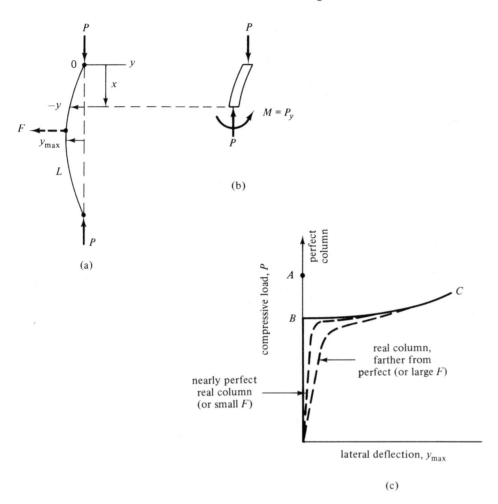

Fig. 11-3. Analysis of critical loads.

column remains bent when the force F is removed, because any part of it can be in equilibrium whether it is straight or slightly curved (Fig. 11-3b); the only difference is that the straight element has no moment on it. Thus, a temporary lateral load may cause a permanent lateral deflection, while the compressive load is constant. The compressive load P corresponding to a finite, permanent lateral displacement caused by a nearly zero lateral load is defined as the critical load. It is marked by B in Fig. 11-3c and is called the *bifurcation point*. The curve rises toward point C because these lateral deflections are stable; they are caused by F, exist after F is removed, but increase further only if P is increased. Naturally, the bifurcation point is lowered by increased lateral forces.

Real columns can buckle without the assistance of lateral forces. The causes of effective eccentricity (unavoidable curvatures, inhomogeneities) discussed above are equivalent to lateral forces. Consequently, real columns have load versus lateral deflection curves that are below the curve $0BC$, which represents ideal behavior in Fig. 11-3c.

The mathematical statement for the critical load, or buckling load, is derived with the aid of Fig. 11-3a and b. A differential equation can be written to relate the external load P and the internal resisting moment M for the element in Fig. 11-3b, as in a beam. M is positive according to the convention and the deflection is negative in the chosen coordinates, so

$$-EI\frac{d^2y}{dx^2} = M = Py, \qquad \frac{d^2y}{dx^2} + \frac{P}{EI}y = 0 \qquad (11\text{-}1)$$

The solution of this equation is

$$y = A\sin\sqrt{\frac{P}{EI}}x + B\cos\sqrt{\frac{P}{EI}}x \qquad (11\text{-}2a)$$

where A and B are constants that can be determined from the boundary conditions:

$$y(0) = y(L) = 0$$

Substituting these in the solution,

$$0 = A\sin 0 + B\cos 0 \qquad (11\text{-}2b)$$

$$0 = A\sin\sqrt{\frac{P}{EI}}L \qquad (11\text{-}2c)$$

From Eq. 11-2b, $B = 0$. From Eq. 11-2c, either $A = 0$, which is a trivial

solution, or $\qquad L\sqrt{\frac{P}{EI}} = n\pi \qquad$ with $n = 0, 1, 2, 3, \ldots$

Since the smallest finite value of the load is the critical load, P_c, n must be taken as unity. Thus, for the column in Fig. 11-3a (without the force F),

$$P_c = \frac{EI\pi^2}{L^2} \qquad (11\text{-}3)$$

Sometimes it is realistic to use $n = 2, 3, \ldots$. These values correspond to higher modes of buckling but have physical significance only if the column is restrained in one or more places besides the ends. For $n = 2$, the lateral deflection is also zero at the center of the column; for $n = 3$, it is zero at $L/3$ and $2L/3$, besides the ends.

Equation 11-3 is usually expressed in one of the following ways (only for elastic buckling):

$$P_c = \frac{\pi^2 EI}{(kL)^2} = \frac{\pi^2 E(Ar^2)}{(kL)^2} = \frac{\pi^2 EA}{k^2(L/r)^2} \qquad (11\text{-}4)$$

where E = modulus of elasticity
      I = smallest moment of inertia of the cross-sectional area about its centroid
      L = length of column
      k = a constant; it depends on the constraints of the column
      A = area of cross section
      r = radius of gyration of cross-sectional area; $r = \sqrt{I/A}$ by definition of the moment of inertia

The ratio L/r is called the *slenderness ratio* of a column. The most common engineering materials would not tend to buckle in an unstable manner if the slenderness ratio is less than about 30. At the other extreme, a ratio of about 200 or higher indicates that the member is very slender and may not be able to support large compressive loads.

    The formula that relates the critical load to the properties of a slender column was first derived by Leonard Euler in 1757, and his name is frequently used in this area.

End Conditions. The critical load for a column depends on how it is constrained. The linear or angular displacements of the ends partially control the extent of bending deformation (and tendency for instability) of the column. Figure 11-4 shows several extreme possibilities of end conditions. The constant k in Eq. 11-4 is

Fig. 11-4. Common end conditions for slender columns.

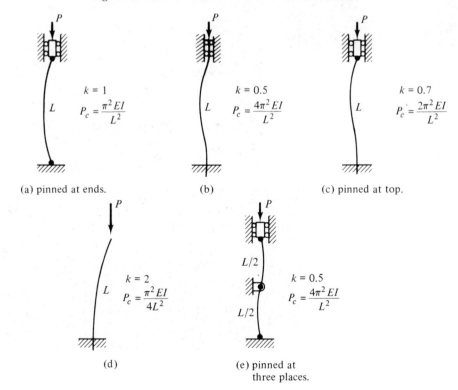

$k = 1$

$P_c = \dfrac{\pi^2 EI}{L^2}$

(a) pinned at ends.

$k = 0.5$

$P_c = \dfrac{4\pi^2 EI}{L^2}$

(b)

$k = 0.7$

$P_c' = \dfrac{2\pi^2 EI}{L^2}$

(c) pinned at top.

$k = 2$

$P_c = \dfrac{\pi^2 EI}{4L^2}$

(d)

$k = 0.5$

$P_c = \dfrac{4\pi^2 EI}{L^2}$

(e) pinned at
three places.

Elevator on long rod of hydraulic piston. *Photograph by R. Sandor.*

Strain gage mounted on the piston rod of a small engine; the strain signal is recorded on a digital oscilloscope. *Photograph by J. Dreger, University of Wisconsin, Madison, Wis.*

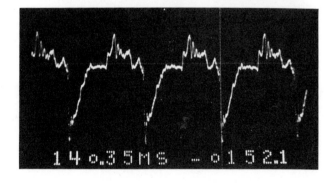

the way of dealing with the end conditions quantitatively. The physical meaning of this constant is that it gives the distance, in terms of the length of the column, between the inflection points where the moment is zero in the column. Sometimes kL is called the *effective length*.

EXAMPLE 11-1

A fiber glass rod is 15 ft long and has a diameter of 0.8 in. A fiber glass tube is also made with the same length and same weight, but with an outside diameter of 1.6 in. Which one has the higher buckling load in the pin-ended configuration? $E = 3 \times 10^6$ psi for both.

Solution

With E and L held constant, the buckling load is proportional to I. The pole with the higher I buckles at the higher load. For the solid rod,

$$I_1 = \pi \frac{(0.4)^4}{4} = 0.0064\pi$$

For the same cross-sectional area, the tube has an inner radius given by

$$\pi(0.4)^2 = \pi(0.8^2 - r_i^2), \qquad r_i = 0.693$$

$$I_2 = \frac{\pi}{4}(0.8^4 - 0.693^4) = 0.0448\pi$$

The hollow tube is seven times stronger than the solid pole of the same length and weight.

EXAMPLE 11-2

An aluminum tube is 5 m long, has a wall thickness of 2 mm, and has an elliptical cross section with outside dimensions of 4 cm $\times$ 3 cm. What is the buckling load in the pin-ended configuration if $E = 70$ GPa?

Solution

The critical quantity is the smallest I, which is I_x according to Fig. 11-5.

$$I_x = \frac{\pi}{4}(ab^3 - a'b'^3) = 7.74 \times 10^{-8} \text{ m}^4$$

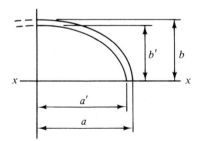

Fig. 11-5. Example 11-2.

From Eq. 11-4, with $kL = L$,

$$P_c = \frac{\pi^2 EI}{L^2} = \frac{\pi^2(7 \times 10^{10})(7.74 \times 10^{-8})}{5^2} = 2.14 \text{ kN}$$

EXAMPLE 11-3

A thin-walled steel tube is 20 ft long, has an outside diameter of 1.4 in., and has a wall thickness of 0.04 in. Both ends are pinned. What are the buckling loads in the first and second mode of buckling (half sine wave and full sine wave, respectively)? $E = 30 \times 10^6$ psi.

Solution

$$kL = L = 20(12) \text{ in.}, \qquad I = \frac{\pi}{4}(R_o^4 - R_i^4) = 0.0395 \text{ in}^4$$

In the first mode, $n = 1$, and

$$P_{c_1} = \frac{\pi^2 EI}{L^2} = \frac{\pi^2(3 \times 10^7)(0.0395)}{240^2} = 203 \text{ lb}$$

In the second mode, $n = 2$, but everything else remains the same, so

$$P_{c_2} = 4P_{c_1} = 812 \text{ lb}$$

EXAMPLE 11-4

What is the change in the buckling load of a hollow tube if an identical tube is joined to it along the full length? Assume pin-ended conditions and the possibility of buckling in two orthogonal planes.

Solution

The important quantity is again I. The original tube has $I_x = I_y = I_o$ (Fig. 11-6a). For the double tube, it is necessary to consider buckling about x' and y' (Fig. 11-6b).

$$I_{x'} = 2I_o, \qquad I_{y'} = 2(I_o + Ar^2) = 2(3I_o) = 6I_o$$

The double tube tends to buckle about the x'-axis, and its buckling load is twice that of the single tube.

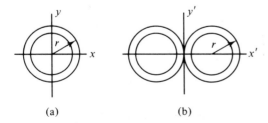

(a)   (b)   **Fig. 11-6.** Example 11-4.

11-2. INELASTIC BUCKLING

If the material of a column is not ideally elastic at all stresses, the Euler equation may not be useful for determining the critical load. The reason is that E in the equation means a linear relationship between stress and strain, whereas the material may be deforming plastically under loads less than the critical load calculated for elastic buckling. This may happen when the slenderness ratio is relatively low, somewhere in the approximate range of 30 to 100. Problems of this kind can be solved by substituting the tangent modulus E_T for the elastic modulus E in Euler's equation (first mode, $n = 1$):

$$P_c = \frac{\pi^2 E_T A}{k^2 (L/r)^2} \tag{11-5}$$

This is sometimes called the generalized Euler equation, or the Engesser equation. Equations 11-4 and 11-5 give the same result for brittle materials, as shown schematically in Fig. 11-7a. Even for ductile materials, the two equations are the same for average stresses that do not cause yielding. In other words, if the column is sufficiently long (as those to the right of the proportional limit in Fig. 11-7b), critical buckling initiates before the material yields. Only the Engesser equation is valid for intermediate columns of ductile materials; these yield before buckling. Short columns tend to flatten instead of buckling.

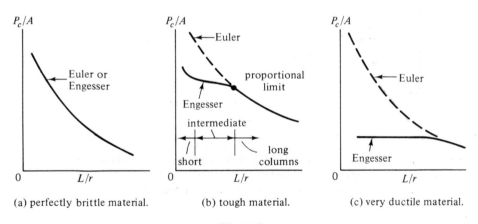

(a) perfectly brittle material.   (b) tough material.   (c) very ductile material.

Fig. 11-7

Unfortunately, the tangent modulus is not a constant for most materials at all stresses. Thus, a trial and error procedure must be applied in using Eq. 11-5. The critical load P_c is the load for which the average stress P_c/A and the tangent modulus satisfy Eq. 11-5 on the basis of the stress-strain curve of the material (E_T must be valid for the given stress).

Another potential complication is that even for a given material and a given stress the tangent modulus may not be a constant. Cycle-dependent hardening or softening of the material may change E_T in time. For example, a compressive load P is applied to a member and it does not buckle. Subsequently, a load equal to P (or maybe even less than P) is applied to the member many times. If the material softens during this process, inelastic buckling may develop in time. Cycle-dependent buckling develops gradually, so it may be detected if a reasonable attempt is made. The bending deformation may increase at an increasing rate from cycle to cycle, however. More will be said about this problem area in Chapter 12.

EXAMPLE 11-5

Sketch the Euler and Engesser equations on the same diagram of P_c/A versus L/r for an ideally elastic-ideally plastic metal. Show the yield strength on the diagram.

Solution

For long columns, the two curves coincide. When σ_y is reached, the curves deviate as shown in Fig. 11-8. The shorter the column, the larger the deviation, or error, if the elastic modulus is used inadvertently.

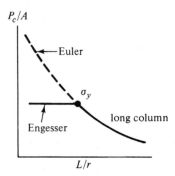

Fig. 11-8. Example 11-5.

EXAMPLE 11-6

An aluminum alloy rod is 5 ft long and 2 in. in diameter. The stress-strain curve for compressive loading is the solid line in Fig. 11-9. Plot stress versus E_T on the same diagram and compute the buckling load for the pin-ended rod.

Solution

The dashed line in Fig. 11-9 is σ versus E_T. The buckling load is found by trial and error. For a pinned rod,

$$P_c = \frac{\pi^2 E_T A}{(L/r)^2}, \qquad \sigma_c = \frac{P_c}{A} = \frac{\pi^2 E_T}{(L/r)^2}$$

$$r = \sqrt{\frac{I}{A}} = \sqrt{\frac{\pi R^4/4}{\pi R^2}} = 0.5 \text{ in.}$$

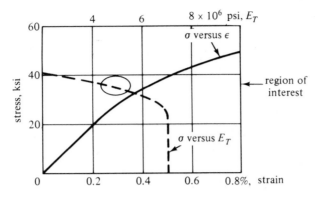

Fig. 11-9. Example 11-6.

From Eq. 11-5,

$$\sigma_c = 6.85 \times 10^{-4} \, E_T$$

It appears (using Fig. 11-9) that this equation is satisfied when $\sigma_c = 35$ ksi and $E_T = 5.1 \times 10^7$ psi. The buckling load is

$$P_c = \sigma_c A = 35,000\pi \text{ lb}$$

EXAMPLE 11-7

The swivel boom of a large truck crane is supported in the forward position during transportation as shown in Fig. 11-10. The vertical supporting member A on the bumper is a 2-m long steel tube with $D_o = 8$ cm and $D_i = 6$ cm. It can be assumed fixed-ended on the bottom and pinned on the top. Some yielding of this member is allowed. The stress-strain curve for the steel is approximated by $\sigma = 800\epsilon^{0.2}$ MPa. What is the buckling load for the tube?

Solution

In this case, since σ is a known function of ϵ, the tangent modulus may be written as a function of ϵ or σ.

$$E_T = \frac{d\sigma}{d\epsilon} = 8 \times 10^8 (0.2)\epsilon^{-0.8} = 1.6 \times 10^8 \left(\frac{\sigma}{8 \times 10^8}\right)^{-4}$$

$$= \frac{6.55 \times 10^{43}}{\sigma^4} \tag{a}$$

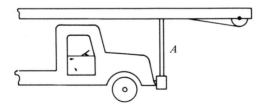

Fig. 11-10. Example 11-7.

Self-propelled hydraulic crane; maximum load of 55 tons and maximum
tip height of 165 ft are possible with the modified hexagonal cross section
of the boom that has high resistance to buckling. *Warner & Swasey Co.,
Solon, Ohio.*

For a pinned-fixed column,

$$P_c = \frac{2\pi^2 E_T A}{(L/r)^2}, \qquad \sigma_c = \frac{P_c}{A} = \frac{2\pi^2 E_T}{(L/r)^2} \qquad \text{(b)}$$

Substituting (b) into (a) gives

$$\sigma_c^5 = \frac{2\pi^2 (6.55 \times 10^{43})}{(L/r)^2}$$

$$r = \sqrt{\frac{I}{A}} = \sqrt{\frac{\pi/4(4^4 - 3^4)}{\pi(4^2 - 3^2)}} = 2.5 \text{ cm} = 0.025 \text{ m}$$

$$\sigma_c = 182 \text{ MPa}$$

The buckling load is

$$P_c = \sigma_c A = 401 \text{ kN}$$

11-3. SIMPLIFIED COLUMN FORMULAS

There have been numerous efforts to avoid the trial and error method of solution of the tangent-modulus formula. These involve simple mathematical functions with which the solid line in Fig. 11-7b can be approximated. Three functions have been widely used for this purpose, each one for columns that have intermediate slenderness ratios. The functions and their names are

Straight line: $\qquad\qquad \dfrac{P_c}{A} = \sigma_o - C\dfrac{kL}{r}$ $\qquad\qquad\qquad$ (11-6)

Parabolic: $\qquad\qquad\quad \dfrac{P_c}{A} = \sigma_o - C\left(\dfrac{kL}{r}\right)^2$ $\qquad\qquad\qquad$ (11-7)

Gordon-Rankine: $\qquad\quad \dfrac{P_c}{A} = \dfrac{\sigma_o}{1 + C(kL/r)^2}$ $\qquad\qquad\quad$ (11-8)

where P_c/A = average buckling stress

$\qquad \sigma_o$ = compressive strength of short block

$\qquad C$ = constant for a given material

$\qquad kL$ = effective length of column depending on the constraints

$\qquad r$ = radius of gyration of cross section

The constants σ_o and C are not necessarily the same in the three equations even for a given material. The reason is that they are adjused on the basis of experimental data to provide a reasonable fit to the Engesser line in Fig. 11-5b; their differences in the three equations are unimportant. Figure 11-11 shows a schematic comparison of these equations. Obviously, none of them is perfect over wide ranges of slenderness ratios.

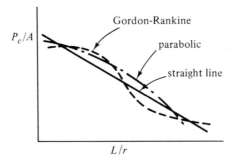

Fig. 11-11

It can be expected that the approximate column formulas will gradually lose much of their former appeal to designers. Recent advances in computing equipment and methods make them less significant than in the past.

11-4. EVALUATION OF A VAULTING POLE

An attempt can be made now to analyze the loading on a vaulting pole and the possible mode of its failure. It is best to start by eliminating those items from consideration that are irrelevant or insignificant in the problem.

If the pole is made of fiber glass, inelastic buckling may be ignored as a possibility. Fiber glass is known to be brittle; its fracture ductility is only a few percent. Assuming that information concerning the cycle-dependent behavior of fiber glass is not available, there is enough qualitative evidence from experience with various objects to indicate that fiber glass does not soften significantly when loaded repeatedly. A column could buckle under its own weight, but this is not expected to happen to a vaulting pole; in fact, the pole can be assumed as weightless for the analysis.

Fig. 11-12. Pole-vaulting configurations.

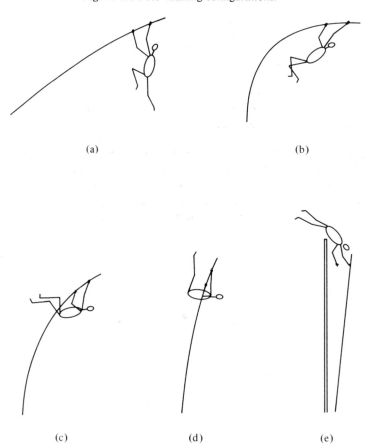

(a)   (b)

(c)   (d)   (e)

The condition at the lower end of the pole (in the "pole box") is similar to a pin joint with fairly low friction. The load is applied somewhere near the other end (for a free-body diagram of the pole, see Appendix A, Section 2), and the condition there is that of a frictionless joint. The load is applied eccentrically to the pole at the time of the takeoff, as seen from the position and direction of motion of the vaulter in Fig. 11-12a. Assume for now, however, that the load is centroidal to the pole.

The critical load for a given pole can be calculated with the assumptions made. For an example, consider a 16.5-ft long pole that has a 1.5-in. outside diameter and a wall thickness of 0.094 in. Its modulus of elasticity is 3×10^6 psi. The results of the calculations are

cross-sectional area, $A = 0.416$ in^2
moment of inertia of cross section, $I = 0.103$ in^4
radius of gyration, $r = \sqrt{I/A} \cong 0.5$ in.
slenderness ratio (assuming an effective length of 15 ft), $L/r = 360$

The critical load is

$$P_c = \frac{\pi^2 E A}{(L/r)^2} \cong 96 \text{ lb}$$

Most pole-vaulters weigh more than this, so it appears that the pole is totally inadequate. Pole vaulting is a dynamic event, however, where mass and acceleration must be considered. This will not change the critical load for the pole, but it will provide a clearer idea of what is done to the pole during a jump.

A qualitative understanding of the dynamic loading of the pole can be obtained by considering the sequence of configurations in Fig. 11-12. In Fig. 11-12a the vaulter (assume he weighs 150 lb) has just left the ground. His initial horizontal velocity (30 ft/sec) is larger than his vertical velocity (20 ft/sec). The pole begins to bend under the moment applied to it, but the pole in turn slows the horizontal motion of the vaulter. It is found that, for a uniform decrease of his horizontal velocity from 30 ft/sec to a few feet per second over a distance of about 15 ft, the force on the pole is about 140 lb, enough to cause buckling. It seems reasonable, however, to assume that his initial deceleration is the largest with a force of about 200 lb that he can apply to the pole with nearly straight arms. The pole begins to bend rapidly under the excessive initial load, but the load is not constant. As the curvature of the pole increases dangerously, the load caused by the horizontal deceleration decreases rapidly. If the pole doesn't break in this configuration, it can begin to straighten. Much of the initial kinetic energy of the vaulter was converted into elastic strain energy in the pole, and this energy is released as the pole straightens, applying an upward force to the vaulter. This force depends on his strength, and it must be less than what he can exert in the essentially hanging position shown in Fig. 11-12b. The force does not have to be 150 lb to increase the jumper's vertical velocity. In Fig. 11-12e the pole is straight and the vaulter is making a final push against the pole with one arm. If he is fairly strong, he can apply a force that is somewhere around 50 lb (it is a difficult, rapidly changing position).

The conclusions based on the qualitative statements above are as follows. The situations shown in Fig. 11-12a and b are similar to elastic buckling. The pole is slender enough, and a load is applied that increases the curvature of the member. The load decreases during this process (this is not typical in structures), however, so that the pole has a chance to straighten. If the pole survives the maximum deformation, there is little chance that it could buckle during the rest of the jump. Figure 11-12c and d show a spring doing work and not buckling in progress. Interestingly, increases in the vaulter's weight, approach velocity, and arm strength all would necessitate the use of a stronger pole.

11-5. PROBLEMS

Sec. 11-1

11-1. Calculate and plot the deflections for a pole such as that in Fig. 11-1. Use at least three loads: W, $2W$, and $3W$.

11-2. Do you obtain true stress if you divide the critical load by the cross-sectional area of a column?

11-3. Consider a spherical flaw in a cylindrical column. Its diameter is one-tenth the diameter of the column. What is the maximum bending moment in the column that can be attributed to having this flaw?

11-4. Consider a column that is loaded as the one in Fig. 11-4b but for which the supports are not perfectly rigid at either end. What is the probable range for the value of k for this column?

11-5 to 11-20. Determine the unknown quantity in each of these problems for the cross section of the column shown. Assume that the members are weightless.

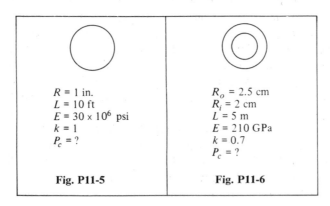

$R = 1$ in.
$L = 10$ ft
$E = 30 \times 10^6$ psi
$k = 1$
$P_c = ?$

Fig. P11-5

$R_o = 2.5$ cm
$R_i = 2$ cm
$L = 5$ m
$E = 210\,\text{GPa}$
$k = 0.7$
$P_c = ?$

Fig. P11-6

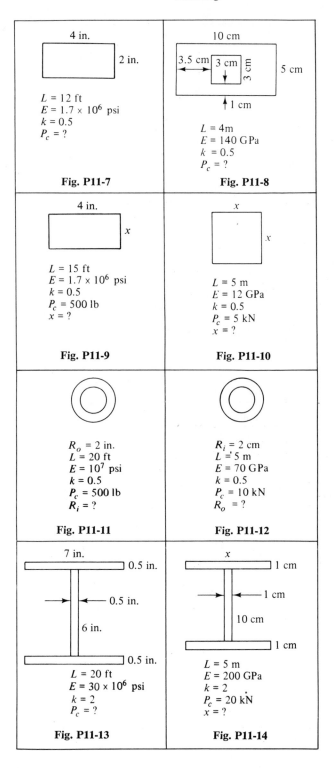

Fig. P11-7

4 in.

2 in.

$L = 12$ ft
$E = 1.7 \times 10^6$ psi
$k = 0.5$
$P_c = ?$

Fig. P11-8

10 cm

3.5 cm   3 cm   3 cm   5 cm

1 cm

$L = 4$ m
$E = 140$ GPa
$k = 0.5$
$P_c = ?$

Fig. P11-9

4 in.

x

$L = 15$ ft
$E = 1.7 \times 10^6$ psi
$k = 0.5$
$P_c = 500$ lb
$x = ?$

Fig. P11-10

x

x

$L = 5$ m
$E = 12$ GPa
$k = 0.5$
$P_c = 5$ kN
$x = ?$

Fig. P11-11

$R_o = 2$ in.
$L = 20$ ft
$E = 10^7$ psi
$k = 0.5$
$P_c = 500$ lb
$R_i = ?$

Fig. P11-12

$R_i = 2$ cm
$L = 5$ m
$E = 70$ GPa
$k = 0.5$
$P_c = 10$ kN
$R_o = ?$

Fig. P11-13

7 in.

0.5 in.

0.5 in.

6 in.

0.5 in.

$L = 20$ ft
$E = 30 \times 10^6$ psi
$k = 2$
$P_c = ?$

Fig. P11-14

x

1 cm

1 cm

10 cm

1 cm

$L = 5$ m
$E = 200$ GPa
$k = 2$
$P_c = 20$ kN
$x = ?$

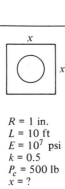

$R = 1$ in.
$L = 10$ ft
$E = 10^7$ psi
$k = 0.5$
$P_c = 500$ lb
$x = ?$

Fig. P11-15

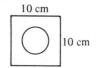

$L = 6$ m
$E = 70$ GPa
$k = 0.5$
$P_c = 3$ kN
$R_{max} = ?$

Fig. P11-16

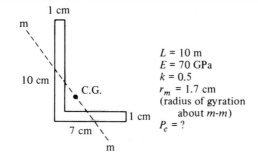

$L = 10$ m
$E = 70$ GPa
$k = 0.5$
$r_m = 1.7$ cm
(radius of gyration
about m-m)
$P_c = ?$

Fig. P11-17

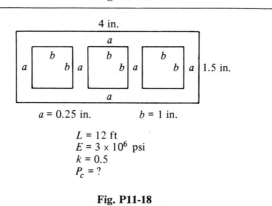

$a = 0.25$ in. $b = 1$ in.

$L = 12$ ft
$E = 3 \times 10^6$ psi
$k = 0.5$
$P_c = ?$

Fig. P11-18

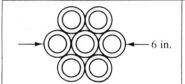

R_o = 1 in.
R_i = 0.8 in
L = 30 ft
E = 30 × 10⁶ psi
k = 2

The tubes are welded together.

P_c = ?

what is the improvement
over a single tube?

Fig. P11-19

R = 1 cm
L = 4 m
E = 200 GPa
k = 2

the rods are welded together.

P_c = ?

what is the improvement
over a single rod?

Fig. P11-20

Sec. 11-2

In Prob. 11-21 to 11-26 the column is 6 cm in diameter and 5 m long. Determine the buckling load.

11-21. $\sigma = 70\epsilon^{0.2}$ ksi

11-22. $\sigma = 300\epsilon^{0.15}$ MPa

11-23. $\sigma = 320\epsilon^{0.1}$ ksi

11-24. $\sigma = 20\epsilon^{0.4}$ MPa

11-25. $\sigma = 30\epsilon^{0.3}$ ksi

11-26. $\sigma = 1600\epsilon^{0.05}$ MPa

Sec. 11-4

11-27. In Section 11-4 it was assumed that the vaulter applied the force to the pole at 15 ft from the lower end. Calculate the slenderness ratios and critical loads for the following effective lengths of the pole: 10 ft, 14 ft, 16 ft. What is the implication from these results for the future when the effective length may have to be 20 or 25 ft? If you think that these lengths are unrealistic, consider the facts that the current record in pole vaulting is about twice what it was 100 years ago, and new records are made almost every month.

11-28. Calculate the approximate maximum stress in the pole from the curvature shown in Fig. 11-12b and the information given in Section 11-4. Compare this stress with that calculated using the Euler equation.

11-29. The wall thickness is essentially constant for vaulting poles. Could this dimension be varied somehow to make a better pole? Assume that the outside diameter should be constant.

12

CONCENTRATIONS OF STRESSES AND TIME-DEPENDENT PHENOMENA

In 1953 the future of British aviation appeared most promising. The elegant, four-engine Comet airplanes were the first jet-propelled civil aircraft in the world, and they were successful. The design and development of the Comets had been pains-taking to assure success. Design had commenced in 1946. The first prototype had flown in 1949. A special school had been established for the training of pilots and crews. A full Certificate of Airworthiness was issued for the Comet aircraft early in 1952, and passenger service commenced in May, 1952. Comet aircraft had flown about 20,000 hours by the end of 1953.

On January 10, 1954, Comet G-ALYP (also called Yoke Peter), which was the first passenger-carrying jet aircraft in the world, crashed in the Mediterranean near the Isle of Elba. All those aboard were killed. Comet aircraft were removed from service at once, and an investigation of the cause of the accident was begun. The investigation included a naval search and recovery of Yoke Peter from water 600 ft deep. Television equipment was used for this purpose for the first time, and this made it possible to recover about 70% of the aircraft.

The Air Safety Board could not establish a definite reason for the accident, but certain modifications were made in Comet aircraft to cover every possibility that could be imagined as a cause of the disaster.   Considering the modifications, new flight tests, and favorable technical advice, the Minister of Transport and Civil Aviation gave permission for the resumption of Comet services. The first passenger flight was on March 23, 1954.

On April 8, 1954, Comet G-ALYY (also called Yoke Yoke) crashed in the Mediterranean near Naples. All those aboard were killed.

An unprecedented investigation began after this accident. The Lord Chancellor appointed a high-level Court of Inquiry. Extensive and expensive new tests were made. For example, a Comet aircraft that had made 1230 flights without any serious problems was subjected to full-scale cabin pressurization tests. The whole fuselage was immersed in a tank of water (Fig. 12-1), and the pressure inside was raised and lowered repeatedly by pumping water in and out of it. After each application of the cabin pressure, fluctuating loads were applied to bend the wings as they may be affected by gusts under normal conditions.

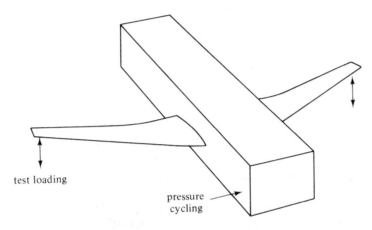

Fig. 12-1. Comet aircraft in testing tank.

It was established with a high degree of certainty that the two disasters were caused by the same kind of structural failure. Aviation around the world benefited from these experiences and new knowledge. New Comet aircraft were designed and built, and these were successful from the technological point of view. British aircraft manufacturers lost much money and prestige after the two catastrophies, however, and have not completely recovered in more than 2 decades. The basic problems that concerned the British designers and investigators in the 1950's still confront designers and will remain significant in the foreseeable future. Many engineers and technical managers have to be alert to the possibility of such problems at all times.

The following sections provide insights into several major problem areas. The list is not exhaustive, and each topic is discussed in an introductory manner. Many of these subjects are still being developed vigorously in numerous laboratories around the world. Some of the potential problems are interrelated, as shown in the final review of the Comet failures.

12-1. STRESS CONCENTRATIONS

It was shown in other chapters that nonuniform stress distributions are quite common. A special and important cause of a nonuniform stress distribution is a discontinuity in the material. Consider Fig. 12-2a for an example of this. Uniformly distributed stresses σ_A are applied at the ends of the member. At level B, the

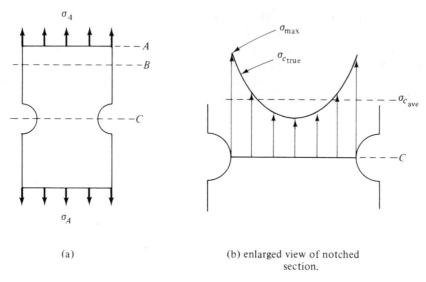

(a)

(b) enlarged view of notched section.

Fig. 12-2. Stress distribution in a notched member.

stresses are also uniform, and $\sigma_B = \sigma_A$. At level C, one expects higher stresses because the area is smaller. If $\text{area}_C = \frac{1}{2}\text{area}_A$, $\sigma_C = 2\sigma_A$, apparently. This is called the *nominal stress* or *average stress* at C. Unfortunately, the stress distribution differs from what is found by comparing the cross-sectional areas. The stresses near the notches are higher than the average, and they are lower than the average in the middle of the member. The stress distribution is shown in Fig. 12-2b for an ideally elastic material.

Knowledge of the average stress is not worthless. It can give an idea of the adequacy of a member, and it is used in some calculations. Failure of a member is most often caused by the largest stresses, however, so the true stress distribution should be known. The true peak stress $\sigma_{\max}$ is given by

$$\sigma_{\max} = K_t \sigma_{\text{ave}} \tag{12-1}$$

where K_t = theoretical stress concentration factor.

The constant K_t does not depend on the material except that it is not valid if there is yielding. K_t is a geometric factor that depends strongly on the radius of curvature of the notch. A small radius means K_t is high. It is for this reason that brittle materials such as glass break so easily after being scratched or nicked superficially. Circular holes of common size (diameter < 2 in.) often have theoretical stress concentration factors of 2 to 3 associated with them. Exact values depend on the other dimensions and on the loading of the particular members.* Figure 12-3 shows examples of these.

*R. E. Peterson, *Stress Concentration Design Factors* (New York: John Wiley & Sons, 1953).

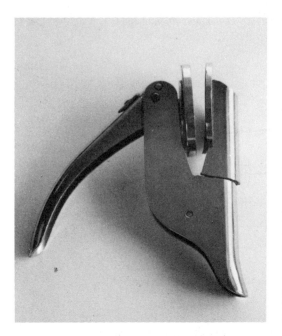 

Broken seal of a professional engineer; the crack started at the V-shaped notch. *Photograph by R. Sandor.*

Broken drawbar of truck train hauling gasoline; it caused the train to overturn and burn on an exit ramp of a highway; the change in geometry and flaws at the same region were stress raisers. Engineering Case Library, Stanford University. *Courtesy Professor Henry O. Fuchs, Stanford University, and Professor Charles O. Smith, University of Nebraska.*

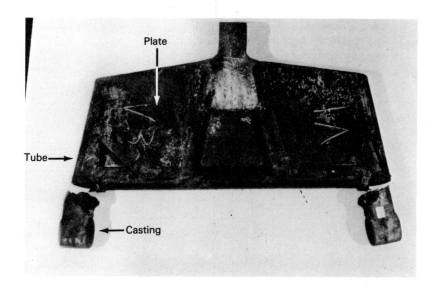

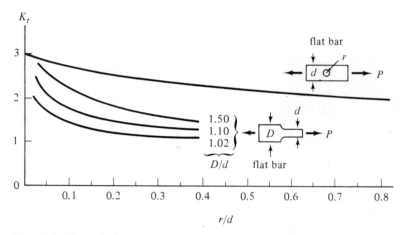

Fig. 12-3. Theoretical stress concentration factors as functions of geometry.

Members made of ductile materials are less affected by geometric disconti-nuities than those made of brittle materials. The reason is that plastic flow at the places of highest stress prevents these stresses from increasing according to the stress concentration factor. To appreciate the difference in behavior between brittle and ductile materials, consider Figs. 12-4 and 12-5. These show stress-strain curves and stepwise increments in the external load P for members similar to that shown in Fig. 12-2. One must assume that K_t changes only negligibly as the load on the member is increased. For the brittle material, a doubling of the load means a doubling of σ_{ave} and also of σ_{max}, according to Eq. 12-1. This is not meant as an argument because this is how K_t is defined. For the ductile material (Fig. 12-5), a doubling of the load causes a doubling of σ_{ave} but only a less than 100% increase of

Fig. 12-4. Stress distributions in a notched brittle member.

$$\sigma_3 = K_t \, (\sigma_{ave} \text{ for } P_3), \text{ etc.}$$

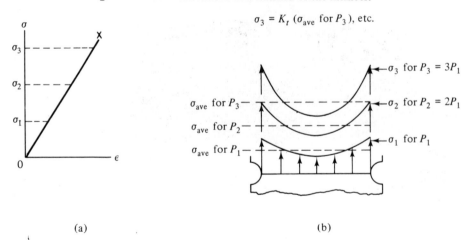

(a)   (b)

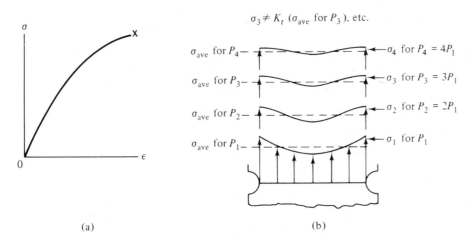

$$\sigma_3 \neq K_t \ (\sigma_{ave} \text{ for } P_3), \text{ etc.}$$

σ_{ave} for P_4 —   ← σ_4 for $P_4 = 4P_1$

σ_{ave} for P_3 —   ← σ_3 for $P_3 = 3P_1$

σ_{ave} for P_2 —   ← σ_2 for $P_2 = 2P_1$

σ_{ave} for P_1 —   ← σ_1 for P_1

(a)   (b)

Fig. 12-5. Stress distributions in a notched ductile member.

Rudder component of sailboat; the fatigue cracks started at the hole in the center. *Photograph by R. Sandor.*

This aluminum mast of a sailboat broke at rivet holes that could have been avoided. *Photograph by B. Boyce, University of Wisconsin, Madison, Wis.*

σ_{max}. Instead of a large increase of σ_{max}, the plastic flow spreads from the roots of the notches toward the center of the member. Each successive, equal increment of the load causes relatively small increases of σ_{max} and relatively large increases of σ_{min}. Sometimes this is not entirely clear at first sight. A careful consideration of force equilibrium and of local and average stresses with respect to Fig. 12-5a may help in trying to understand the situation.

The theoretical stress concentration factor cannot be used to find the true maximum stress in a notched, ductile member. It can be used, however, to make qualitative statements regarding the occurrence of plastic deformations. The behavior of ductile materials necessitates the definition of two new concentration factors, one for stress and one for strain. Both of these are similar to the definition of K_t. The true stress concentration factor, K_σ, is defined as

$$K_\sigma = \frac{\sigma_{max}}{\sigma_{ave}} \tag{12-2}$$

The true strain concentration factor, K_ϵ, is

$$K_\epsilon = \frac{\epsilon_{max}}{\epsilon_{ave}} \tag{12-3}$$

where $\epsilon_{ave} = \sigma_{ave}/E$.

In ductile members the stress concentration is relatively mild. This leads many people to think that such members are not weakened by notches. In some cases this is true, but the generalization is not justified. The price paid for having a mild concentration of stresses is a severe concentration of strains. This is more advantageous than the high concentration of stresses in brittle members, but it does not eliminate the possibility of fracture. For example, fatigue failures are caused by the alternating plastic strains. It is common practice to talk about stress concentrations in both brittle *and* ductile materials, even though in the case of the latter *strain concentration* would be the more correct terminology.

Factors of Safety. Stress concentrations are nebulous because they include the effects of internal flaws. The determination of the locations and severities of flaws (which are all but unavoidable) is very difficult and expensive. These are among the reasons why designers rely on factors of safety. For example, it is calculated that the maximum load on a member should not exceed 6000 lb in order to avoid yielding in it. Using a factor of safety of 2, one would say that the working load must be 3000 lb. For a factor of safety of 3, the working load is 2000 lb, and so on.

The magnitudes of factors of safety are determined, somewhat arbitrarily, on the bases of experience and of the intended purpose of each member. Relatively large factors of safety may be used for a bridge or machinery in permanent installation; they must be quite small for moving vehicles. Any factor of safety can be called a factor of ignorance. The challenge of reducing this ignorance will remain for a long time.

Component of fatigue testing machine broken by fatigue; note the particularly severe stress concentration by two close holes and by the sharp change in each hole. *Photograph by B. Boyce, University of Wisconsin, Madison, Wis.*

EXAMPLE 12-1

What are the true stress and strain concentration factors for the material whose stress-strain curve is shown in Fig. 12-4?

Solution

The entire stress-strain curve is linear (the material is perfectly brittle), so the theoretical stress concentration factor is valid for stresses *and* strains:

$$K_\sigma = K_\epsilon = K_t$$

EXAMPLE 12-2

A flat bar is 10 cm wide and 1 cm thick and has a 2-cm diameter hole in the center. What is the maximum axial force if the stress should not exceed 100 MPa? Assume the material is brittle.

Solution

K_t must be determined first. From Fig. 12-3,

$$\frac{r}{d} = 0.1, \qquad K_t \simeq 2.7$$

$$\sigma_{max} = K_t \sigma_{ave}$$

where σ_{ave} is on the minimum area.

$$P_{max} = \sigma_{ave} A_{min} = \frac{\sigma_{max}}{K_t} A_{min} = 29.6 \text{ kN}$$

12-2. RESIDUAL STRESSES

Solid materials are never entirely free of internal stress, not even when external loads are not acting on them. It is common practice to call all internal stresses that are present in the absence of external loading *residual stresses*. These stresses may be generated during the solidification of the material, during electroplating, during welding, and during machining and forming operations. For a demonstration of the latter, hammer at one point in the middle of an originally flat piece of thin metal. Note that it becomes curved and remains so in the absence of bending moments. The curvature is the combined result of plastic deformation and residual stresses in the metal.

Plastic deformation is the most important to consider with respect to residual stresses. Solidification, machining, or welding seldom occur during the service life of a member, but it may be deformed plastically many times, and perhaps not the same way each time. The details of the role of plastic flow and other causes in generating or eliminating residual stresses and the effects of these stresses cannot be discussed here,* but it is worthwhile to remember the following qualitative statements:

1. If a member has no significant residual stresses in it, uneven plastic deformation will leave residual stresses.

2. Plastic deformation will alter preexisting residual stresses and leave a new distribution and magnitudes of these stresses.

3. If a sequence of different plastic deformations is applied to a member, the last deformation tends to govern the distribution and magnitudes of the residual stresses. A possible exception to this is when the last deformation involves very little plastic flow in contrast to a preceding deformation.

4. At a given point in the material, a plastic deformation in tension gives rise to a compressive residual stress after the external load is removed from

---

*See G. E. Dieter, Jr., *Mechanical Metallurgy* (New York: McGraw-Hill Book Company, 1961); B. I. Sandor, *Fundamentals of Cyclic Stress and Strain* (Madison, Wis.: The University of Wisconsin Press, 1972).

the member. The reverse is also true. A plastic deformation in compression gives rise to a tensile residual stress. These may not seem to be correct, but they are.

5. A residual stress of a given sign (tension or compression) cannot exist alone in a member. If there is tension, there is also compression, and vice versa. This is necessary for the equilibrium of internal forces and moments on any cross section of the member. The magnitudes of the stresses with opposite signs and the areas affected by them do not have to be the same, however.

6. Stresses caused by external loads are superimposed on the residual stresses. For example, the residual stresses in a notched member are shown in Fig. 12-6a. Next, a uniform stress σ_a is applied at the ends of the member. Assume that $K_t = 1.5$ is valid after σ_a is applied. The resulting approximate stress distribution at the minimum section is shown in Fig. 12-6b.

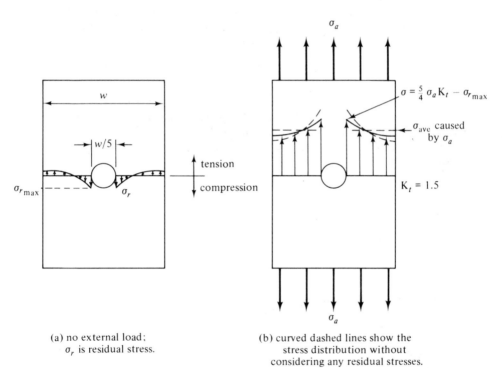

(a) no external load;
σ_r is residual stress.

(b) curved dashed lines show the
stress distribution without
considering any residual stresses.

Fig. 12-6. Superposition of residual stresses and those caused by external loading.

7. Compressive residual stresses are generally beneficial and tensile residual stresses are harmful. More is said about these in Section 12-4.

8. Factors of safety are often used to take care of the uncertainties of the distribution and magnitudes of residual stresses.

9. Residual stresses may decrease in time, especially if the member is cyclically loaded. This is discussed in Sections 12-3 and 12-4. Stress relief is accomplished rapidly by heating the member; this is discussed in Section 12-3.

As imagined from the statements above, the area of residual stresses is quite challenging. There is an ever-increasing need for people who have some knowledge, experience, and new ideas relevant to this subject.

EXAMPLE 12-3

Assume that the residual stresses shown in Fig. 12-6a were caused by a linear stress distribution applied at the ends of the plate. Describe the external stress distribution as well as possible.

Solution

The symmetry of the residual stresses indicates that the external stresses must have been uniformly distributed on the ends. Since the notch roots are in residual compression, the external stresses must have been in tension.

EXAMPLE 12-4

A strong wire A is clad with a more ductile metal, B. The stress-strain curves for these are shown in Fig. 12-7. An axial tensile load is applied to the composite wire, and the cladding metal yields a little. Describe the stresses in the wire after the tensile load is removed. Assume that the wire was entirely free of stresses prior to the tensile loading.

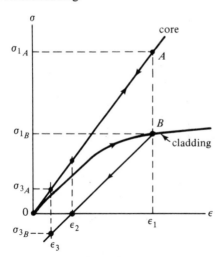

Fig. 12-7. Example 12-4.

Solution

It is reasonable to assume that the tensile load causes the same strain ϵ_1 in the two metals. The peak stresses corresponding to this strain are σ_{1_A} and σ_{1_B}. Imagine that the load is slowly removed. At the strain of ϵ_3 (which is still identical in the two metals), the stress becomes zero in the cladding, but it is still tensile in the core. Obviously, the external load cannot be zero for this state of stress. A further decrease of the strain occurs as the load is completely removed. At the strain ϵ_3 (somewhere between 0 and ϵ_2) the core is still in tension (σ_{3_A}) but the cladding is in compression (σ_{3_B}), and the internal forces can be in equilibrium when the external load becomes zero. The magnitudes of σ_{3_A} and σ_{3_B} depend on the relative cross-sectional areas of the core and the cladding and on the extent of yielding in the cladding. Severe yielding causes large residual stresses, as seen by increasing ϵ_1 in Fig. 12-7.

12-3.   CREEP AND STRESS RELAXATION

Creep is a time-dependent deformation that can be demonstrated easily. Take a piece of solder wire or thin strip of lead and suspend a mass with it as shown in Fig. 12-8. The mass should be large (say, 5 kg) but not large enough to break the wire at once. There will be an instantaneous elastic deformation (strain $= \epsilon_0$) of the wire when the weight is applied. After this, the weight will sink gradually as the wire stretches. The gradual deformation is inelastic. If the weight is removed, only the initial elastic deformation is recovered. The gradual creep of the wire leads to fracture of the wire if enough time is allowed.

The creep strain varies with time, as shown in Fig. 12-9. The creep rate $\dot{\epsilon}$ is minimum in secondary creep (also called steady state), although this is not always distinguishable as an extensive stage. It is the minimum creep rate that is important

Fig. 12-8. Creep experiment.

Fig. 12-9. Creep curve.

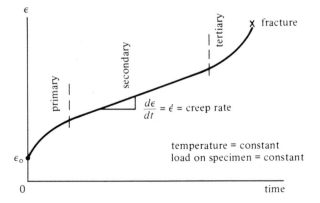

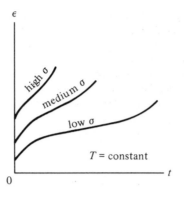

Fig. 12-10. Effect of stress on creep rate.

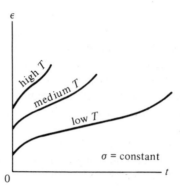

Fig. 12-11. Effect of temperature on creep rate.

in design. The creep rate of a given material depends on the applied load (which is normally constant throughout each test), as shown in Fig. 12-10.

The simple creep test discussed above is not successful with a steel wire in place of the solder wire. One can apply a load that is large enough to break the steel wire, but even a slightly smaller load than that will not cause obvious creep. Steel wire at high temperature does creep, however, and creep curves similar to those in Fig. 12-9 and 12-10 can be obtained. It is found that temperature has an effect in creep similar to that of stress: The creep rate increases with temperature (Fig. 12-11). The two effects are different in that a low stress causes noticeable creep when the temperature is high, but the reverse is not true. A high stress causes insignificant creep when the temperature is low.

It is necessary to clarify what is meant by high and low temperatures with respect to creep. A ratio T_H is defined, which is sometimes called the *homologous* temperature (even though it has no units), as

$$T_H = \frac{T}{T_m} \tag{12-4}$$

where T = actual temperature (°K)
    T_m = melting point temperature (°K)

Creep can be expected in metals when $T_H \gtrsim 0.5$. Thus, the results of the simple creep tests are not surprising. Room temperature is relatively high for lead and solder, but it is quite low for steel.

An interesting aspect of creep is that the creep strength at high temperature is usually higher for a metal with large grains than for one with fine grains (room temperature metal-working processes such as rolling and drawing produce fine grains). This is in contrast to the behavior at lower temperature where the finer-grained metals have the higher strength in many cases.

The development of creep-resistant alloys, their testing, and proper design with th m are particularly important for jet engines, nuclear reactors, and rockets.

Stress relaxation is a complementary phenomenon to creep. Creep occurs under stress control with strain as the dependent variable. Stress relaxation is under strain control with stress as the dependent variable. For an understanding of the latter, imagine that a wire is attached firmly to a perfectly rigid support (A) as in Fig. 12-12a. The wire here is a little short of reaching another perfectly rigid support (B). Next, the free end of the wire is gripped, it is stretched toward support B, and finally it is attached to B firmly (Fig. 12-12b). This causes a tensile strain ϵ_0 and a corresponding stress σ_0 in the wire. The strain remains constant in the wire at all times, but the stress is free to change according to the properties of the material and the environmental conditions.

The tendency is for the stress to decrease in time as shown in Fig. 12-13. The rate of decrease depends on the initial deformation, as shown in the figure. Both

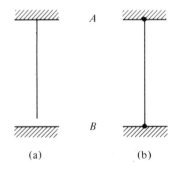

(a)   (b)

Fig. 12-12. Model for stress relaxation.

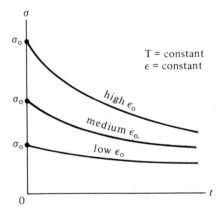

T = constant
ϵ = constant

high ϵ_0

medium ϵ_0.

low ϵ_0

Fig. 12-13. Stress relaxation curves for various initial stresses.

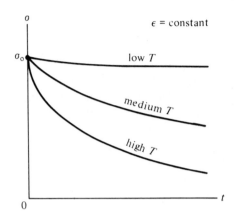

Fig. 12-14. Stress relaxation curves for various temperatures.

tensile and compressive initial stresses tend to decrease toward the state of zero stress. Stress relaxation can be expected in a metal when $T_H \gtrsim 0.5$. Figure 12-14 shows schematically the rates of stress relaxation for a given metal at different temperatures when all specimens have the same initial stress.

One could wonder why a naturally decreasing stress is worth talking about. The practical significance of stress relaxation is that it affects residual stresses. A small region with a residual stress is constrained by the material surrounding it. Thus, it is essentially in strain control. If the temperature in the region is high enough, the residual stress decreases in time. Naturally, this may be beneficial or harmful, depending on the sign of the residual stress.

EXAMPLE 12-5

What would happen to the plate shown in Fig. 12-6a if it were heated at $T_H = 0.75$?

Solution

There is no external load on the plate, so it would not creep. The residual stresses would relax toward zero everywhere in the plate.

EXAMPLE 12-6

What would happen to the plate shown in Fig. 12-6b if it were heated at $T_H = 0.75$?

Solution

This plate would creep under the tensile load. The creep rate would be highest near the hole, but not quite so high as in a homogeneous plate of the same material under the same stress and temperature. The reason is that material at the minimum section but far from the hole acts as a constraint for the material in the critical regions.

12-4. FATIGUE

Fatigue, which means failure caused by stresses less than the ultimate strength, is one of the most common causes of material failures. It is best for the engineer to assume that all materials could fail by fatigue. One could distinguish static fatigue and cyclic fatigue. In the first case the loads are not applied repeatedly; failure occurs in time because a crack may grow (corrosion) to a critical size. Static fatigue will not be discussed here because it occurs very rarely in its pure form. Cyclic fatigue means that the loads or deformations are imposed on a member more than once. This kind of fatigue deserves and requires much attention because the subject is complex.

Exhaust manifold of automobile fatigued by thermal stresses. *Photograph by B. Boyce, University of Wisconsin, Madison, Wis.*

In the simplest form of fatigue loading the stress or strain imposed on the material varies repetitively in a predictable way. Figure 12-15 shows a complete cycle of a sine function with the common terminology used in the fatigue area. Constant-amplitude loading is relatively simple but it is important both in research and in practical situations. Random loading (Fig. 12-16) is quite common in practice, but a detailed discussion of it exceeds the scope of this book. Note the interesting result that stress and strain in a given member are seemingly unrelated. This can be explained by considering plastic deformations.

The most basic diagram used in fatigue design is the stress amplitude versus fatigue life plot, often referred to as *S-N* plot. Such a plot is obtained by testing a number of identical specimens under identical conditions. The stress amplitude is the only parameter that can vary from test to test, but even that is constant during

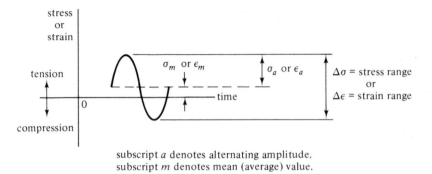

subscript *a* denotes alternating amplitude.
subscript *m* denotes mean (average) value.

Fig. 12-15. Terminology for cyclic loading.

Fig. 12-16. Repeated blocks of random loads on ductile steel.

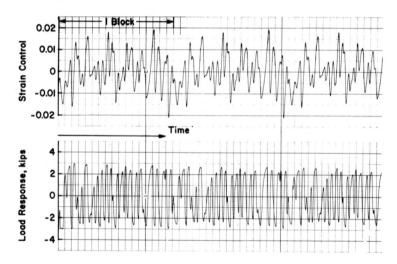

a given test. The first *S-N* plots were obtained for steels. For these metals the plot has a characteristic bend and leveling off at low stresses, as shown in Fig. 12-17. The lowest stress amplitude at which specimens fail in the longest tests (several million cycles) is called the *fatigue limit*. For almost a hundred years there was a preoccupation with the fatigue limit. Many designers thought that stresses above this limit should not be allowed. This philosophy cannot be held now for two reasons. Many important metals such as aluminum alloys have no distinct fatigue limit. The intended lives of many members are too short to allow efficient design on the basis of the fatigue limit, even if there is one. The alternative concept is called the *fatigue strength*. This is the stress amplitude that causes failure in an arbitrary number of

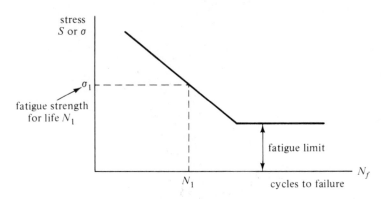

Fig. 12-17. Stress amplitude versus fatigue life for steels.

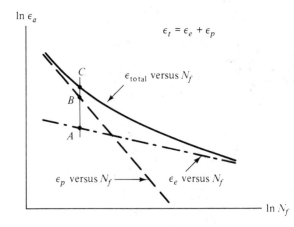

Fig. 12-18. Strain amplitudes versus fatigue life.

cycles. Thus, the appropriate number of cycles should always be given with the values of the fatigue strength.

A variation of the *S-N* plot is the strain amplitude versus fatigue life diagram. A typical example of this is shown in Fig. 12-18. The dashed lines show the elastic strain amplitude versus life and the plastic strain amplitude versus life plots. The total strain amplitude versus life plot is the sum of the component strains versus the appropriate lives. The summation is in the vertical direction; as an example, $A + B$ gives the value at C (note the natural log axis). Such diagrams are the bases for obtaining several important fatigue properties. Fatigue tests employing strain control are indispensable in the low-cycle (short life) region of lives, which is generally considered as less than 10^5 cycles to failure. The reason for preferring strain control will be explained later in this section.

The basic *S-N* or ϵ-*N* diagrams are obtained with the mean stress σ_m equal to zero throughout each test. Situations where σ_m is not equal to zero are extremely important in practice. A mean stress is often imposed by the external loading on a member. Furthermore, residual stresses are mean stresses on which the externally

Corrosion fatigue test; the specimen is inside the chamber; crack growth is monitored with a traveling microscope. *J. M. Barsom*, *"Corrosion Fatigue of High-Yield-Strength Steels," Closed Loop, 3, No. 3 (1972), 2–7.*

induced stresses are superimposed. It is found in general that a tensile mean stress reduces the fatigue life and a compressive mean stress increases it at a given alternating loading amplitude. The *S-N* plots shift as shown schematically in Fig. 12-19.

Large fully reversed loads may cause substantial plastic deformation in each cycle. An important feature of these deformations is that their directions are reversed twice in each cycle. For example, there is plastic deformation in tension during the loading from O to A in Fig. 12-20. From A to B the material is unloaded, and there is essentially no change in the plastic strain in this process. From B to C

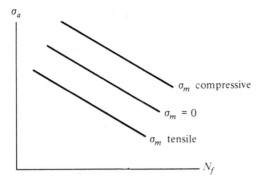

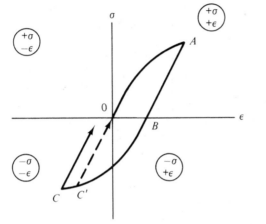

σ_m compressive

$\sigma_m = 0$

σ_m tensile

N_f

Fig. 12-19. Effect of mean stress on the fatigue life.

$\overrightarrow{OA}$—loading
$\overrightarrow{AB}$—unloading
$\overrightarrow{BC}$—loading
$\overrightarrow{C'O}$—unloading

Fig. 12-20. Reversed plastic deformations in cyclic loading; hysteresis loop.

the material is loaded in compression and there is compressive plastic strain. Going clockwise from point C, the material is unloaded again. Completely closed stress-strain loops (such as unloading from C' to O in Fig. 12-20) are called *hysteresis loops*. The area inside the loop is the work done per cycle in a unit volume of the material. This work is converted into heat.

The concept of the hysteresis loop has been helpful in recognizing that all combinations of tensile (+) and compressive (−) stresses and strains are possible. As shown in Fig. 12-20, it is possible to have

tensile stress with tensile strain	$(+\sigma, +\epsilon)$
tensile stress with compressive strain	$(+\sigma, -\epsilon)$
compressive stress with tensile strain	$(-\sigma, +\epsilon)$
compressive stress with compressive strain	$(-\sigma, -\epsilon)$

The implications of these are important. Since there is no such thing as a stress gage, stresses are experimentally determined from the results of strain measurements.

The strain measurement does not provide the sign of the stress, without ambiguity, however, and this may be important in subsequent loading.

The hysteresis loop and progressive changes of it are excellent pictorial views of the fatigue process. First of all, the width of the loop shows the extent of plastic deformation in a half cycle (horizontal distance OB in Fig. 12-20). This is related to the damage done, and thus it indicates the life range. A fat loop means short life; a loop that has no apparent width (just a straight, slanted line on stress versus strain) means long life. One problem with the latter is that current technology does not enable the recording of hysteresis loops for small volumes of material. The loops that can be recorded are averages of the material's response. Thus, the loop may have no measurable width indicating long life. Within the volume monitored, there may be a small, critical region that experiences considerable cyclic plasticity that makes the life of the whole part relatively short.

Cycle-dependent changes in the size and position of a hysteresis loop provide additional information about the fatigue process. Gradual softening of a material can be observed in one of two extreme situations. Under stress control (the stress amplitude is the same throughout the life) the hysteresis loops become fatter as cyclic loading progresses. Under strain control the hysteresis loops become shorter in the vertical direction (σ-axis); a smaller and smaller stress can stretch the material to the prescribed strain limits. Cycle-dependent hardening has the opposite effects under the two extreme control conditions.

The cyclic stress-strain curves that can be plotted for softening or hardening materials have been introduced already in Section 4-4. It can be added here that one way of obtaining a cyclic stress-strain curve is by plotting hysteresis loops from different tests of identical specimens. Figure 12-21 illustrates this method. Each loop in the diagram is at half-life of a specimen tested under a given control condi-

Fig. 12-21. Cyclic stress-strain curve obtained from different fatigue tests.

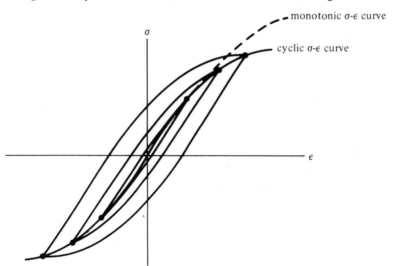

tion. The largest loops are for the most severe loading and shortest lives. The curve connecting the tips of these so-called stable hysteresis loops (at half-life they change relatively little) is the cyclic stress-strain curve. The monotonic stress-strain curve for a hypothetical cyclically softening material is shown in the same diagram for comparison.

The position of a hysteresis loop may change along the stress or strain axes during cyclic loading. The first of these occurs under strain control and indicates the *relaxation of a mean stress* (Fig. 12-22). The general rule is that the rate of relaxation is directly proportional to the magnitudes of the initial mean stress and the cyclic plastic strains. Thus, a high mean stress (or residual stress) cannot be maintained for many cycles of large deformations. A loop shifting along the strain axis occurs under stress control and it indicates *cyclic creep* (Fig. 12-23). Cyclic creep in the compressive direction (to the left in Fig. 12-23) is sometimes called

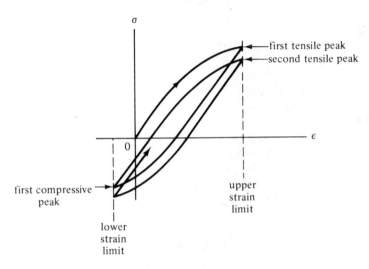

Fig. 12-22. Relaxation of mean stress in cyclic loading.

Fig. 12-23. Cyclic creep in tension.

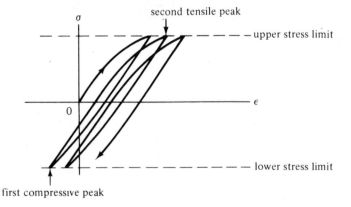

cycle-dependent buckling. The cause of cyclic creep is a mean stress (even a small one) or some inhomogeneity in the material. Since these are difficult to avoid, stress control is undesirable in low-cycle fatigue tests. Strain control is preferred because the possible changes in mean stress cannot lead to runaway strains. Cyclic creep deserves consideration in practice because in some cases it may not be possible to detect it in the early part of the service life of a member.

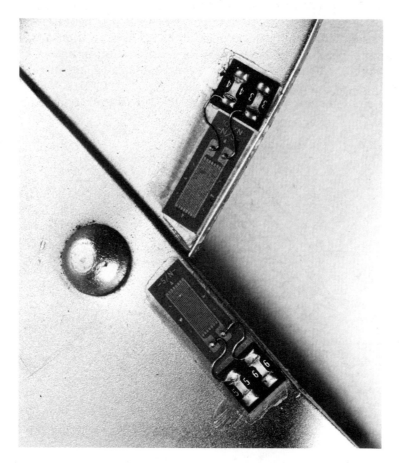

Fatigue life gages mounted at a critical region of an automobile fan blade; these gages are similar to strain gages but change resistance permanently as the life of the member is gradually exhausted, allowing an estimate of the remaining life. *Micro-Measurements Division, Vishay Intertechnology, Inc., Romulus, Mich.*

There is much more that could be said about fatigue. This brief introduction to the subject is ended by the following statements concerning the general cycle-dependent changes in the mechanical properties of metals (more is known about metals in this respect than about other materials).

1.   Cyclic loading does not affect the modulus of elasticity significantly. Changes in E are less than 10%.

2.   The flow properties (those related to plastic deformation) are all potentially changed by cyclic loading. These properties are the hardness, yield strength (sharp yielding or offset yield strength), ultimate strength, and the strain hardening exponent. For metals, the latter will change only slightly in the range of 0.1 to 0.2. An initial value of n below this range indicates that the metal is capable of cycle-dependent softening; n will increase gradually with progressive cyclic loading. An initially high n (annealed metal) generally means cycle-dependent hardening.

3.   The fracture properties may change because of cyclic deformations, but not so much as the flow properties. These properties are the true fracture strength, true fracture ductility, and the percent reduction of area.

EXAMPLE 12-7

The total strain amplitude versus life plot in Fig. 12-24 was obtained by testing a number of specimens, each one at a different strain amplitude, which was kept constant during that particular test. The elastic strain amplitude versus life line was determined by plotting σ_a/E versus N_f from all the tests (the stress amplitudes

Fig. 12-24. Strain amplitude versus fatigue life plots—Example 12-7. Notes: There are two reversals in each cycle of loading; amplitude is from mean to peak.

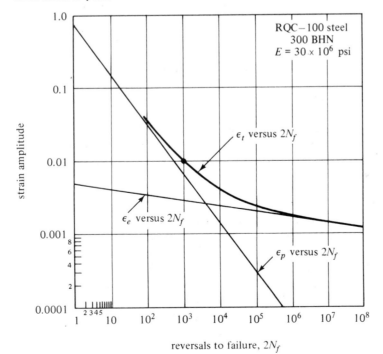

reversals to failure, $2N_f$

σ_a were measured independently in each test). Thus, the elastic strain amplitude versus life plot is essentially an *S-N* diagram. The plastic strain amplitude versus life line was determined by subtracting the elastic strain from the total strain at the same life. A sample calculation involving the data from one test is given here.

At the total strain amplitude (this is the controlled quantity) of 0.01 the life was $2N_f = 10^3$ reversals or 500 cycles. The average stress amplitude measured during this test was $\sigma_a = 85,000$ psi. The calculated elastic strain amplitude is

$$\epsilon_e = \frac{\sigma_a}{E} = \frac{85,000}{30 \times 10^6} = 0.0028 \qquad \text{at } 2N_f = 10^3$$

The calculated plastic strain amplitude is $\epsilon_p = \epsilon_t - \epsilon_e = 0.01 - 0.0028 = 0.0072$ at $2N_f = 10^3$. The lines drawn for the total strain, elastic strain, and plastic strain in Fig. 12-24 do not necessarily go through each appropriate point of the measured or calculated data. This is because of unavoidable scatter in the data. The lines should be drawn smoothly to represent the material's behavior over a large range of lives and not necessarily to go through all the points of the data.

EXAMPLE 12-8

Assume that for a certain metal the yield strength is the same in tension and compression and that the behavior is not altered by several reversals of the load (cyclically stable material). The 0.2% offset yield strength is 200 MPa, and it is reached by gradual yielding. Consider a loading sequence that consists of tensile loading to 200 MPa, unloading, reversed loading to 220 MPa, unloading, and reversed loading to 30 MPa. The last load is held constant. What are the final signs of the stress and strain?

Solution

Figure 12-20 can be used as a model for this problem. Assume that point *A* represents the tensile stress of 200 MPa. The first unloading leads to point *B*. A reversed loading to 200 MPa would lead to point *C′*; 220 MPa may fall to the left of point *C* on the extension of the *BC* curve. Unloading from here results in zero stress and a negative (compressive) strain. The final reversed loading to 30 MPa must result in a positive (tensile) stress and a negative (compressive) strain at any given point in the metal.

12-5. IMPACT LOADING

Dynamic applications of loads frequently must be considered in the analysis of machines, vehicles, and even stationary structures. The simplest model for these is a mass falling on a spring or a mass falling with a spring, as shown in Fig. 12-25. In either case, h is the height of fall before contact and d is the maximum deformation of the spring. The kinetic energy of the mass is zero at the beginning of the fall and when the spring is deformed by the amount d. The change in potential energy of the mass in this fall is $W(h + d)$, and it must be equal to the energy stored in the spring at the largest deformation,

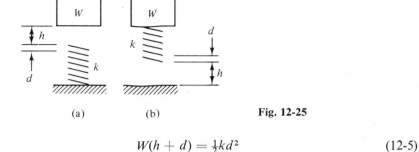

(a)   (b)   **Fig. 12-25**

$$W(h + d) = \tfrac{1}{2}kd^2 \qquad (12\text{-}5)$$

where k is the spring constant.

The same deformation, d, can be produced by applying a static load, P, which must be larger than W. The magnitude of this load is

$$P = kd \qquad (12\text{-}6)$$

The work done on the spring by P is

$$U = \tfrac{1}{2}Pd$$

Assuming that the spring responds linearly to the loads, the deformations and corresponding stresses in the spring are proportional to the loads:

$$\frac{P_1}{P_2} = \frac{d_1}{d_2} = \frac{\sigma_1}{\sigma_2} \qquad (12\text{-}7)$$

where the subscripts denote different levels of loading. The last equalities are also valid with the following substitutions:

$P_1 = P = kd,$ the static equivalent of the dynamic load
$d_1 = d,$ the deformation caused by the falling body
$\sigma_1 = \sigma,$ the stress caused by the falling body
$P_2 = W,$ the static weight of the falling body
$d_2 = d_{st},$ the deformation caused by W applied slowly
$\sigma_2 = \sigma_{st},$ the stress caused by W applied slowly

Equations 12-5 and 12-7 are combined to express the dynamic deformation in terms of the static deformation and the height of fall:

$$d^2 - 2dd_{st} - 2hd_{st} = 0$$

for which $$d = d_{st} + \sqrt{d_{st}^2 + 2hd_{st}} = d_{st}\left(1 + \sqrt{1 + \frac{2h}{d_{st}}}\right) \qquad (12\text{-}8)$$

The dynamic and static stresses can be related using a similar procedure (based on the stresses in Eq. 12-7):

$$\sigma = \sigma_{st}\left(1 + \sqrt{1 + \frac{2h}{d_{st}}}\right) \tag{12-9}$$

The quantity in parentheses in Eq. 12-8 and 12-9 is called the *impact factor* because it shows the magnification in deflection and stress when a load is applied dynamically instead of statically. The real impact factor is somewhat smaller than the calculated one because some energy is always dissipated by friction during the fall and deceleration of the body. This includes the internal friction during plastic flow at the points of contact between bodies. Other errors are caused by neglecting the inertia and possible inelasticity of the spring.

A special value of the impact factor is worth noting. When the load is applied suddenly, without involving the prior fall of a body, $h = 0$, so

$$d = 2d_{st} \quad \text{and} \quad \sigma = 2\sigma_{st} \tag{12-10}$$

This shows the merit in applying loads slowly.

EXAMPLE 12-9

Assume that the coil spring in the landing gear of an airplane deflects 5 in. when the plane stands on the ground. During landing, even a gentle touchdown, a given wheel may be the only one in contact, so the load on it may exceed the static load. Assume that this increase is by 120%. Furthermore, gusts or pilot error may cause a hard landing. Assume that this is equivalent to a free fall from a height of 5 ft. What is the magnification of deformations and loads compared to the static situation?

Solution

Denote the normal rest load by P_{st} and the corresponding deflection by e_{st}. The worst load and deflection are $2.2P_{st}$ and $2.2e_{st}$. A hard landing causes a further

magnification of $\qquad 1 + \sqrt{1 + \frac{2(5)(12)}{5}} = 6$

The largest deformation and load would tend to be

$\qquad e = (2.2)(6)e_{st} = 13.2e_{st} = 66$ in.   (it is likely that the spring
$\qquad\qquad\qquad\qquad\qquad\qquad\qquad\qquad\qquad$ would bottom out)

$\qquad P = 13.2P_{st}$ (things may break)

EXAMPLE 12-10

A jet engine weighs 50 kN and is located on the wing at a distance of 7 m from the fuselage. What is the magnification of stresses at the root of the wing during a hard landing, which is equivalent to a free fall from a height of 2 m? Assume that the wing is weightless and that the static deflection caused by the engine where it is mounted is 5 cm.

Solution

The wing is a cantilever beam. The deflection of the wing at the engine is proportional to the load applied (engine weight). Since the load-deflection relation is linear, Eq. 12-8 applies:

$$d = d_{st}\left(1 + \sqrt{1 + \frac{2(2)}{0.05}}\right) = 10d_{st}$$

The engine deflection, the strains throughout the wing, and the stresses at the wing root are all 10 times as large as in the static situation.

12-6. REVIEW OF THE COMET FAILURES

The structure of an airplane is complex and it has many possible sources of failure. Factors of safety cannot be used liberally if the airplane is to fly economically (or, to fly at all). The use of generous factors of safety can be self-defeating, anyway: The increase in weight of one part may require increases in weight of several other components. The net result may be an overall increase in weight with little or no change in the factors of safety. Thus, even a properly designed airplane has low factors of safety.

The investigation of the cause of a crash such as the Comets' appears to be almost hopeless at first. The number of purposely made stress concentrations alone is in the thousands. The whole structure has an indescribable state of residual stresses, and this changes in time in unknown ways. The maximum stresses cannot be known at all the critical locations.

The major clues were that the same time-dependent phenomenon may have caused both Comet crashes and that both aircraft fell from high altitudes. Repeated pressurization of the fuselage in each flight satisfied these conditions, but this was considered in the original design and was not found to be critical. This was reconsidered (among other items) after the two crashes, and new tests were made. The results of the tests and observations made on the recovered parts of Yoke Peter confirmed beyond reasonable doubt that the failure was caused by fatigue of the

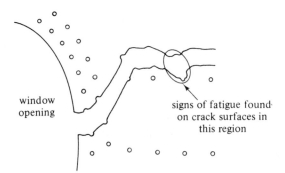

window opening

signs of fatigue found on crack surfaces in this region

Fig. 12-26. Sketch of failure of skin of a Comet aircraft.

fuselage from repeated pressurization (the gage pressure for Comets was about 50% greater than that in general use at the time).

The origin of each failure was in the skin of the cabin at a window, as shown in Fig. 12-26. The window is a stress concentration and so are the numerous rivet holes and bolt holes around it. The fatigue failure occurring in about 3000 cycles of pressurization is an example of low-cycle fatigue. This is remarkable considering that many parts such as engine components of the same aircraft must have endured millions of cycles of loading.

12-7. PROBLEMS

Sec. 12-1

12-1.   Assume it is unavoidable to have an elliptical hole in a flat tension member. Which of the two orientations of the hole shown in Fig. P12-1 is best and by how much? Guess how the answer may be different if the minimum cross-sectional areas of the member are the same for the two orientations of the same ellipse (assume the stress remains the same at the ends of the member).

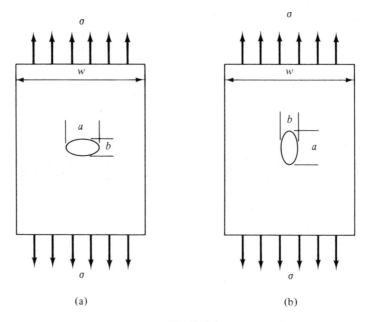

Fig. P12-1

12-2.   Is there any specific relationship between the areas under the average stress and the true stress distributions in Fig. 12-2b?

12-3.   Is the theoretical stress concentration factor independent of the load applied to a member if the material is (a) ideally elastic? (b) very ductile?

12-4.　Sketch the sequence of stress distributions similar to those in Fig. 12-5b but for a low carbon steel that yields sharply.

12-5.　Somebody proposes that $K_t = \sqrt{K_\sigma K_\epsilon}$. Discuss the validity of this formula for (a) brittle materials; (b) ductile materials.

12-6 to 12-11.　Determine the maximum average stress and the peak stress for each of the members shown. Use Fig. 12-3 for each problem. If you do not have enough information, do the best with what you have.

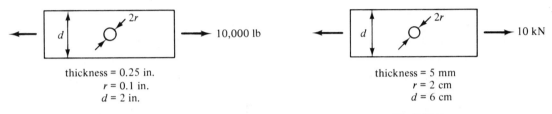

thickness = 0.25 in. r = 0.1 in. d = 2 in. **Fig. P12-6**	thickness = 5 mm r = 2 cm d = 6 cm **Fig. P12-7**

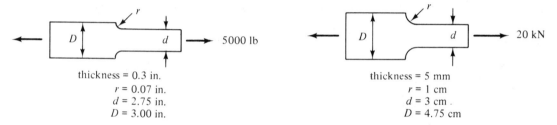

thickness = 0.3 in. r = 0.07 in. d = 2.75 in. D = 3.00 in. **Fig. P12-8**	thickness = 5 mm r = 1 cm d = 3 cm . D = 4.75 cm **Fig. P12-9**

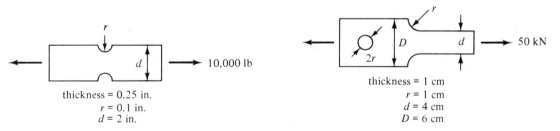

thickness = 0.25 in. r = 0.1 in. d = 2 in. **Fig. P12-10**	thickness = 1 cm r = 1 cm d = 4 cm D = 6 cm **Fig. P12-11**

Sec. 12-2

12-12.　Make as good an estimation as you can of the force equilibrium for the stress distribution shown in Fig. 12-6a. In other words, can the member be in equilibrium as shown?

12-13. Consider a thick wire and a very fine wire made of the same metal. Which wire may have the largest residual stress in it?

12-14. What is the number of equilibrium conditions that must be checked to make sure that a certain distribution of residual stresses is reasonable?

*12-15. Consider a ductile bar with a rectangular cross section that is bent so that the outer layers of the bar yield both in tension and in compression. Sketch a reasonable distribution of residual stresses on a cross-sectional area to show the state of stresses after all external loads are removed.

Sec. 12-3

12-16. Would you expect the stress to be constant throughout the creep test shown in Fig. 12-8?

12-17. The initial elastic strain ϵ_0 in creep (Fig. 12-9) is loosely called instantaneous strain. Certainly, instantaneous strain cannot mean that it occurs in zero time. How much time do you think is necessary to produce the elastic strain? Would this time vary from material to material?

12-18. The creep curve in Fig. 12-9 is for a constant load. Sketch creep curves to show what could be expected if the load is suddenly (a) increased during the test; (b) decreased during the test.

12-19. Is it possible to use degrees Celsius (°C) for T and T_m in Eq. 12-4?

12-20. How would you demonstrate stress relaxation with a minimum of equipment?

12-21. Sketch the results after some heating of the member shown in Fig. 12-6a and b.

12-22. Assume you have an aluminum component and a steel component that you want to relieve of internal stresses. What is the approximate minimum temperature in each case that is necessary to accomplish this?

12-23. Assume you want to obtain the stress-strain curve for lead. Are there any special problems in such a test that you should consider on the basis of this section?

Sec. 12-4

12-24. Describe in reasonable detail how many ways you can get the following stresses and strains simultaneously at a given point in a material:
(a) tensile stress and tensile strain;
(b) tensile stress and compressive strain;
(c) compressive stress and compressive strain;
(d) compressive stress and tensile strain.

12-25. Under what conditions can the sign and magnitude of a stress be determined from a measured strain? When is this not possible?

12-26. Draw a hysteresis loop that shows stress versus plastic strain in a complete cycle of loading.

12-27. How many fatigue strengths does a metal have?

12-28. Plot three successive hysteresis loops for a metal that hardens cycle-dependently. Start with a very fat hysteresis loop, and use (a) stress control; (b) strain control.

*12-29 to 12-32. The data in each of the plots were obtained in tests where the mean stress was zero. Plot the *S-N* diagrams (stress amplitude versus life) from the available data. Indicate qualitatively the effects of tensile and compressive mean stresses on the *S-N* diagrams.

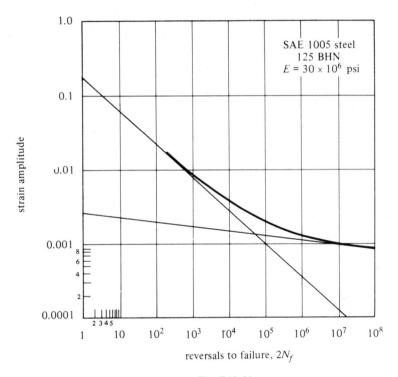

Fig. P12-29

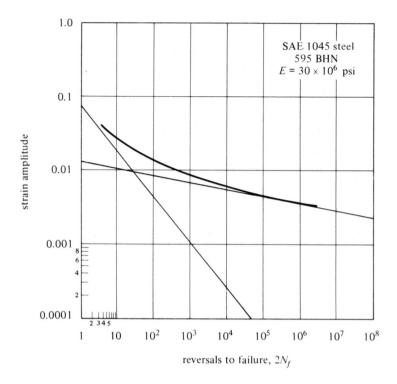

Fig. P12-30

Fig. P12-31

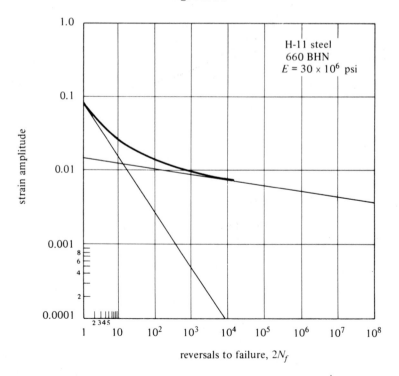

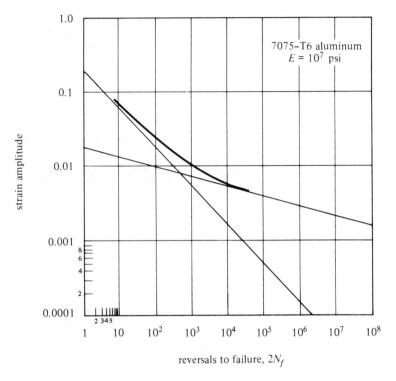

Fig. P12-32

*12-33. In Example 12-7 and in Probs. 12-29 to 12-32, determine the slopes and the strain and stress amplitudes at one reversal from the elastic strain amplitude versus life lines. Compare the appropriate values for the different metals and try to generalize from the results.

Sec. 12-5

> *A body of weight W is dropped from a height h on a spring that has a constant k. Determine the unknown quantity in Probs. 12-34 to 12-39.*

12-34. $W = 100$ lb, $h = 2$ in., $k = 500$ lb/in. What is the maximum deflection of the spring and the impact factor?

12-35. $W = 1$ kN, $h = 5$ cm, $k = 1.2$ kN/cm. What is the maximum deflection of the spring and the impact factor?

12-36. $W = 200$ lb, $h = 5$ in., $d_{max} = 1$ in., $k = ?$

12-37. $W = 5$ kN, $h = 10$ cm, $d_{max} = 2$ cm, $k = ?$

12-38. $k = 1000$ lb/in., $h = 5$ in., $d_{max} = 0.4$ in. $W = ?$

12-39. $k = 5$ kN/cm, $h = 4$ cm, $d_{max} = 0.3$ cm, $W = ?$

13

FAILURE CRITERIA AND DESIGN CONCEPTS

A failure criterion is a concise statement of the most significant variables that are thought to be involved in the failure of a solid object. Such statements (also called theories of failure) are used in attempts to prevent failures or to predict the set of conditions that causes a certain kind of failure. Many different failure criteria have been proposed because even the definition of failure is not unique. For example, a certain amount of elastic strain may be considered failure in a machine designed with tight tolerances, whereas even more strain than that may be allowed in a different situation. No single theory of failure should be expected to be entirely satisfactory in a variety of design problems.

Failure criteria in strength of materials can be grouped in two broad categories. The *classical theories of failure* deal with fundamental statements regarding critical values of stress, strain, or strain energy. These theories are useful in situations where the critical quantity can be measured or predicted accurately throughout the service life of each member. This means that they are most useful for unnotched, flaw-free materials whose properties do not change in time (or change in known ways). In other words, the classical theories are not very helpful in the majority of practical problems. Unfortunately, the *failure criteria in the category of modern design concepts* have not yet been able to fill all the void that was left by the classical theories. These criteria, which must include at least the effects of notches, are still in the process of evolution, but they have already assumed commanding roles in comparison with the classical theories. The uneven emphasis on the various theories presented in this chapter reflects the changing state of affairs in design concepts and research efforts in strength of materials.

13-1. CLASSICAL THEORIES OF FAILURE

The simplest of these theories are nothing more than formal statements of the most obvious concepts of strength of a material. Some of the others are difficult to derive. Not all the classical theories proposed will be presented here, and those presented will not necessarily be given in sufficient detail to appreciate them fully.

Maximum Normal Stress (or Rankine) Theory. According to this theory, failure occurs at some point in a member when the maximum principal normal stress at that point reaches the strength level that is considered critical (either the yield strength or the ultimate strength). This theory is most reasonable when used to predict the fracture of brittle materials. Its main limitations are that it does not take into account possible differences between tensile and compressive strengths, the possibility that strength may depend on the orientation of the principal planes (anisotropic materials), or that shear strength may be the critical quantity.

Maximum Shear Stress (or Tresca) Theory. This theory states that yielding occurs when the maximum shear stress in a member is equal to τ_{ys}^*, the yield stress in pure shear (torsional loading). In terms of principal normal stresses σ_1, σ_2, and σ_3 (for example, $\sigma_1 > \sigma_2 > \sigma_3$), the yield criterion is given as

$$\tau_{max} = \frac{\sigma_1 - \sigma_3}{2} = \tau_{ys} \qquad (13\text{-}1)$$

The theory is most applicable to ductile materials in which shear strength may be the limiting quantity as far as overall strength is concerned. It is applicable only to isotropic materials.

Distortion Energy (or von Mises, or Octahedral Shear Stress) Theory. This gives a relationship between the principal stresses σ_1, σ_2, and σ_3 (triaxial state of stress) and the yield strength in uniaxial tension, σ_{ys}. It is based on the concept that yielding will occur in an element subjected to triaxial stresses when the distortion energy is equal to that in an axially loaded element at the beginning of yielding. It is not necessary to present the somewhat complex details of the derivation here.[†] The mathematical statement of the theory is

$$(\sigma_1 - \sigma_2)^2 + (\sigma_2 - \sigma_3)^2 + (\sigma_3 - \sigma_1)^2 = 2\sigma_{ys}^2 \qquad (13\text{-}2)$$

---

[*]Throughout this chapter the subscripts *ys* are used to denote yield strength (the material property) to distinguish it from the applied stress in the *y* direction (denoted by the subscript *y*). Commonly, and elsewhere in this text, both are denoted only by the subscript *y*. The distinction is helpful in solving problems in this Chapter.

[†]See E. P. Popov, *Introduction to Mechanics of Solids* (Englewood Cliffs, N.J.: Prentice-Hall, Inc., 1968).

The equation is useful for predicting yielding in ductile metals that have no flaws or notches. Localized yielding in regions containing discontinuities (which is important in the fracture of practical members) is a relatively complex phenomenon because it is governed by the internal constraints in the neighborhood of the discontinuity, as shown in Section 13-2. In other words, the failure theory's usefulness depends on how well the stresses in critical regions are known, and this is often a difficult problem.

Summary of the Three Classical Theories. A comparison of the three theories given here can be made if the statements are presented graphically, as for a hypothetical ductile metal shown in Fig. 13-1. It is assumed that the metal has identical strengths in tension and compression and that there is a possibility of having biaxial stresses.

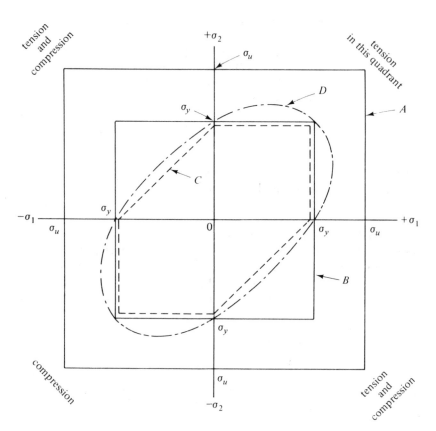

Fig. 13-1. Schematic comparison of three classical failure theories.
A—maximum normal stress theory; σ_u critical.
B—maximum normal stress theory; σ_{ys} critical.
C—maximum shear stress theory.
D—distortion energy theory.

If normal stresses up to the ultimate strength are allowed, square A bounds the region of the maximum reasonable stresses in the design. Any arbitrary factors of safety can be used simply to scale down the square. Yielding must be expected if the factor of safety is small. Square B bounds the region of allowable stresses if the maximum normal stress theory is used to prevent yielding.

The dashed line C bounds the region of allowable principal normal stresses if the maximum shear stress is the failure criterion. Note that the combination of the signs of the normal stresses becomes important according to this theory.

The dotted line D represents the limitations imposed on the allowable principal normal stresses when the distortion energy is used as a failure criterion (the safe region is within the oval shape). Note again that the combination of the signs of the normal stresses is important.

The experimental data fall between the dashed and the dotted lines in Fig. 13-1 for some ductile metals. Many people believe that the data are often closest to the values predicted by using the distortion energy theory. According to this, $\tau_{ys} = 0.577\sigma_{ys}$. Actually, many metals have yield properties far from this value predicted using the theory. The range of values measured is $\tau_{ys} = 0.25$ to $0.75\sigma_{ys}$. It should be emphasized that the classical theories should be used with caution in the design of notched members and in cases involving time-dependent phenomena.

EXAMPLE 13-1

A ductile metal has a yield strength of 300 MPa. Find the factors of safety for the given state of stress using the three classical theories just discussed. Assume that yielding means failure.

Given: $\sigma_x = 150$ MPa, $\sigma_y = -100$ MPa. (There are no other applied stresses.)

Solution

The center of Mohr's circle is at

$$\frac{\sigma_x + \sigma_y}{2} = 25$$

and its radius is

$$r = \frac{\sigma_x - \sigma_y}{2} = 125$$

The principal stresses are

$$\sigma_1 = \text{center} + \text{radius} = 150 \text{ MPa}$$
$$\sigma_2 = \text{center} - \text{radius} = -100 \text{ MPa}$$
$$\tau_{max} = \text{radius} = 125 \text{ MPa}$$

According to the maximum normal stress theory, the factor of safety, N, is

$$N = \frac{\sigma_{ys}}{\sigma_1} = \frac{300}{150} = 2$$

The maximum shear stress theory gives

$$N = \frac{\tau_{ys}}{\tau_{max}} = \frac{150}{125} = 1.2$$

Note that $\tau_{ys} = \sigma_{ys}/2$ was used in finding N.

The distortion energy theory gives, with $\sigma_3 = 0$,

$$(\sigma_1 - \sigma_2)^2 + \sigma_2^2 + \sigma_1^2 = 2\sigma_{ys}^2, \qquad \sigma_1^2 - \sigma_1\sigma_2 + \sigma_2^2 = \sigma_{ys}^2$$

Yielding occurs when this equation is satisfied. The factor of safety is

$$N = \frac{\sigma_{ys}}{\sqrt{\sigma_1^2 - \sigma_1\sigma_2 + \sigma_2^2}} = \frac{300}{218} = 1.37$$

Note that a similar procedure should be used for any state of stress: The principal stresses must be determined first.

13-2. MODERN FAILURE CRITERIA

The most important failure criteria in modern design are those that take into account the effects of geometric discontinuities. Relatively few people realize that it is not sufficient to consider the stress concentration factors in dealing with this problem. The sharpness of a notch is important, but it has been found that the length or depth of the notch (whether it is machined or a natural flaw) and the thickness of the member also play important roles in the initiation of fractures. These aspects of fracture mechanics are not known in sufficient detail to satisfy all the requirements of the most rigorous designers, but the subject has been developing very rapidly in recent years. The time has come when it is imperative to introduce some of the basic concepts of modern fracture mechanics to all future designers in strength of materials. A whole course would be needed just to give an extensive introduction to fracture mechanics. Thus, the material given here is intended only to show some simple formulas and general tendencies of behavior in order to alert the reader to potential problems and to available rational ways of dealing with them.

13-3. THE GRIFFITH CRITERION

The era of modern fracture mechanics began in 1920 when Griffith proposed his famous theory of fracture. The theory states that a crack will propagate when the decrease in elastic strain energy is at least equal to the energy required to create the new crack surfaces. In other words, there must be enough elastic strain energy in a solid at the onset of fracture to be converted into other forms of energy that are associated with the fracture process. These other forms of energy include

(a) surface energy of new cracks (this is similar in concept to the so-called surface tension of liquids),
(b) heat from plastic deformation near the crack,
(c) kinetic energy of parts that move as a result of the fracture,
(d) sound, etc.

Obviously, some of these energies may be quite negligible in certain cases. For example, a brittle solid dissipates essentially no heat during fracture, whereas the energy required to create the crack surfaces is absolutely unavoidable (and there is also some sound). The surface energy requirement is also unavoidable in the case of a ductile material, but its magnitude may be a thousandth of the plastic work.

A very simple derivation of the Griffith equation can be given here to show the significant variables in the fracture of a brittle solid. Consider a crack of length $2c$ in a large, flat plate of thickness t (Fig. 13-2a). There are no stresses in the material acting at the two crack faces in directions perpendicular to those faces.

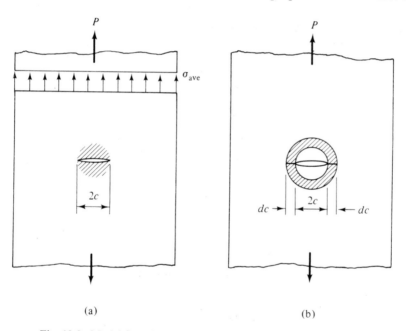

(a)   (b)

Fig. 13-2. Model for release of strain energy in crack propagation.

The stresses increase gradually in the shaded region from zero stress near the crack to about σ_{ave} at the outer edges of the shaded circle. Assume that the stress is zero everywhere in the shaded circle, however. Imagine next that the crack grows a small distance dc at both ends. Make the crude approximation that this crack propagation results in the stress suddenly decreasing from σ_{ave} to zero within the shaded ring in Fig. 13-2b. Thus, elastic strain energy is released in the volume indicated by the ring, and energy was required to create four new crack surfaces, each surface with an area of *dct*. If all other forms of energy involved in the process are negligible, the two energies can be related approximately as follows.

Assuming that the stress in the ring was uniform and equal to σ_{ave} prior to the cracking by 2*dc*, the total elastic strain energy released during the crack growth is

$$dW_E = \frac{\sigma_{ave}^2}{2E} \times (2c\pi tdc) \qquad (13\text{-}3)$$

$$\underset{\substack{\text{strain} \\ \text{energy} \\ \text{per unit} \\ \text{volume}}}{} \qquad \underset{\substack{\text{volume} \\ \text{of} \\ \text{ring}}}{}$$

The incremental increase in the surface energy of the plate during the crack growth is

$$dW_S = 4dct\gamma \qquad (13\text{-}4)$$

where γ is the surface energy per unit area, a constant for the material.

According to the Griffith criterion,

$$\frac{dW_S}{dc} \leq \frac{dW_E}{dc} \qquad (13\text{-}5)$$

$$\underset{\substack{\text{surface energy} \\ \text{increase per} \\ \text{unit extension} \\ \text{of the crack}}}{} \qquad \underset{\substack{\text{strain energy} \\ \text{released per} \\ \text{unit extension} \\ \text{of the crack}}}{}$$

Thus,
$$4\gamma t = \frac{\sigma_{ave}^2}{2E} 2c\pi t$$

so the average stress that can be considered critical is

$$\sigma_{critical} = 2\sqrt{\frac{\gamma E}{\pi c}} \qquad (13\text{-}6)$$

A more precise analysis than this has led Griffith to conclude that

$$\sigma_{critical} = \sqrt{\frac{2\gamma E}{\pi c}} \qquad (13\text{-}7)$$

Equation 13-7 is significant mainly because it shows that the length of a flaw or notch is an important variable in the initiation of unstable fracture. It also shows why a brittle fracture is catastrophic (unstoppable). Once the load is large enough to start the crack, even a rapidly decreasing load that keeps σ_{ave} a constant over the remaining cross-sectional area could not stop the crack from growing. The minimum stress necessary to keep the crack growing decreases with the square root of the crack length, as shown in Fig. 13-3.

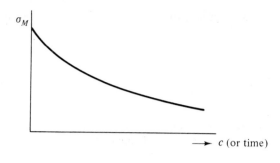

Fig. 13-3. Minimum stress sufficient to propagate a crack.

$\longrightarrow c$ (or time)

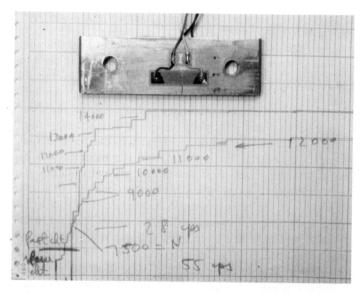

Crack propagation gage on specimen cut from flawed bearing cap of large marine Diesel engine; test record shows acceleration of crack growth rate for constant amplitude loading. *Photograph by J. Dreger, University of Wisconsin, Madison, Wis.*

The Griffith equation indicates that nondestructive searches for flaws are highly desirable. For any given stress σ_{ave} chosen in the design there is a critical flaw size. The nondestructive searching technique must be at least sensitive enough for the detection of every flaw whose size is equal to or greater than the critical size.

EXAMPLE 13-2

A brittle material has $E = 15 \times 10^6$ psi and $\gamma = 6 \times 10^{-3}$ in.-lb/in². What are the critical flaw sizes c_c corresponding to the proposed average operational stresses of $\sigma_1 = 5$ ksi, $\sigma_2 = 10$ ksi, and $\sigma_3 = 20$ ksi?

Solution

From Eq. 13-7,

$$c_{c_1} = \frac{2\gamma E}{\pi \sigma_1^2} = 2.3 \times 10^{-3} \text{ in.}$$

$$c_{c_2} = 5.7 \times 10^{-4} \text{ in.}$$

$$c_{c_3} = 1.4 \times 10^{-5} \text{ in.}$$

Obviously, even the largest of these is quite small, so a careful inspection procedure should be specified. The smallest critical flaw is practically impossible to detect, and failure under σ_3 is rather certain.

13-4. EMBRITTLEMENT CAUSED BY NOTCHES

Engineers frequently assume that ductility alone is required to have reliability against brittle fractures. This is wrong. Ductility is necessary but not sufficient to assure high toughness. The reason is that yielding must occur to take advantage of the available ductility, and yielding occurs only if the shear stress is high enough. There are conditions when the normal stresses are high in a material, but the shear stresses are low (review three-dimensional principal stresses and absolute maximum shear stress). In such cases the member cannot yield even if the material has intrinsic ductility (observed in a uniaxial tension test), and the member fails by brittle fracture when the maximum normal stress reaches the critical value.

There are two important aspects of embrittlement by triaxial states of stress. First, the signs and magnitudes of the three principal stresses do not have to be identical, resulting in zero shear stress, in order to obtain brittle fracture in an intrinsically ductile material. There may be a finite shear stress (but low because of the triaxiality) at the onset of fracture. Second, the external loading on the member does not have to be three-dimensional to cause a triaxial state of stress. Uniaxial loading can cause such stresses when the member has a discontinuity. This phenomenon is somewhat complex and has to be explained in detail.

It is advantageous to consider a simple model at first, such as that shown in Fig. 13-4. Sections A and B are identical, rigid members. The thin section C is firmly attached to A and B; it is smaller in cross-sectional area than they are; it has an identical Poisson's ratio; and it has a lower elastic modulus (the last two are not required features of the model). A small uniaxial load P is applied to the model. The average stresses in the y-direction are largest in part C:

$$\sigma_{Cy} > \sigma_{Ay} = \sigma_{By}$$

The three parts tend to contract in the x- and z-directions because of the Poisson effect. Section C wants to contract more than the other two parts do, however. The three parts constrain each other in the neighborhood of C and develop stresses as follows:

(a) C cannot contract as much as it would under σ_{Cy} in the absence of A and B; it feels tensile stresses in the x- and z-directions.

(b) C in turn applies compressive stresses to A and B in the x- and z-directions.

Autoclave assembled in a 500-kip load frame for testing pressure vessel steels at elevated temperature and pressure; (b) and (c) notched specimens for different components of a nuclear reactor. *E. Kiss, T. L. Gerber, and J. D. Heald, "Evaluation of Fatigue Crack Growth in a Simulated Light Water Reactor Environment,"* Closed Loop, **4**, *No. 2 (1974), 2–9.*

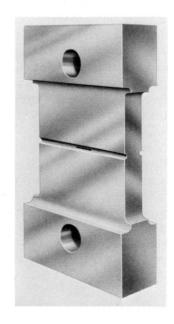

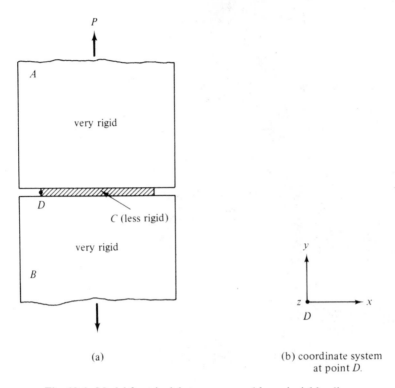

P

A

very rigid

D

C (less rigid)

very rigid

B

(a)

y

z x

D

(b) coordinate system
at point D.

Fig. 13-4. Model for triaxial stresses caused by uniaxial loading.

Thus, C (which is the critical region anyway) has a high tensile σ_y stress and also tensile σ_x and σ_z stresses. The latter are not so high as the originally imposed stress σ_y in the same part. The triaxial state of tensile stresses reduces the maximum shear stress in C and enhances the possibility of brittle fracture there even if this material is ductile.

The same concepts can be applied to a notched member made of a single homogeneous material. For example, the plate shown in Fig. 13-5 has a uniform material and a constant thickness t; a coordinate system is put at the center of the root of the notch. The average stress in region A is higher than that in region B, so region A wants to contract more than region B does. The mutual constraint of these regions (and the region below A) results in triaxial tensile stresses in region A. These stresses are not constant over the minimum section; their distributions must be considered one by one.

The stress distributions shown in Fig. 13-6 are applicable to the plane of the notch, the x-z-plane from O to C (Fig. 13-5). The σ_y distribution is imposed by the load P on the notched member and should be self-explanatory (Fig. 13-6a). It is clear that there must be tensile stresses in the x-direction because of the constraint, but σ_x must be zero at O and C, which are on the free surfaces of the plate. Thus, the magnitudes of σ_x vary approximately as in Fig. 13-6b. The unsymmetric

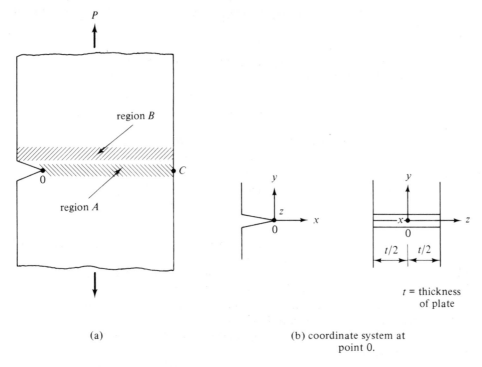

(a)

(b) coordinate system at
point 0.

t = thickness
of plate

Fig. 13-5. Model for mutual constraint in a notched member.

Fig. 13-6. Patterns of triaxial stresses in a notched member under unixial loading.

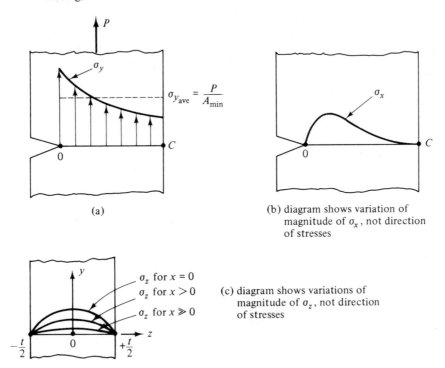

$$\sigma_{y_{ave}} = \frac{P}{A_{min}}$$

(a)

(b) diagram shows variation of
magnitude of σ_x, not direction
of stresses

σ_z for $x = 0$
σ_z for $x > 0$
σ_z for $x \gg 0$

(c) diagram shows variations of
magnitude of σ_z, not direction
of stresses

distribution is reasonable because the differences (and mutual constraint) between regions A and B are large near the notch root and negligible far from the notch. Similar considerations are applied in sketching the variation in the magnitudes of σ_z, with the exception that the distribution is expected to be symmetric about the x-axis (Fig. 13-6c). σ_z must be zero at $\pm(t/2)$ because of the free surfaces there. The three sample distributions drawn in Fig. 13-6c are based on the fact that σ_z can be largest at $x = 0$ where the constraint caused by region B on A is the most severe. The constraint and the resulting maximum value of σ_z decreases as the distance from the notch root increases. The three sketches in Fig. 13-6 show that the most severe triaxiality of stresses should be expected at $y = 0$ and $z = 0$, but at x slightly larger than zero.

13-5. THE THICKNESS PROBLEM

It has been found that the severity of a triaxial state of stress caused by uniaxial loading in a notched member depends significantly on the thickness of that member. This can be explained by considering the stresses in the z-direction in thin and thick notched members as shown in Fig. 13-7. Assume that the straight lines A-B and C-D are the notch roots in the two members. The notch geometries and $\sigma_{y_{ave}}$ are identical in both cases.

It is clear that σ_z must be zero at points A, B, C, and D (Fig. 13-7). There must be some σ_z inside both members because of the constraints caused by the notches, and these stresses should be largest at $z = 0$ (Fig. 13-6c). Next, it must be argued that there should be a limit to the gradient of σ_z with respect to z. Very steep gradients in internal stresses may occur when the temperature gradient is large, but here the temperature is uniform throughout each plate. Let us say that the maximum gradient in σ_z is indicated by 45° lines in Fig. 13-7 (the exact magnitude of these slopes is not important here). Any assumed limitation on the stress

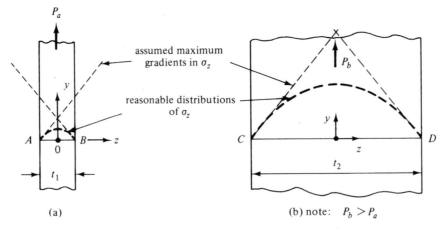

(a)

(b) note: $P_b > P_a$

Fig. 13-7. Model for the thickness effect in notched members.

gradients creates a limitation on the maximum values of σ_z. In the present cases, the absolute maximum values possible are such that

$$\sigma_{z_b} = \frac{t_2}{t_1}\sigma_{z_a}$$

The reasonable distributions of σ_z as shown in Fig. 13-7 indicate that a very thin member may have negligible σ_z even at $z = 0$, whereas it is best to expect large σ_z at $z = 0$ in thick members. These possibilities have important implications. A thin member is likely to have only a biaxial state of stress even if it is notched, so if it is ductile, it can yield [$\tau_{max} = (\sigma_1 - 0)/2$ or $(\sigma_2 - 0)/2$]. The same metal with the same notch geometry acts more and more brittle as the thickness is increased since the presence of a sizable third principal stress reduces τ_{max}. Some of the practical considerations of this problem area are discussed in Section 13-6.

The most important conclusion from the discussion above is that, for yielding to occur, ductility is a *necessary* but *not* a *sufficient* condition. The sufficient condition is that the absolute maximum shear stress is larger than the yield strength τ_{ys}.

EXAMPLE 13-3

A ductile metal has strength properties $\tau_{ys} = 20$ ksi, $\sigma_{ys} = 40$ ksi, and $\sigma_u = 65$ ksi. It is to be used in two different thicknesses, and each plate has an edge notch as in Fig. 13-5. The following maximum stresses are obtained by computer analysis (these subscripts denote axes as in Fig. 13-5):

thin plate	thick plate
$\sigma_x = 15$ ksi	$\sigma_x = 25$ ksi
$\sigma_y = 55$ ksi	$\sigma_y = 55$ ksi
$\sigma_z = 0.5$ ksi	$\sigma_z = 21$ ksi

What are the chances of failure in the two plates if the external load may increase by 20%?

Solution

It is recognized, on the basis of Fig. 13-6, that the three maximum stresses do not occur at the same place in a notched member. They occur in regions very close to each other, however, so it will be assumed that they are present at the same point. This assumption makes the results of the analysis a little conservative, which is acceptable.

The maximum shear stress in the thin plate is above τ_{ys} for the original normal stresses; thus, there is yielding. A 20% increase in the load does not result in a 20% increase in σ_y (from 55 to 66 ksi). Rather, the peak stress can be expected to increase a few kips per square inch while the minimum stress in the y-direction increases by more than 20% (stress redistribution upon yielding; $\sigma_{y_{ave}}$ increases 20%). The thin plate is not expected to fracture for the given increase in load.

Rock specimen is inserted into ultra-stiff, servo-controlled load frame; crack length, specimen thickness, and crack orientation of rocks give information for shale oil recovery, spark drilling, and mine design; (b) three-point-bend specimens of rock; large specimen is equipped with LVDT displacement transducer. *Richard A. Schmidt*, "*Fracture Toughness Testing of Rock*," Closed Loop, **5**, *No. 2 (1975), 2–12.*

The maximum shear stress in the thick plate is

$$\tau_{max} = \frac{55 - 21}{2} = 17 \text{ ksi} < \tau_{ys}$$

Thus, the plate is not able to yield. Any increase in the external load will cause proportional increases in the stresses (according to linear elastic behavior, with K_t fully applicable). For a 20% increase in the load, the principal stresses become

$$\sigma_x \longrightarrow 30 \text{ ksi}, \qquad \sigma_y \longrightarrow 66 \text{ ksi} > \sigma_u, \qquad \sigma_z \longrightarrow 25.2 \text{ ksi}$$

The maximum shear stress is

$$\tau_{max} = \frac{66 - 25.2}{2} = 20.4 \text{ ksi} > \tau_{ys}$$

Yielding has become possible, however, only when one of the principal stresses reached the ultimate strength of the metal. This is a very dangerous situation because catastrophic brittle failure is likely before yielding could reduce the severity of the stress concentration.

13-6. FRACTURE TOUGHNESS OF NOTCHED MEMBERS AS MATERIAL PROPERTY (K_{Ic})

Concepts such as those in Sections 13-4 and 13-5 are helpful in qualitative descriptions of the phenomena involved, but they create difficult problems for the designer. The basic material properties (E, σ_{ys}, σ_u) may be known, but what should be done if a given material is used in various thicknesses and has many different notches? The basic properties cannot be used to predict the effects of any given notch and thickness, and the results of tests on specimens with a certain geometry may not be applicable to specimens with different geometries. There is no completely satisfactory solution to these problems at this time. A method has evolved, however, to deal with certain kinds of problems within the area. This is based on the concept of the stress intensity factor.

The stress intensity factor was defined by G. R. Irwin* as

$$K = \sigma_{ave}\sqrt{\pi c} \tag{13-8}$$

where c = half crack length. A quantity related to this factor is defined as

$$G = \frac{K^2}{E}$$

which is called the crack extension force (its units are lb/in. or in.-lb/in.²).

*A. S. Tetelman and A. J. McEvily, Jr., *Fracture of Structural Materials* (New York: John Wiley & Sons, Inc., 1967), p. 49.

It can be shown that K and G have meanings with respect to the Griffith equation; this is left as an exercise for the reader (show that $G = 2\gamma$).

Three modes of deformations are distinguished for notched members. These are shown schematically in Fig. 13-8. The opening mode is the simplest and it is

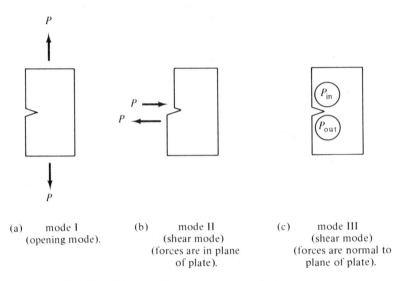

(a)	mode I	(b)	mode II	(c)	mode III
	(opening mode).		(shear mode)		(shear mode)
			(forces are in plane		(forces are normal to
			of plate).		plane of plate).

Fig. 13-8. Modes of deformation of notched members.

used in analyses and tests most frequently. K and G are normally given with the appropriate subscript that indicates the mode (K_I, G_I, etc.). The equation that gives the stress intensity factor depends on the mode of deformation and the geometry. Equation 13-8 is the simplest of these (opening mode), but it has the basic feature of all equations of stress intensity factors. This feature is $\sigma \sqrt{c}$.

The stress intensity factor can be considered as analogous to a stress-strain curve as shown in Fig. 13-9. When $\sigma = 0$, $K = 0$ by definition. For an ideally elastic material, K increases with σ essentially linearly up to K_c (small deviations

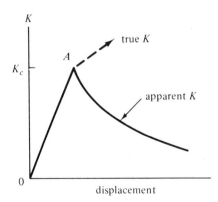

Fig. 13-9. Variation of stress intensity factor with crack opening displacement (or overall deformation of member).

from linearity may occur because of bending during the opening of the notch). K_c represents a critical event similar to yielding in a simple tension test: The notch or flaw suddenly begins to grow and complete fracture occurs. It appears that K decreases rapidly after point A on the curve. This is because the force measured in laboratory tests may decrease and still keep the crack growing. Naturally, the true stress intensity factor may be increasing during the crack propagation since the crack length and σ_{ave} both are increasing in most cases.

K_c is commonly called the *fracture toughness* of the material. G_c is called the *crack resistance force* of the material (or toughness or work done in initiating unstable fracture at the notch). Unfortunately, these critical values depend on the geometry of the member. The general tendency with respect to thickness is shown in Fig. 13-10. At very small thicknesses the curve drops rapidly because the volume of material in which plastic work is done becomes small. In most engineering problems such thin members are not involved.

An interesting feature of Fig. 13-10 is that the curve seems to approach asymptotically a lower limiting value. The recognition of this has led to the use of the limiting value as a pseudo-material property in design. The idea is this: If brittle fracture is to be avoided in a notched member, find the limiting value in laboratory tests of the same material, and use it to determine the safe stresses for various notch or defect sizes that must be considered in the design.

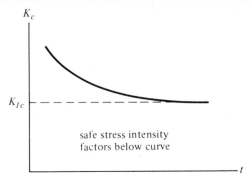

Fig. 13-10

The most commonly used lower limiting value of K_c is called K_{Ic}, the plane-strain fracture toughness. It is determined by making the sharpest crack possible (a fatigue crack of a given length created by many cycles at low stress amplitudes) in a sufficiently thick specimen ($t \approx 1$ in. or more). A tension test is performed on this cracked specimen, and the stress intensity at the first noticeable crack extension is taken as the limiting value, K_{Ic}. In design this is considered the worst possible stress intensity (sharpest possible crack, most severe triaxiality of stresses). Obviously, certain criteria must be satisfied before one has confidence that the measured K_{Ic} is the lowest that can be obtained for the given material.* On the other hand, in some cases the use of K_{Ic} is too conservative.

*1976 Annual Book of ASTM Standards, Part 10 (Philadelphia: American Society for Testing and Materials, 1976).

Insertion of an irradiated fracture toughness specimen into the grips of a testing machine in a shielded "hot cell," using a remotely controlled manipulator. *Earl B. Schwenk, Jr., "Fracture Mechanics Testing of Irradiated Steel Specimens,"* Closed Loop, **3**, *No. 8 (1971), 2–7.*

Further Reading About Modern Failure Criteria. The reader is encouraged to check the most recent ASTM Standards for descriptions of standardized or recommended mechanical test methods. ASTM Special Technical Publications cover useful recent developments in mechanical testing.

EXAMPLE 13-4

4340 steel is chosen for a certain structural member. It is known that for this metal $\sigma_{ys} = 1.5$ kN/mm², $\sigma_u = 1.85$ kN/mm², and $K_{I_c} = 1.5$ kN/mm$^{3/2}$. What is the largest crack that can be tolerated in this steel if the maximum average operating stress is 60% of the ultimate strength? The critical stress is

$$\sigma_c = \frac{K_{I_c}}{\sqrt{\pi c}} = 0.6\sigma_u$$

and the largest allowable crack is

$$2c = \frac{2K_{I_c}^2}{(0.6\sigma_u)^2\pi} = 1.1 \text{ mm long}$$

13-7. PROBLEMS

Sec. 13-1

Use the classical theories to find the factors of safety for the given states of stresses in Probs. 13-1 to 13-12. σ_{ys}, τ_{ys}, and σ_u are material properties; σ_x, σ_y, σ_z, or τ are the applied stresses (there are no other applied stresses). Do calculations both for yielding and ultimate failure where applicable.

13-1. $\sigma_{ys} = 40$ ksi, $\tau_{ys} = 20$ ksi, $\sigma_u = 65$ ksi, $\sigma_x = \sigma_y = 20$ ksi

13-2. $\sigma_{ys} = 300$ MPa, $\tau_{ys} = 150$ MPa, $\sigma_u = 450$ MPa, $\sigma_x = 150$ MPa,
$\sigma_y = -150$ MPa

13-3. $\sigma_{ys} = 100$ ksi, $\tau_{ys} = 50$ ksi, $\sigma_u = 130$ ksi, $\sigma_x = \sigma_y = -80$ ksi

13-4. $\ $ $\sigma_{ys} = 700$ MPa, $\tau_{ys} = 350$ MPa, $\sigma_u = 1$ GPa, $\sigma_x = 650$ MPa,
$\sigma_y = 750$ MPa

13-5. $\sigma_{ys} = \sigma_u = 300$ ksi, $\sigma_x = \sigma_y = 250$ ksi

13-6. $\sigma_{ys} = \sigma_u = 1.4$ GPa, $\tau = 700$ MPa

13-7. $\sigma_{ys} = 80$ ksi, $\tau_{ys} = 40$ ksi, $\sigma_u = 140$ ksi, $\sigma_x = 40$ ksi, $\tau = 40$ ksi

13-8. $\sigma_{ys} = 560$ MPa, $\tau_{ys} = 280$ MPa, $\sigma_u = 1$ GPa, $\sigma_x = -280$ MPa,
$\tau = 280$ MPa

13-9. $\sigma_{ys} = 60$ ksi, $\tau_{ys} = 30$ ksi, $\sigma_u = 100$ ksi, $\sigma_x = 2\sigma_y = 3\sigma_z = 60$ ksi

13-10. $\sigma_{ys} = \sigma_u = 1.8$ GPa, $\sigma_x = 2\sigma_y = 3\sigma_z = -1$ GPa

13-11. $\sigma_{ys} = 60$ ksi, $\tau_{ys} = 30$ ksi, $\sigma_u = 110$ ksi, $\sigma_x = 10$ ksi, $\sigma_y = 20$ ksi,
$\tau = \pm 25$ ksi
(The shear stresses are reversed cyclically while the normal stresses are constant.)

13-12. $\sigma_{ys} = 400$ MPa, $\tau_{ys} = 200$ MPa, $\sigma_u = 600$ MPa, $\sigma_x = 2\sigma_y = 3\sigma_z = -120$ MPa, $\tau = \pm 100$ MPa on the *y*- and *z*-faces
(The shear stresses are reversed cyclically.)

13-13. Could you use the data from some fracture tests to determine the surface energy of a solid? If yes, what kind of tests would be desirable? If no, what is the main problem?

13-14. Plot the minimum average stress (versus crack length) necessary to keep the crack propagating in a brittle material for (a) two different initial flaw sizes; (b) two different initial critical stresses.

13-15. A brittle material has $E = 20 \times 10^6$ psi and $\gamma = 7.5 \times 10^{-3}$ in.-lb/in.^{2}. Assume that flaws smaller than 0.01 in. long cannot be detected. What is the maximum allowable average stress if the factor of safety must be at least 1.5?

13-16. A brittle material has $E = 150$ MPa and $\gamma = 2$ N·m/m^2. What is the requirement in flaw detection if the maximum average stress must be 30 MPa?

Sec. 13-4, 13-5

Determine the possibility of brittle or ductile fracture in Probs. 13-17 to 13-22. The members are edge-notched, and the applied uniaxial load may increase by 30% in each case. σ_{ys}, τ_{ys}, and σ_u are material properties; σ_x, σ_y, and σ_z are the maximum calculated stresses (directions of axes are as in Fig. 13-5).

13-17. $\sigma_{ys} = 46$ ksi, $\quad \tau_{ys} = 23$ ksi, $\quad \sigma_u = 70$ ksi, $\quad \sigma_x = 20$ ksi, $\quad \sigma_y = 50$ ksi, $\sigma_z = 1$ ksi

13-18. $\sigma_{ys} = 400$ MPa, $\quad \tau_{ys} = 200$ MPa, $\quad \sigma_u = 600$ MPa, $\quad \sigma_x = 80$ MPa, $\sigma_y = 350$ MPa, $\quad \sigma_z = 60$ MPa

13-19. $\sigma_{ys} = 100$ ksi, $\quad \tau_{ys} = 50$ ksi, $\quad \sigma_u = 140$ ksi, $\quad \sigma_x = 30$ ksi, $\sigma_y = 120$ ksi, $\sigma_z = 20$ ksi

13-20. $\sigma_{ys} = 200$ MPa, $\quad \tau_{ys} = 100$ MPa, $\quad \sigma_u = 280$ MPa, $\quad \sigma_x = 60$ MPa, $\sigma_y = 250$ MPa, $\quad \sigma_z = 5$ MPa

13-21. $\sigma_{ys} = 80$ ksi, $\quad \tau_{ys} = 40$ ksi, $\quad \sigma_u = 120$ ksi, $\quad \sigma_x = 15$ ksi, $\quad \sigma_y = 100$ ksi, $\sigma_z = 2$ ksi

13-22. $\sigma_{ys} = 1.2$ GPa, $\quad \tau_{ys} = 600$ MPa, $\quad \sigma_u = 1.25$ GPa, $\quad \sigma_x = 200$ MPa, $\sigma_y = 950$ MPa, $\sigma_z = 30$ MPa

Determine the ranges of allowable yield and ultimate strengths to avoid brittle fracture in edge-notched plates for which the maximum normal stresses are predicted as in Probs. 13-23 to 13-26. The loading is uniaxial.

13-23. $\sigma_x = 15$ ksi, $\quad \sigma_y = 50$ ksi, $\quad \sigma_z = 2$ ksi

13-24. $\sigma_x = 120$ MPa, $\quad \sigma_y = 500$ MPa, $\quad \sigma_z = 20$ MPa

13-25. $\sigma_x = 18$ ksi, $\sigma_y = 50$ ksi, $\sigma_z = 15$ ksi

13-26. $\sigma_x = 180$ MPa, $\sigma_y = 500$ MPa, $\sigma_z = 120$ MPa

13-27. $\sigma_x = 30$ ksi, $\sigma_y = 100$ ksi, $\sigma_z = 1$ ksi

13-28. $\sigma_x = 200$ MPa, $\sigma_y = 800$ MPa, $\sigma_z = 180$ MPa

Sec. 13-6

Determine the largest allowable flaw size in Probs. 13-29 to 13-32. The first row gives the material and its properties.

13-29. 4340 steel: $\sigma_{ys} = 210$ ksi, $\sigma_u = 260$ ksi, $K_{I_c} = 40$ ksi $\sqrt{\text{in.}}$
Stress to be applied = 200 ksi.

13-30. 4340 steel: $\sigma_{ys} = 1.45$ GPa, $\sigma_u = 1.8$ GPa, $K_{I_c} = 45$ MPa $\sqrt{\text{m}}$
Stress to be applied = 200 MPa.

13-31. 7075-T6 aluminum alloy: $\sigma_{ys} = 70$ ksi, $\sigma_u = 80$ ksi, $K_{I_c} = 30$ ksi $\sqrt{\text{in.}}$
Stress to be applied = 10 ksi.

13-32. 7075-T6 aluminum alloy: $\sigma_{ys} = 500$ MPa, $\sigma_u = 550$ MPa,
$K_{I_c} = 30$ MN/m$^{3/2}$
Stress to be applied = 300 MPa.

Determine the maximum allowable average stress in Probs. 13-33 to 13-36. The first row gives the material and its properties.

13-33. 5 Cr-Mo-V steel: $\sigma_u = 250$ ksi, $K_{I_c} = 40$ ksi $\sqrt{\text{in.}}$
Smallest detectable flaw = 0.01 in.

13-34. Maraging 300 steel: $\sigma_u = 1.85$ GPa, $K_{I_c} = 90$ MPa $\sqrt{\text{m}}$
Largest allowable flaw = 1 cm.

13-35. Ti-6Al-6V-2Sn titanium alloy: $\sigma_{ys} = 175$ ksi, $K_{I_c} = 32$ ksi $\sqrt{\text{in.}}$
Largest allowable flaw = 1 in.

13-36. 18 Ni maraging steel: $\sigma_{ys} = 1.9$ GPa, $\sigma_u = 2$ GPa, $K_{I_c} = 57$ MN/m$^{3/2}$
Smallest detectable flaw = 3 mm.

14

COMBINED STRESSES

Some members are subjected to various combinations of axial, bending, and torsional loads. Problems like these can be solved by appropriate superposition of the stresses that are calculated for each of the loads at the same point in the material. There are no new methods of analysis that have to be considered at this stage. A little experience is helpful in selecting the most reasonable points in the member where the stresses should be determined and in doing the superpositions. The following examples provide experience for solving a large variety of problems that involve more than one kind of loading.

14-1. COMBINATIONS OF NORMAL STRESSES

A common problem is when axial and bending loads are acting simultaneously on a member. The key to solving such problems is that the normal stress at any given point is the sum of the normal stresses caused by bending and axial loading. Each of these is calculated while ignoring the other.

EXAMPLE 14-1

Consider a bicycle pedal as shown in Fig. 14-1. Assume that the given loading is the worst possible. What is the maximum tensile stress at the section *A-B* if the area there is 0.5 in. $\times$ 0.5 in.?

Solution

The statically equivalent system that is best to consider has a downward force of 200 lb at point *C* and a clockwise moment of 800 in.-lb about *C*. The normal

Drill dynamometer used to relate drill performance to drilling variables; this is similar to a load cell with strain gages; capacity of device shown: 2200 lb axial thrust, 220 ft-lb torque. *Courtesy Professor Marvin DeVries, University of Wisconsin, Madison, Wis.*

Section through coal pulverizer, showing 7-in. diameter shaft that fractured, repaired itself by friction welding, and fractured a second time. *By permission, from* Metals Handbook, *Volume 10. Copyright American Society for Metals, 1975.*

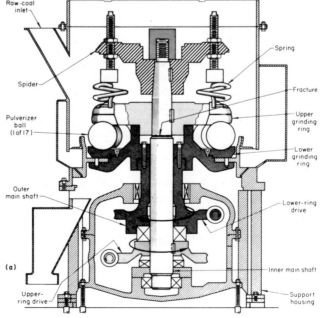

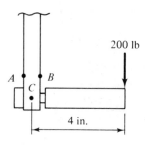

Fig. 14-1. Example 14-1.

stress at both *A* and *B* is

$$\sigma_{\text{axial}} = \frac{P}{A} = \frac{200}{(0.5)^2} = 800 \text{ psi} \qquad \text{(tensile)}$$

The normal stress at *A* caused by bending is

$$\sigma_{A\,\text{bending}} = \frac{Mc}{I} = \frac{800(0.25)}{\frac{1}{12}(0.5)^4} = 38,400 \qquad \text{(compressive)}$$

while at point *B* it is

$$\sigma_{B\,\text{bending}} = 38,400 \qquad \text{(tensile)}$$

The total stresses at *A* and *B* are

$$\sigma_A = 800 - 38,400 = -37,600 \text{ psi} \qquad \text{(compressive)}$$
$$\sigma_B = 800 + 38,400 = 39,200 \text{ psi} \qquad \text{(tensile)}$$

The maximum tensile stress is 39,200 psi at point *B*, provided the material does not yield.

EXAMPLE 14-2

A water tower is designed for a town in an earthquake zone. The lenticular container's center is 40 m above ground. The leg is 1 m in diameter. The full container's weight is 10 MN. In comparison, the leg's weight is negligible. The maximum horizontal acceleration during a severe earthquake is 3 g's. Is it sufficient to have a 2-cm wall thickness in the leg if the allowable tensile or compressive stresses are 800 MPa? Ignore the effect of wind.

Solution

Treating the problem as one of statics, the forces shown in Fig. 14-2 must be considered. The axial stresses are the same everywhere:

$$\sigma_{\text{axial}} = \frac{-10^7}{2\pi(0.5)(0.02)} = -159 \text{ MPa} \qquad \text{(compressive)}$$

The stresses caused by bending are largest at *A* and *B*. With $M_{\text{max}} = 1.2$ GN·m,

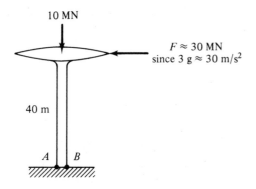

Fig. 14-2. Example 14-2.

$$\sigma_{A\,bending} = \frac{Mc}{I} = -\frac{(-1.2 \times 10^9)(0.5)}{\pi(0.5)^3(0.02)} = -76.4\ \text{GPa} \quad (\text{compressive})$$

$$\sigma_{B\,bending} = 76.4\ \text{GPa} \quad (\text{tensile})$$

The total stresses are

$$\sigma_A = -76.6\ \text{GPa} \quad (\text{compressive})$$

$$\sigma_B = 76.2\ \text{GPa} \quad (\text{tensile})$$

Clearly, the 2-cm wall thickness is not adequate.

> Simulation of seismic events on structures; the spectrum response is for a large graphite column of a nuclear reactor. *Charles Berriaud and Yves Tigeot*, "*Paraseismic Testing of Model Structures*," Closed Loop, *3, No. 6* (*1973*), *17–21*.

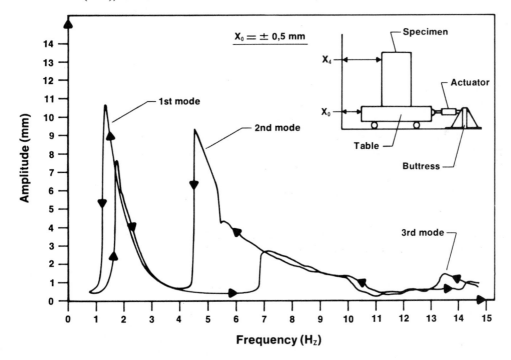

14-2. COMBINATIONS OF SHEAR STRESSES

Sometimes the shear stresses caused by more than one kind of loading must be considered. They always must be superimposed to have a single resultant shear stress on any given small plane area.

EXAMPLE 14-3

A strong filament composite is considered for the shaft of a new golf club. Since such composites are relatively weak in shear, the normal stresses caused by bending are ignored and only shear is considered. Outline a procedure in as much detail as possible to determine the maximum shear stress caused by the equivalent static load P in the shaft (Fig. 14-3a). Consider only direct shear by P and torsion; d is the distance between the shaft's center line and the line of action of P.

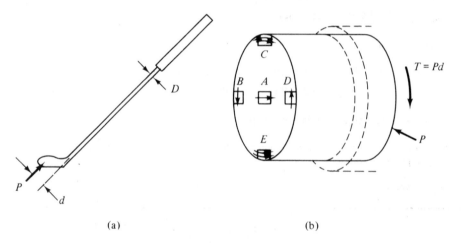

(a)   (b)

Fig. 14-3. Example 14-3.

Solution

Figure 14-3b shows a statically equivalent system of loading transferred to the hosel (the junction of the shaft and the head of the club). There is a shear force P and a torque $T = Pd$. Also shown is a cross section of the shaft with the shear stresses transmitted by the rest of the shaft to maintain equilibrium. One component of the shear stress is always counteracting P, and its magnitude is

$$\tau_P = \frac{PQ}{It}$$

It is maximum at elements A, C, and E and is zero at B and D (because Q vanishes). The other component of the shear stress always opposes the torque:

$$\tau_T = \frac{Tc}{J} = \frac{Pdc}{J} \qquad \text{at } B, C, D, \text{ and } E \text{ (it is zero at point } A)$$

Element E is the most severely stressed:

$$\tau_E = \frac{PQ}{It} + \frac{Pdc}{J} = \frac{16P}{3\pi D^2} + \frac{16Pd}{\pi D^3}$$

The major assumptions are that the shaft is solid and behaves as an elastic, homogeneous material and that the dynamic loading can be replaced by an equivalent static loading (ignoring shock waves). In the prediction of failure the orientation of the filaments must also be considered.

EXAMPLE 14-4

The vibration-isolator springs of an instrument are helical coils with $R = 1$ cm and $d = 1$ mm according to Fig. 14-4a. P is 10 N on each spring. What is the required strength of the spring metal? Generalize the answer to make it useful for analyzing other similar springs.

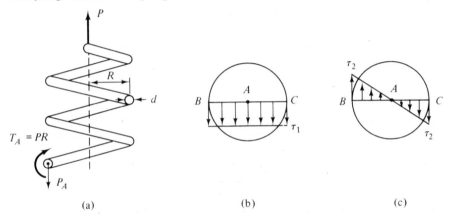

Fig. 14-4. Example 14-4.

Solution

Consider a partial free-body diagram of the spring as in Fig. 14-4a. A is a plane perpendicular to the wire anywhere along the coil. For equilibrium, there is a direct shear of $P_A = P$ and a torque of $T_A = PR$ acting on section A. Assume that P_A is the integral of a uniformly distributed shear stress τ_1 on A; τ_1 is shown for the line BC in Fig. 14-4b. T_A causes a linearly varying shear stress with a maximum of τ_2 as shown in Fig. 14-4c (assuming entirely elastic behavior).

The resultant shear stress at any point is always the vector sum of the local values of the two distributions. This is easiest to do for the line BC: $\tau_B = \tau_1 - \tau_2$, and $\tau_C = \tau_1 + \tau_2$. At any other place, τ_1 is still perpendicular to BC, but τ_2 (or lesser stresses from the nonuniform distribution) is not. It is found that τ_C (which is nearest to the center of the coil) is the largest resultant on the section A. Thus

$$\tau_{\max} = \tau_1 + \tau_2 = \frac{4P}{\pi d^2} + \frac{16PR}{\pi d^3} = \frac{16PR}{\pi d^3}\left(1 + \frac{d}{4R}\right) \qquad (14\text{-}1)$$

This formula is based on the torsion formula for straight members, so it is increasingly in error as d/R increases (more tightly wound spring). In the torsion of curved members it is necessary to consider the difference in fiber lengths between the inside and outside fibers (analogous to curved beams). A correction for the curvature was developed by A. M. Wahl, and this result is

$$\tau_{max} = \frac{16PR}{\pi d^3}\left(\frac{4m-1}{4m-4} + \frac{0.615}{m}\right) \tag{14-2}$$

where $m = 2R/d$, called the *spring index*.

For the given spring, Eq. 14-1 gives

$$\tau_{max} = \frac{16(10N)(0.01m)}{\pi(0.001m)^3}(1.025) = 522 \text{ MPa}$$

From Eq. 14-2, with $m = 20$,

$$\tau_{max} = 545 \text{ MPa}$$

High-strength spring materials are available to work at such a stress level or even higher.

EXAMPLE 14-5

A broken tubular member of a farm machine is repaired simply by shrink-fitting another tube over the broken pieces. The sleeve has an inside diameter of 3 cm, and it covers each broken end of the original member over a length of 4 cm. A torsion test is performed and it is found that a 1-kN force acting on a moment arm of 0.8 m does not twist the member in the sleeve. What is the tensile force that the repaired member can be expected to survive? What is the tensile force that can be applied if a torque equal to half the test torque is applied simultaneously?

Solution

The load is transmitted by friction, and the shear stress is assumed to be constant over the area of contact between the sleeve and the tube. The torque of $T = 800$ N·m causes a circumferential shear stress:

$$\tau_c = \frac{T}{rA} = \frac{800}{(0.015)2\pi(0.015)(0.04)} = 14.15 \text{ MPa}$$

If there is no torque but there is an axial force P, the axial shear stress τ_a is taken as equal to τ_c (τ_c originally obtained in the absence of P), so

$$P = \tau_c A = 5.33 \times 10^4 \text{ N} \qquad (T = 0)$$

For a torque of $T_1 = 400$ N·m, $\tau_{c_1} = 7.07$ MPa. An elemental area dA (Fig. 14-5) can have a force dF acting on it:

$$dF = \tau_{max}dA = (14.15 \text{ MPa})dA$$

Characteristic and endurance testing of spring with servo-controlled equipment; the hydraulic actuator and load cell are on the top. *J. P. Norman, "Testing at British Railways Research Department,"* Closed Loop, **2**, *No. 6 (1970), 2–7.*

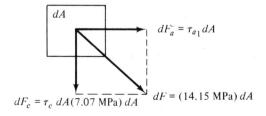

$dF_a = \tau_{a_1} dA$

$dF_c = \tau_c dA(7.07 \text{ MPa}) dA$ $dF = (14.15 \text{ MPa}) dA$

Fig. 14-5. Example 14-5.

The allowable tensile force P_1 in the presence of T_1 is obtained after determining the allowable axial shear stress, τ_{a_1}.

$$dF_a = \sqrt{dF^2 - dF_c^2} = \sqrt{(14.15 \text{ MPa})^2(dA)^2 - (7.07 \text{ MPa})^2 dA^2}$$
$$= (12.26 \text{ MPa})dA = \tau_{a_1}dA$$
$$P_1 = \tau_{a_1}A = (12.26)2\pi(0.015)(0.04) = 46.2 \text{ kN}$$

14-3. COMBINED AXIAL, BENDING, AND TORSIONAL LOADING

The methods used in Sections 14-1 and 14-2 can be readily extended to solve problems involving three simultaneous loads.

EXAMPLE 14-6

Consider an eyebolt A of the radio tower mentioned in Prob. 5-18. The shaft of the bolt is 2.5 in. in diameter. Assume that the following loads are acting simultaneously on the bolt:

Axial: 100,000 lb in tension, caused by the pretension of the guy plus a strong wind.

Bending: 15,000 in.-lb, caused by misalignment of the bolt head on the insulator assembly.

Torsion: 10,000 in.-lb, caused by the tendency of the stranded guy wires to unwind under the axial load.

Determine the maximum normal stress in the bolt.

Solution

The individual stresses caused by these loads must be determined first. Assuming elastic behavior and that the cross-sectional area of the bolt is 5 in², the following are the maximum stresses:

$$\sigma_A = \frac{P}{A} = \frac{100,000}{5} = 20 \text{ ksi}$$

$$\sigma_B = \frac{Mc}{I} = \frac{15,000(1.25)}{(\pi/4)(1.25)^4} = 9.8 \text{ ksi}$$

$$\tau_T = \frac{Tr}{J} = \frac{10,000(1.25)}{(\pi/2)(1.25)^4} = 3.3 \text{ ksi}$$

It should be noted that σ_A is constant over a cross-sectional area, σ_B is valid only for a small region at the distance of the radius away from the neutral axis, and τ_T is valid at every point on the outer surface of the bolt's shaft. Thus, the most severe state of stress is where the bending causes the largest stress. The problem is then simplified to a small rectangular element being stressed by a uniaxial normal stress of $\sigma_A + \sigma_B = 29.8$ ksi and a shear stress of 3.3 ksi. The center of Mohr's circle is at 14.9 ksi, and its radius is 15.3 ksi. The maximum tensile stress is 30.2 ksi.

This analysis does not take into account that there is at least one region of stress concentration in the bolt, in the fillet area under the head. This must be considered separately for the three kinds of loading.

14-4. COMBINED LOADS INCLUDING INTERNAL PRESSURE

Two interesting examples are presented here to show the wide range of problems where internal pressure plays a role in combination with other loads. The complexities of some of these far exceed those of the conventional pressure vessels.

EXAMPLE 14-7

Gigantic superconducting magnets are planned for the storage of electric energy in two special situations. Magnets with about 1-MW·h capacity are needed in conjunction with pulsed fusion reactors to smooth out the large fluctuations of energy that occur at the rate of about 1 Hz. Other magnets could be used to take care of short-range variations in customer demand (such as day to night). A 10-GW·h magnet is designed, for example, to deliver electricity for up to 10 hours to an industrialized area with 5 million inhabitants. Such a magnet would be about 45 m high, with a coil radius of 150 m.

Conceptual design of a superconductive energy storage facility.

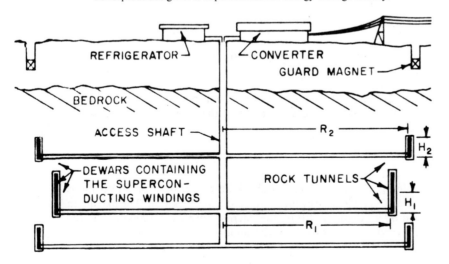

The planned conductor is a composite structure as shown in the cross-sectional view in Fig. 14-6a. There are two major loads on it. The maximum compressive stress in the stack of conductors is 30 MPa. There is a large magnetic force acting to expand the rings of conductors. This magnetic force can be represented as an internal pressure (Fig. 14-6b); its magnitude is 10 MPa. Since yielding in the conductor must be prevented, the absolute maximum shear stress in it must be known. Assume that a ring is similar to a thin-walled pressure vessel without end plates.

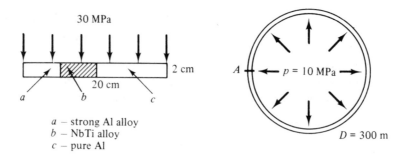

(a) cross section of ring at point *A* in top view.   (b) top view of ring.

Fig. 14-6. Example 14-7.

Solution

The magnetic pressure creates a circumferential stress in a ring, which is calculated as for a pressure vessel:

$$\sigma_c = \frac{pD}{2t} = \frac{10^7(150)}{0.2} = 7.5 \text{ GPa}$$

The magnetic field causes no stress perpendicular to the plane of the ring, so from Mohr's circle

$$\tau_{max} = \frac{\sigma_1 - \sigma_2}{2} = \frac{7500 - (-30)}{2} = 3.765 \text{ GPa}$$

Both σ_c and τ_{max} are much too high for the required materials, so special steps must be taken in the design to reduce the effect of the magnetic pressure, which cannot be altered.

EXAMPLE 14-8

Each landing gear (may be called a shock absorber) of a large airplane is designed to have a 20-in. diameter cylinder with a 0.5-in. thick wall. The maximum compressive load on a cylinder is 10^6 lb, the internal pressure is 1000 psi, and the bending moment is 10^5 ft-lb. The three maximum loads may occur simultaneously.

Is the cylinder properly designed if its tensile strength is 180 ksi and its shear strength is 100 ksi? Assume that it has to survive only one of these worst possible combinations of loads. What happens if the internal pressure is lost?

Solution

The axial compression causes a stress of $\sigma_a = 31{,}830$ psi everywhere in the cylinder. The bending causes a compressive stress at an arbitrary point A:

$$\sigma_{b_A} = \frac{Mc}{I} = \frac{(12 \times 10^5)(10)}{\pi(10)^3(0.5)} = 7640 \text{ psi} \qquad \text{(compressive)}$$

The bending causes a tensile stress of equal magnitude at point B, which is opposite to point A on the cylinder. The internal pressure causes a tensile axial (longitudinal) stress at both A and B:

$$\sigma_{l_A} = \sigma_{l_B} = \frac{pD}{4t} = \frac{10^3(20)}{4(0.5)} = 10{,}000 \text{ psi}$$

It also causes a circumferential stress at both A and B:

$$\sigma_{c_A} = \sigma_{c_B} = \frac{pD}{2t} = 20{,}000 \text{ psi}$$

Adding all stresses together results in the two elements shown in Fig. 14-7. Element A is loaded more severely. Since there is no shear stress, these are principal stresses.

$$\sigma_1 = 20{,}000 \text{ psi}, \qquad \sigma_2 = -29{,}470 \text{ psi}, \qquad \sigma_3 = 0, \qquad \tau_{max} = 24{,}740 \text{ psi}$$

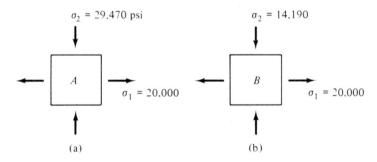

Fig. 14-7. Example 14-8.

The axial stress becomes larger if the internal pressure is lost:

$$\sigma_1 = 0, \qquad \sigma_2 = -39{,}470 \text{ psi}, \qquad \sigma_3 = 0, \qquad \tau_{max} = 19{,}740 \text{ psi}$$

The landing gear is conservatively designed for the most severe loading.

14-5. PROBLEMS

Sec. 14-1

Determine the maximum normal stresses in Probs. 14-1 to 14-8 by adding the axial load F to the bending load shown.

14-1. Figure P7-5; $F = 1000$ lb   **14-2.** Figure P7-8; $F = 4$ kN

14-3. Figure P7-19; $F = -500$ lb   **14-4.** Figure P7-22; $F = -20$ N

14-5. Figure P7-27; $F = 10,000$ lb   **14-6.** Figure P7-36; $F = -2$ kN

*14-7. Figure P7-42; $F = 30$ kN   ***14-8.** Figure P7-43; $F = -1000$ lb

Sec. 14-2

Determine the maximum transverse shear stress in Probs. 14-9 to 14-13 by adding the torque T (twisting the member about its axis) to the bending load shown. T is clockwise as the member is viewed from the free end.

14-9. Figure P7-7; $T = 600$ ft-lb   **14-10.** Figure P7-8; $T = 300$ N·m

14-11. Figure P7-21; $T = 1000$ ft-lb   **14-12.** Figure P7-22; $T = 0.05$ N·m

14-13. Figure P7-43; $T = 1000$ ft-lb

Sec. 14-3

Determine the maximum tensile stress in Probs. 14-14 to 14-18 by adding the axial load F to the given bending and torsional loads. T is clockwise as the member is viewed from the free end.

14-14. Figure P7-7; $T = 600$ ft-lb $F = 1000$ lb

14-15. Figure P7-8; $T = 300$ N·m $F = 6$ kN

14-16. Figure P7-21; $T = 1000$ ft-lb $F = -300$ lb

14-17. Figure P7-22; $T = 0.05$ N·m $F = -10$ N

14-18. Figure P7-43; $T = 1000$ ft-lb $F = 20,000$ lb

Sec. 14-4

In Probs. 14-19 to 14-24 a closed tube of diameter D and wall thickness t is subjected to an internal pressure p, external axial force F, bending moment M, and torque T. Determine the maximum normal stress.

*14-19. $D = 2$ in., $t = 0.05$ in., $p = 100$ psi, $F = 200$ lb, $M = 30$ ft-lb, $T = 5$ ft-lb

*14-20. $D = 6$ cm, $t = 1$ mm, $p = 100$ kPa, $F = 500$ N, $M = 80$ N·m, $T = 40$ N·m

*14-21. $D = 8$ in., $t = 0.1$ in., $p = 200$ psi, $F = -300$ lb, $M = 20$ ft-lb,
 $T = 30$ ft-lb

*14-22. $D = 10$ cm, $t = 2$ mm, $p = 300$ kPa, $F = -500$ N, $M = 60$ N·m,
 $T = 40$ N·m

*14-23. $D = 6$ in., $t = 0.03$ in., $p = 50$ psi, $F = \pm 100$ lb, $M = 60$ ft-lb,
 $T = 30$ ft-lb

*14-24. $D = 2$ cm, $t = 0.5$ mm, $p = 20$ kPa, $F = \pm 20$ N, $M = 0.6$ N·m,
 $T = 0.1$ N·m

15

PROJECT PROBLEMS

You should be proud if you have learned most of the concepts presented in the preceding chapters and if you can perform the basic computations in the various problem areas. Think of it this way: You have learned several things that could be useful to a carpenter, a plumber, a farmer, a garage mechanic; you could also be on your way as a designer or inventor of new devices, machines, vehicles, and structures.

Of course, there is much more that can be learned or must be learned in some cases than is presented here. The important thing is that you should start standing on your own feet and accept the challenge of solving real problems. This involves much more than using a few formulas and coming up with *the* right answer to each question. Real problems are seldom of this kind. Life would be rather simple (and rather boring for many of us) if there were always a single correct solution to each problem. The excitement and challenge of working on real problems is that they require intelligent assumptions, approximations, critical thinking, judgment, and choice.

You should realize that perfection in solving real problems is a very elusive thing. One does not suddenly become an expert in strength of materials or in any other area. The problems in this chapter allow you to practice your general skills in solving problems, with the emphasis on strength of materials. It is not expected that anybody alive could just breeze through all of these without any stumbling and fumbling and making mistakes. The idea is to work on a few of these. Some of the questions will be easy to answer, to be sure, others could be difficult for anybody. Nevertheless, much can be done with what you should know at this stage.

Many of the project problems are sufficiently taxing as they are; however, more could be done with them. The students should strive to define their own questions and problems (and solve them) relevant to a project under consideration. The ability to come up with reasonable questions is precious.

Imagine that you are an assistant of Archimedes and are developing the equipment that he dreamed up for the destruction of Roman ships (Fig. 15-1). There are many technical problems to solve, but for the time being you are concerned with the ropes. Soon it becomes clear that this is a complex matter by itself. The ships are estimated to weigh 100 tons (200,000 lb), which is quite a weight to handle in a hurry and under fire. There are many ropes of different sizes available, but only a small one has been tested so far. This rope, 0.5 in. in diameter, has an average strength of 2000 lb. There are several questions that come to mind:

1. What is the minimum strength of a rope that could be used to cause substantial damage to a Roman ship?
2. Is it best to aim for a single, heavy rope or several smaller ropes with separate hooks?
3. Could the size of the pulley be a factor in rating a rope's effective strength?

Fig. 15-1. Artist's concept of a machine invented by Archimedes for the defense of Syracuse.

If yes, how could one choose the most appropriate pulley for a given rope or vice versa?

4. How would age and repeated loading affect the strength of a rope? Should the size of the rope be considered also in these respects?

5. Whether one or more ropes are used at a time, the tensions in them must not exceed certain predetermined levels. What factor of safety is reasonable for the ropes?

6. Some people consider ropes old-fashioned. They talk about using copper (annealed, soft; or cold-worked, hard) or bronze or iron. Could these be superior to ropes in some form? If yes, how much ductility is desirable?

7. The elastic modulus of the metals is much higher than that of the ropes. Is this desirable when trying to cause damage to the enemy's ships?

8. Proof tests involve the application of loads that exceed somewhat the expected service loads to members. The idea is that the member can go into service if it did not fail during the proof test; the probability of failure in service is minimized. Are proof tests desirable for the ropes or chains or wires used in the defense of Syracuse? What is the general value of proof tests?

9. If a pulley with a given diameter has been chosen for a certain rope, can the same pulley be used safely for an iron wire that has the same tensile strength (force) as the rope?

10. Analyze the tensile strength of an arbitrary, circular link of chain.

PROBLEM B

Imagine that you are Belisatius, the chief engineer appointed by Ptolemaios IV Philopator who reigned in Egypt from 221 to 205 B.C. Under orders from your king, you are designing the largest war galley ever conceived by man. The ship is to be 420 ft long and 60 ft wide and rowed by 4000 men (free, paid workers). The challenge of this design is fantastic. For example, it is necessary to invent the dry dock to handle the monster ship. New problems come up every day. A huge problem is the propulsion with oars. The largest oars in the upper bank must be 57 ft long with 40 men pulling each one (this is why some people call the ship a *fortier*). The schematic arrangement of a large oar is shown in Fig. 15-2. It is of

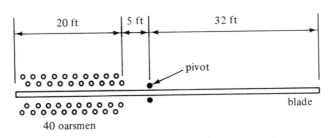

Fig. 15-2. Schematics of largest oar in war galley.

utmost importance that each oar have the largest possible ratio of strength to weight. Many questions arise concerning the large oars:

1. What is the most severe loading on an oar?
2. To find the first, approximate idea of size, assume that the oar is a uniform cylinder. What is the required minimum diameter if the tensile strength of the wood is 8000 psi?
3. Would it be reasonable to have a nonuniform cross-sectional area? If yes, what would be an acceptable variation of diameter along the length of the oar?
4. Is a circular cross section the best?
5. It may be a good idea to use counterweights of lead to help the oarsmen. How would a lead counterweight in an oar affect the answers to the preceding questions?
6. Wood is strong in tension along the grains but relatively weak in shear parallel to the grains. There is shear in torsion, so the possibility of torsion of large oars must be considered. If there is torsion, estimate the largest shear stress in the oar mentioned in Question 2.
7. Is there sufficient reason to try to make at least a part of each oar hollow?
8. Is it a good idea to make laminated oars by nailing together thin boards from different trees?
9. Could copper, bronze, or iron be used with wood somehow to make superior composite oars?
10. Could metal be used somehow to prestress a wooden oar so that the stresses in the wood during pulling are minimized?

PROBLEM C

The horizon scanner in a satellite tracking station failed unexpectedly because of a fractured torsional pendulum. The member was made from a complex alloy, similar to those used in watch springs (40Co-20Cr-15Ni-7Mo-2Mn-0.15C-0.04Be-Fe), for which the vendor had claimed $\sigma_u = 2.2$ GPa, $\sigma_y = 1.65$ GPa, $\tau_u = 1.25$ GPa, $\tau_{ys} = 900$ MPa. It was a solid cylindrical filament (fixed at the top; a small disk is at the bottom) with a diameter of 0.51 mm and a length of 6 cm. The relative twist of the ends was $\pm 15°$ from the rest position at the rate of 4 Hz. Could the pendulum have failed if the axial load was zero? If not, what is the axial load that would make the torsional oscillations critical after a year of continuous operation? Assume that $\tau_{max} = 400$ MPa can be repeated 10^8 times until failure occurs (torsion alone).

PROBLEM D

Suppose you have started a small business of making radio and television antennas. The major claim you plan to make for your products is their light weight combined with strength and unsurpassed durability under all possible environmental conditions. Calculations and tests are necessary to be able to back your claim of excellent products in a competitive market. The first antenna you want to work on is the simple dipole shown in Fig. 15-3. The two metal rods (*L*) are held in a horizontal position by the insulating block *A*. Assume that small

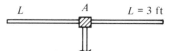

Fig. **15-3.** Dipole antenna.

deflections of the rods, their cross-sectional shapes, and the material within their metallic surface do not affect the transmission of signals. The following questions are raised during the analysis of the dipole antenna:

1. What is the simplest free-body diagram that can be used?
2. What is the simplest cross-sectional shape of the rods that is worth considering?
3. Estimate the maximum stress and deflection of the rods in a 100-mph wind. Choose a material that seems reasonable and assume a simple geometry.
4. Redo Question 3 with the loading on the antenna caused by ice weighing 10 lb/ft.
5. What is the worst possible combined effect of wind and ice on the antenna using the given numbers?
6. What is the effect of a 2-lb bird landing on the antenna? Consider only the maximum tensile stress.
7. What is the best cross-sectional area of the rods, considering the worst possible loading on the antenna? Answer this
(a) disregarding the potential problems and cost of making the rods.
(b) keeping in mind the problems and cost of fabrication.
8. Can materials of different mechanical properties be combined to produce a superior antenna?

PROBLEM E

Suppose a friend has a 20-ft motorboat that he would like to convert into a hydrofoil boat. He asks you to help him, but at first you don't have a clear idea of such boats (very few students of engineering do, for some reason). He explains that a hydrofoil boat is a boat on wings. There are various ways of arranging the wings. Some of these are shown in Fig. 15-4 and the photographs. The hull sits in the water when the boat is not moving. The boat rises gradually as its speed increases, and eventually the hull is completely out of the water. The wings that lift the boat remain below the surface and can be very small: Water is 900 times denser than air, so a wing area has 900 times the lift it would have in air. Now you are becoming quite interested in the project and accept the following tasks to help your friend:

Hydrofoil boats of various design.
Baker Manufacturing Co., Evansville, Wis.

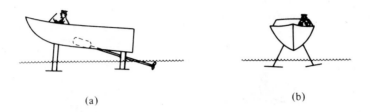

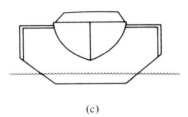

Fig. 15-4. Hydrofoil boats.

1.   Design a minimally adequate solid drive shaft if the motor can deliver 135 hp at 1000 rpm. Power equals torque times the angular velocity:

$$\text{Power} = \text{ft-lb rad/min}$$
$$1 \text{ hp} = 33,000 \text{ ft-lb/min}$$

Consider torsion only. The maximum shear stress is 20 ksi.

2.   What is the cross section of a long shaft for minimum weight if the maximum diameter is 3 in. and the maximum shear stress is 20 ksi? Consider torsion only.

3.   Will the axial thrust of the propeller make the solutions to Questions 1 and 2 conservative or not?

4.   What kind of stress would be caused by a poorly balanced propeller?

PROBLEM F

Suppose your school has obtained a used wind tunnel. It is a good wind tunnel, but there is no instrumentation with it. You want to put models of airplanes and cars in the tunnel and measure the lift and drag forces for them. Design instrumentation using strain gages that will enable you to measure each force in the range of 0 to 30 N. Assume that the strain gages should not be strained to more than 3%.

Model of racing car mounted in instrumented wind tunnel. *Photograph by J. Dreger, University of Wisconsin, Madison, Wis.*

PROBLEM G

Suppose you are an amateur radio operator and want to install a tower with an antenna. You want to maximize the ratio of the tower's height to its cost. Here are several questions that you may want to answer before embarking on the construction:

1. Consider a smooth, cylindrical tube for the tower. How much extra height could be gained by attaching three guy wires to the tower at its middle? Thus, compare the relative heights of the towers in Fig. 15-5a and b. The material and cross-sectional shapes are the same in the two cases.

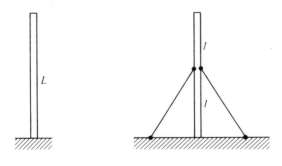

(a)   (b)   **Fig. 15-5.** Antenna towers.

2.   The adding of guy wires as in Fig. 15-5b would allow changing the end condition at the bottom of the tower. Would you change the end condition from that shown in Fig. 15-5a?

3.   Where is it best to attach the single set of guy wires to the tower? Assume that wind causes the major load, which is uniformly distributed.

4.   Could a thin-walled, cylindrical tube be strengthened temporarily by pressurizing the air inside it? If yes, extra strength could be provided during strong winds and ice storms. How would you analyze the problem if internal pressure had to be included?

PROBLEM H

An artificial heart valve of flexible plastic material must be designed for an uninterrupted service life of 80 years. It is essentially a thin, circular disk that is 2.5 cm in diameter and 1 mm thick and firmly attached over a 1-cm length on one side. Assume that the valve is a cantilever attached to an ideally rigid wall. The pressure that causes the valve to open is uniformly distributed over the disk. The valve rises to a maximum of about 1 cm at its free end (opposite to its fixed side) during each heartbeat with respect to its plane configuration. The only reliable test data for the material are the following: in a tension test, the maximum stress is 35 MPa, the maximum strain is 0.08; in a fatigue test, a peak strain of 0.01 causes failure in 1200 cycles. Can this heart valve be installed in a young patient with some confidence that it will not fail?

PROBLEM I

The best creative ideas of artists, engineers, and scientists are often so abstract that much further thought and labor may be necessary before the ideas can materialize. The creative process can also occur in the opposite direction, however: One may contemplate a readily available object until a new purpose for it becomes evident. Consider an exercise of creativity of the latter kind. You contemplate strain gages. You know how they are used normally, but you are thinking of new possibilities. Suppose that one of the ideas is to make a new kind of burglar alarm using strain gages. Can this be done? If yes, what is the rough outline of your system? If not, why is there no chance to make such an alarm? Assume that with the best strain gages and instrumentation the smallest strain that you can measure is 10 μin./in. The maximum strain that can be imposed on a standard gage is 3%.

PROBLEM J

A new test fixture is needed in a precision machine shop. It should consist of a spring-loaded dial gage extensometer mounted on an L-shaped fixture. The long arm of the fixture should be 5 ft long; the short arm, 5 in. The extensometer is to be mounted at the free end of the short arm. The resolution of the gage is 0.0001 in., so it is desired to limit the maximum deflection of the fixture at the gage to 0.00005 in. Design a fixture that satisfies this condition if the largest load on the dial gage is 0.5 lb and it acts perpendicular to the plane of the fixture (Fig. 15-6). The material suggested for the fixture is steel with $E = 30 \times 10^6$ psi and $G = 12 \times 10^6$ psi. Assume that the base of the fixture is firmly connected to an ideally rigid structure.

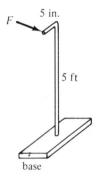

Fig. 15-6. Test fixture.

PROBLEM K

In the future there may be severe shortages of important raw materials. Imagine that an assignment you receive is to reduce the amount of material used in the spokes of bicycle wheels. The saving expected in each bicycle is very small, of course, but there are millions of bicycles. Describe your best ideas relevant to this problem. Estimate the maximum percentage of weight of spokes in a wheel that can be saved, if any.

This is a much more complex problem than it appears to most people. Use simplifications such as assuming smooth roads, ideally rigid wheel rims, and constant speeds. Do take a careful look at standard spokes, and establish the load spectrum on them for a complete revolution of the wheel. Consider possible changes in the geometry of spokes and changing the material on the basis of Table 4 in Appendix C.

PROBLEM L

Imagine that you are an assistant of Leonardo da Vinci while he is working on the design of his celebrated equestrian statue. The horse is to be in a rearing position, and your assignment from the master is to make sure that the critical parts, the lower legs (Fig. 15-7), are sufficiently strong. The sculpture is to weigh about 200,000 lb. The minimum cross-sectional area of a leg can be approximated by an ellipse, 4 in. × 8 in. There is much to do for those who want to be followers of Leonardo da Vinci:

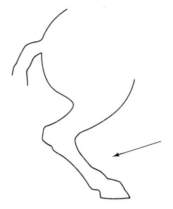

Fig. 15-7. Critical part of equestrian statue.

Only surviving model of Leonardo da Vinci's equestrian statue that is believed to have been made by him (Fine Arts Museum, Budapest). *Photograph by R. Sandor.*

1. The master wants his apprentices to demonstrate a natural ability in their art and science. What are the important mechanical properties of bronze, and what are the approximate (guessed) values of these without relying on test data?

2.   What is the simplest test that will be satisfactory to determine the properties of bronze?

3.   What is the least severe load that can be expected at the minimum section of a leg? How much more severe could the load be than this if practical situations are considered?

4.   Will a solid bronze leg be more than strong enough? If yes, how much bronze could be saved in the cross section?

5.   The statue should be designed to stand for thousands of years. What kind of environmental and time-dependent effects must be considered?

6.   What is the simplest improvement that can be made over a solid bronze leg? The external dimensions of the leg are fixed.

7.   The full load will be applied to the legs slowly and cautiously as the supporting structure of the statue is removed gradually. Could the results of measurements during this process indicate that perhaps not everything is as planned? The technology of measurements in the 15th century is to be considered.

8.   If the results of measurements do indicate that something is becoming critical, are there any countermeasures with which one could avoid starting the casting all over?

9.   Could certain residual stresses at the critical parts of the legs be particularly advantageous or undesirable?

10.   Large castings are difficult to make without having some internal flaws. What is the effect of possible random flaws in the legs? Is there any chance of detecting internal flaws using 15th-century technology?

PROBLEM M

Suppose you are involved with the design of a tool that will be taken on a space mission. It does not matter what material the tool is made of, but its weight must be minimal. The shape of the tool is given in Fig. 15-8. The maximum load is

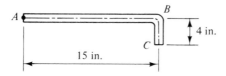

Fig. 15-8. Tool for space mission.

50 lb applied at *A*, perpendicular to the plane *ABC*. Consider two approaches in the design of the tool. First, assume that the cross section is the same from *A* to *C* and that a single material is used; this is the simplest, cheapest tool. Second, use any idea whatever to reduce the weight. Compare the results with those of the simplest tool. Finally, propose a method to prevent an overly eager astronaut from applying a force of more than 50 lb to the tool.

Imagine that you are managing a materials development and research laboratory. One day you receive an interesting letter under the following letterhead:

ACE ATOMS UNLIMITED
1990 Gamma Road
Breeders Brook, Florida

Dear . . . . :

You have been recommended to us as a consultant who may be able to solve a puzzling problem that we have. The exact nature of our work must remain a secret, but in general it involves subjecting large masses of material to radiation for a short period of time. The material to be irradiated is lowered into the pile in 500-lb batches on a steel alloy wire. The wire wraps around a 10-in. diameter drum that is rotated by remote control. The change of batches takes a relatively short time; the full load is on the wire 80% of the time, day and night.

The manufacturer of the wire provided the following information for this material at the operating temperature, which is about 1000°F.

Table of properties at 1000°F

Modulus of elasticity	20×10^6 psi
0.2% offset yield strength	90,000 psi
Ultimate strength	120,000 psi
Percent elongation	20%
Creep life at stress of 60,000 psi	1000 hr

Failure of the wire causes considerable expense and loss of production since the pile must be shut down and allowed to "cool off" before the repairs can be made. Our technical staff includes a number of physicists and chemists who understand basic composition of matter but do not have much experience with materials engineering. They believe that past failures of the wires were caused by creep at high temperature, prolonged loads, and excessive deformation, which was observed on the broken wires.

We cannot arbitrarily increase the amount of material in the equipment that is irradiated, but we have increased the diameter of the wire from 0.100 to 0.150 in. (same metal) to decrease the operating stress and hence increase the life.

The fact is that before this change we operated for about 30 days without a wire failure, and now we have failures in 5 days! Needless to say, we are exceedingly disturbed about this. Could you help us solve this problem? We would like to have answers to the following questions, but we would also appreciate any other relevant information or ideas.

1.   What is the basic problem?

2.   Can we learn anything by having one of our staff scientists look at the wires under a microscope?

3.   We expected the original wire (dia. = 0.1 in.) to last longer than 30 days

on the basis of information supplied by the manufacturer. What went wrong? Could the manufacturer have given us data for a different wire by mistake, or was something improper done here?

4.   Why did the larger wire have even shorter life than the 0.1-in. wire? What could we expect if we tried a 0.2-in. wire?

5.   Should we go back to using the 0.1-in. wire?

6.   This sounds ridiculous to almost everybody here, but what could we expect if we tried a wire smaller than 0.1 in. in diameter?

7.   Would a little cooling of the wire help substantially? We are not sure how we would do this without putting much more equipment into the pile, but we want to know the tendencies of behavior of the wire.

8.   Maybe we should be more specific when we order wires. What is a reasonable list of properties or conditions that we should give in writing?

9.   Are there any simple ways in which we could check the wire before putting it in the pile to make sure that it conforms to the specifications?

10.   Suppose we want to have life tests of samples of the wire under simulated operating conditions, but we want to accelerate the tests. How can we reduce testing time and obtain meaningful results?

We shall appreciate it if you can work on this problem immediately.

Hoping to hear from you soon,

William J. Bullmoose

General William J. Bullmoose, President

PROBLEM O

Suppose you are designing a small sailboat. Describe your plan for a 5-m mast in detail. Try to minimize the weight of the mast and pay particular attention to how it is to be attached to the hull. Consider environmental and time-dependent effects when you choose a material.

PROBLEM P

Imagine that you are a Peace Corps Volunteer stationed in a remote village. Suppose that a severe flood is threatening the villagers and the many refugees who fled there already from other places. Evacuation is a must, but the flood waters have just washed over the small footbridge that could have led to safety. A new bridge would have to be 30 ft long, as shown in the cross section of the stream bed (Fig. 15-9a).

A quick search produces two pieces of 4-in. diameter poles of the right length, and they are installed 2 ft apart with thin boards on them. You recommend that only one small person crosses at a time while you calculate the load-carrying ability of the bridge (if it can't take at least one person, the bridge is no good; so, might as well use it right away, but there is no point in risking fracture).

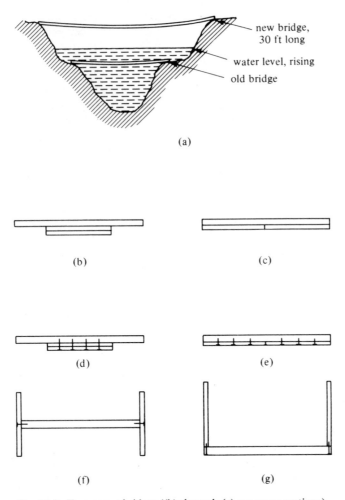

(a)

(b)   (c)

(d)   (e)

(f)   (g)

Fig. 15-9. Emergency bridges ((b) through (g) are cross sections).

A model of the problem has to be used that is valid enough and simple enough. The people moving across are moving, concentrated loads for the bridge. One idea is to determine the maximum stress in the poles if a person were at the center.

The maximum bending moment is about 15,000 in.-lb for a 165-lb person. The maximum stress in each pole is 1200 psi from Mc/I. You recall that the tensile strength of wood is about 8000 psi, so things are all right until more poles are found. But no, there is a commotion . . . one of the poles broke, and a boy fell in the water. The boy and the two halves of the pole were pulled out . . . and three more new poles have been found, too. Now, back to those numbers quickly. There is an infinite number of possible questions about emergency bridges; here are a few.

1.   What went wrong? What was the maximum stress in the poles without anybody on the bridge? Wood weighs about 0.02 lb/in³.

2.   How many people can be allowed at a time on the bridge made of four poles?

3.   To use the broken pole most effectively, where should the pieces be nailed to the others?

4.   What was the maximum deflection of the first bridge

(a)   without anybody on it?

(b)   just prior to failure? (The boy weighed about 100 lb and he was near the center of the bridge.)

5.   Should a sick man be carried on the back of another man or on a light stretcher by two men? The bridge has four poles now.

6.   Three 30-ft long flat boards have been found, too. One is 2 in. thick and 2 ft wide, and there are two 1 in. × 12 in. boards. Which of the two configurations of simply stacking these boards (cross sections shown in Fig. 15-9b and c) is best for a new bridge? Are both of them safe?

7.   Consider four configurations of nailing the three boards together as in Fig. 15-9d through g (cross sections). Which is best?

8.   After finding the best configuration for using the three boards, determine the minimum allowable spacing between people as they cross the bridge.

9.   Describe a procedure for determining the minimum number of nails (3-mm dia., $\tau_{max} = 25$ ksi) that could be used for a given configuration of the three boards. Remember that in some places nails shouldn't be wasted even in an emergency.

PROBLEM Q

Suppose an artist friend comes to you for assistance on a large project of his. He has been commissioned by a university to do an outdoor sculpture for the campus quadrangle. He has already done much work on the project and now reached the point where detailed calculations are necessary. The project and the information provided by the sculptor are described in the following.

The intended sculpture is a welded steel bird of mathematically simple form, depicted at the instant of taking off from its foundation (Fig. 15-10). The body and wings are of thin metal for lively flexibility in wind.

The metal selected is COR-TEN steel, a high-strength, low-alloy steel. It is also called a weathering steel because of its high resistance to corrosion. In normal environments it requires no paint or other protective coating. The thickness of weathering steel that erodes away in 20 years outdoors is only 0.002 to 0.004 in. It takes a few months to form its natural warm, brown patina. The wet and dry cycles in normal atmospheric exposure facilitate the development of its thick, nonporous oxide coating that inhibits further corrosion. It can be welded using all standard methods in the shop and in the field.

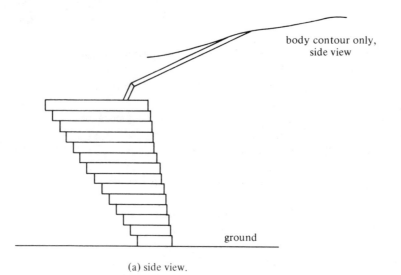

body contour only,
side view

ground

(a) side view.

(b) wing contour, front view.

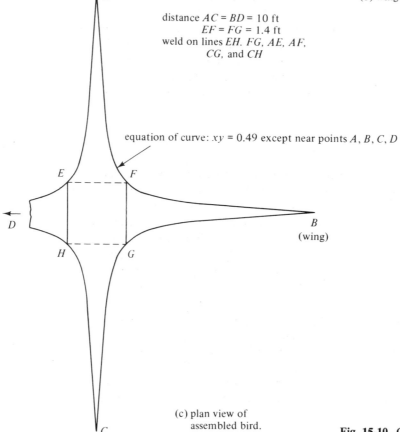

distance $AC = BD = 10$ ft
$EF = FG = 1.4$ ft
weld on lines EH. FG, AE, AF,
CG, and CH

equation of curve: $xy = 0.49$ except near points A, B, C, D

A

E F

D B
       (wing)

H G

C

(c) plan view of
assembled bird.

Fig. 15-10. Outdoor sculpture.

NOTE: a single wing piece ($BGHDEFB$)
is sandwiched between two body
pieces ($AFGCHEA$).

The mechanical properties of the steel are

Minimum yield strength	50,000 psi
Minimum tensile strength	70,000 psi
Modulus of elasticity	30×10^6 psi
Endurance limit	42,000 psi
Elongation in 2 in.	22%

The steel is available in the following forms that are relevant to the project:

Angles:	$1 \times 1 \times \frac{1}{8}$ (in.),	$W = 0.80$ lb/ft,	$I = 0.02$ in⁴
	$1\frac{1}{2} \times 1\frac{1}{2} \times \frac{1}{8}$	1.23	0.08
	$1\frac{1}{2} \times 1\frac{1}{2} \times \frac{1}{4}$	2.46	0.15
	$2 \times 2 \times \frac{1}{8}$	1.65	0.19
	$2 \times 2 \times \frac{1}{4}$	3.30	0.35
	$2\frac{1}{2} \times 2\frac{1}{2} \times \frac{1}{4}$	4.16	0.70
	$3 \times 3 \times \frac{1}{4}$	4.90	1.20

NOTE: The moment of inertia is about an axis parallel to one side and through the centroid.

I Beams:	3-in. depth,	$W = 5.7$ lb/ft,	$I = 2.5$ in⁴
	4	10.5	7.1
	5	11.0	13.0

Sheets:	0.1793 in. thick,	$W = 7.5$ lb/ft²
	0.1345	5.625
	0.1046	4.375
	0.0598	2.5
	0.0359	1.5

Your friend asks you to design the body and legs of the steel bird so that it will stand up against the elements and all but the most vicious and technologically equipped pranksters. It must be kept in mind, however, that a slender sculpture of this kind is the most elegant and appealing.

Assume the following maximum loads:

(a) Three people hanging where the legs are attached to the body.
(b) Ice on the body and wings, uniform thickness, $t = 1$ ft.
(c) Drag force on body and wings in strong wind, 30 lb.
(d) Lift force on body and wings in strong wind, 100 lb, up or down.

PROBLEM R

Suppose you go to an exhibition of hang-gliders and quickly become an enthusiast of the sport. You decide to design and build your own hang-glider, a simple Rogallo wing (Fig. 15-11). You look over the numerous Rogallo wings on exhibit carefully to learn about all their appealing features. One thing in particular

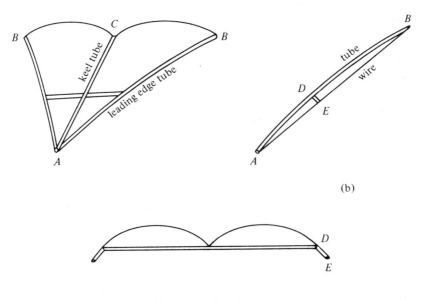

(b)

(c) cross section of wing.

Fig. 15-11. Rogallo wing.

attracts your attention. In most cases, the aluminum tubing AB is bent slightly with the aid of a peg DE and a stranded steel wire as shown in Fig. 15-11b. The peg DE is oriented downward and outward with respect to the boom AC, as shown in Fig. 15-11c.

The wing that you would like to build should be as large and light as possible. Your first major concern is the structural element AB. Can you improve on the simple aluminum tubing that you have observed? Obviously, you may use any tricks in the design as long as they affect the overall size and weight of the wing favorably.

PROBLEM S

An inventive but somewhat lazy engineer makes an addition to his stereo to control the volume remotely. It is a simple device, a 3-m long, straight copper wire with a diameter of 3 mm that is laid horizontally on a bookshelf to hide it from view. The wire is attached firmly as an axial extension to the shaft of the volume-controlling potentiometer. The knob is on the other end of the wire next to the armchair.

The device just doesn't work satisfactorily. There is an annoying problem with overshoot in the sound volume. The increase or decrease in volume is always larger than the intended change, no matter how carefully the knob on the end of the wire is turned. The following questions are relevant to the problem and to possible variations of the scheme of mechanical remote control:

1.   Why is there an overshoot when the volume is adjusted?
2.   Is there a solution to the problem without changing much of anything?
3.   Is the material or the geometry of the wire somehow involved in the problem?
4.   Would it be helpful to impose an axial tension or compression on the wire? If yes, how much?
5.   What might be a perfect solution to the problem and still have mechanical remote control? The distance to the potentiometer should remain 3m.

PROBLEM T

Suppose a painful experience in a dentist's chair has turned your thoughts to the potential problems of teeth. You have heard a lot about chemical effects on teeth (don't chew this, don't drink that), but there seems to be a general lack of information about mechanical effects, even though these must be important. So, you go to the medical library and read what you can find about teeth. It soon becomes obvious that not enough is known about teeth from the point of view of strength of materials, but what is known is quite fascinating. The structure of a tooth is complex both at the macroscopic and microscopic levels (the latter is particularly amazing). You find that the maximum force a human jaw can apply is about 300 lb. The following average data are available for enamel and dentin (material under the thin layer of enamel):

	Enamel	*Dentin*
Modulus of elasticity	5×10^6 psi	1.7×10^6 psi
Proportional limit in compression	30,000 psi	20,000 psi
Compressive strength	40,000 psi	40,000 psi
Proportional limit in tension	Not available	≈ 5000 psi
Tensile strength	Not available	5000 psi
Thermal conductivity	9 mJ$\cdot$s$^{-1}\cdot$cm$^{-1}\cdot$C^{-1}	6 mJ$\cdot$s$^{-1}\cdot$cm$^{-1}\cdot$C^{-1}
Density	2.8 g/cm^3	1.96 g/cm^3

You also find some stress versus strain data (Fig. 15-12) for tooth tissues and several restorative materials. Unfortunately, the tabulated and graphical results are not necessarily in agreement (since there is no standard tooth).

A little thinking about the information found leads to an appreciation of the challenging problems that arise in dealing with human teeth. Here are a few basic questions that come to an engineer's mind:

1.   What kinds of loads does a tooth encounter and roughly how many times during a normal life?
2.   What is the expected maximum stress in a tooth under normal conditions?

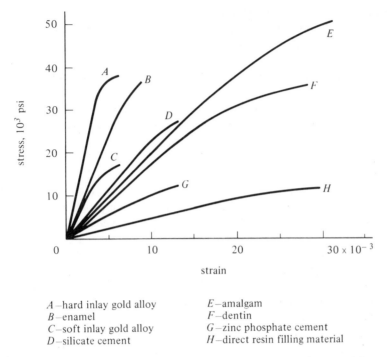

Fig. 15-12. Stress-strain curves of tooth tissues and restorative materials.

3. What are the mechanical or other physical properties of teeth that should be known besides the ones found already?

4. What are the possible mechanical effects of an unfilled cavity?

5. What are the possible mechanical effects of a cavity filled with any of the restorative materials whose stress versus strain curve is available? Consider a (a) shallow cavity; (b) deep cavity.

6. What are the important time-dependent changes that one could expect in the various restorative materials?

7. If one could produce a material after specifying the desirable physical properties, what are the values of those properties for an ideal material intended to fill a deep cavity?

8. What are the possible effects of temperature changes during the normal life of a tooth?

PROBLEM U

Suppose you devote some of your spare time to assist a group of conservationists in finding and evaluating the potential technical problems of a proposed oil pipeline in Alaska. The design criteria and specifications for the pipeline have been completed and made available to all interested parties by the North Alaska

Pipeline System (NAPS).* The information provided leads to many questions concerning the probability of failure of the pipeline. Your job is to come up with reasonable but hard-hitting questions and specific statements regarding the proposed design. Your group will seek a permanent injunction to prevent construction of the pipeline until all doubts about its reliability have been removed. In essence, the information you have to work with is the following:

Length of pipeline: 800 miles

Size of pipe: 40-ft segments, 48-in. diameter, 0.5-in. wall thickness

Construction: welded; the uninsulated pipe is buried in permafrost wherever possible; aboveground, supports are spaced 30 ft apart (pipe is insulated); the supports act as pins for movement in the horizontal plane; the line is essentially a fixed-end column in the vertical plane between supports; anchors are provided to hold the weight component on slopes; construction is at temperatures above 10°F

Operating conditions: oil temperature at the origin station = 145°F; oil temperature farther down the line = 30°F in the winter, 145°F in the summer; normal internal pressure = 900 psi; start-up pressure after a shutdown allowing the pipe to cool to 0°F = 1100 psi

Properties of steel pipe: minimum yield strength = 60,000 psi; pipe remains ductile at temperatures below freezing; a strain of 0.4% is well below the ultimate strain in tension; buckling of pipe is initiated only at strains exceeding 0.4%

Weight: 1000 lb per linear foot of pipe full of oil

Seismic action: earthquake shocks may impose a lateral load on the pipe equal to a maximum of 200 lb per linear foot

NAPS claims that the pipeline can be built and operated successfully, but here are a few questions that come to mind:

1.   Are there any glaring omissions or errors in the design by NAPS?
2.   What is the most severe set of conditions for a peak stress in the pipe at a given point that can be expected?
3.   Where and what is the maximum stress in the pipe caused by its own weight?
4.   What is the worst possible effect of a soil subsidence of 1 ft in 100 ft? Calculate the maximum stress caused by this alone.
5.   What is the largest possible stress caused by a temperature change? Consider both heating and cooling.
6.   Calculate the expected stresses caused by internal pressure.
7.   Is torsion of the pipe possible? If yes, estimate the maximum stresses caused by it.
8.   What would happen if one support (of those at every 30 ft) was lost under a portion of the pipe that is aboveground?

*Not a real name.

9. What is the largest stress caused by an earthquake on the basis of the information given by NAPS? Is this the largest possible stress caused by an earthquake?

10. What about stress concentrations (weldments have many flaws), residual stresses, and time-dependent effects on the metal? Try to be specific on the basis of answers to other questions.

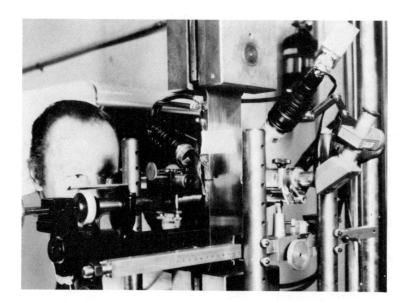

Test of line-pipe steel in air for reference data; two traveling microscopes and direct measurements after fracture are used to determine the positions and curvatures of the crack front throughout the test. *O. Vosikovsky, "Fatigue-Crack Growth in an X-65 Line-Pipe Steel at Low Cyclic Frequencies in Aqueous Environments,"* Closed Loop, **6**, *No. 1 (1976), 2–12.*

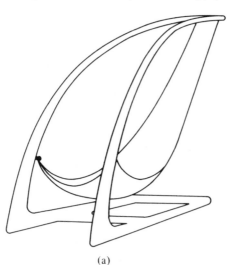

(a)

Fig. 15-13. Views of a springy chair.

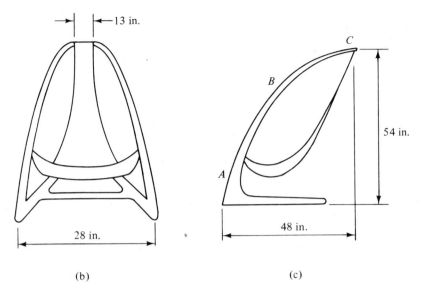

(b) (c)

Fig. 15-13. (Continued)

PROBLEM V

Suppose you are designing a modern chair that has bold lines and exciting springiness (Fig. 15-13). The frame is wood, segments of which may be glued together. The seat is leather. You prefer to use no metal at all. The two main members should have circular cross sections, tapered toward the top as shown. Aside from artistic considerations, what are the minimum diameters at *A*, *B*, and *C* if the maximum stresses should not exceed 35 MPa? The maximum load is the static equivalent of a 120-kg person.

PROBLEM W

Suppose you want to enter a competition for the design of small windmills. The designs are judged mainly on the power generated and cost of the devices. Their practicality must be demonstrated by presenting full-scale prototypes or detailed calculations, preferably both.

It is obvious that most contestants will be able to build and show working devices that are reasonably light in weight and low in cost. You think that the best way to beat the other competitors is to demonstrate the durability of your windmill besides its other features. The other competitors probably will not even think of that, or if they do, they would not know how to handle it. Of course, you realize that this is not a child's game. How could you convince the judges in a few minutes that your device will last at least 20 years?

There are some standard parts such as bearings that present no problems, that's what handbooks are for. But there are major parts that must be designed from scratch. The tower must be as tall as possible because the velocity of the wind increases with distance from the ground, and the propeller blades must be large and light. There is steady force from the wind on all of these ... there are gusts of wind ... there may be vibrations induced by the wind ... there may be ice causing imbalances ....

Describe the causes of the largest normal stresses and how they may vary over a long time for the following parts:

(a) Tower (assume a four-legged structure)
(b) Propeller blade
(c) Propeller shaft

PROBLEM X

Suppose you have taken up karate. One thing that has always been interesting to watch and in which you would like to excel is the breaking of wooden boards with a single blow of a bare hand. This involves muscular strength, speed, technique, showmanship, and (last but not least) the strength of the wood. There are many questions that are relevant to the whole problem, but there is one that seems fundamental from the technical point of view. Should you work with a single board at a time or a stack of boards with a total thickness of the single board (Fig. 15-14)?

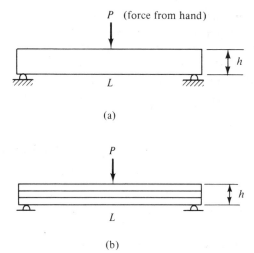

Fig. 15-14. Boards for karate demonstration.

Assume that the lengths (L) and widths (b) of the boards are the same in the two cases. Only one kind of wood is used with the fibers always parallel to the longitudinal axis of the board. After you settle the basic question, consider all aspects of the problem and describe how you could have the most impressive demonstration of breaking boards. You may use tricks, but assume that the nearest spectators are 10 ft from you and would not be easily deceived.

Imagine that you are an independent explorer and prospector on the Moon. You would like to build a small portable cabin in an area remote from the settlements. All materials except rocks and dust are very expensive. The costs of materials are fairly directly proportional to their weights. Describe and justify the design of your ideal cabin in as much detail as you can.

Imagine that you are an active member of the Experimental Aircraft Association. Your latest project is a slow-flying, single-seater aircraft. Its main feature is its light weight: The total weight with payload is 3 kN. Each wing is 4 m long, and the aerodynamic lift force is uniformly distributed over the length. The proposed main structural components in each wing are two aluminum tubes with $L = 3$ m (the wing tip region can be relatively weak), $D_o = 10$ cm, $\sigma_{max} = 400$ MPa. What is the required wall thickness of the tubes if the worst loading is equivalent to three times the static load? There is a suggestion to stiffen the tubes by internal pressurization. Is this a good idea?

APPENDIX **A**

REVIEW OF STATICS
PRINCIPLES

1. NEWTON'S LAW OF ACTION AND REACTION

The forces between two interacting bodies are equal in magnitude, opposite in direction, and collinear. A similar law can be stated for moments of forces.

EXAMPLES:

(a) A man of weight W is standing. This means that a force W is acting downward on the floor, and the same force is acting upward on his feet.

(b) A car is pulling a trailer at a constant speed to the left. This means that a certain force F is acting on the drawbar of the trailer to the left. The same force is acting on the hitch (on the car) to the right.

(c) An airplane cruises at a constant speed in level flight. The weight of each wing is W; the lift force on it is L; the drag force on it is D. There are no engines on the wings. Consider the wings and the fuselage as individual bodies that were connected in the final assembly.

During the flight mentioned the wings apply a total upward force of magnitude $2(L - W)$ to the fuselage. The fuselage applies a downward force of magnitude $V = L - W$ to each wing.

Each wing applies a backward directed force D to the fuselage, and the fuselage applies a forward force D to each wing.

There are also moments at the roots of the wings because the wings are cantilever beams. Looking at the left wing, there are two moments acting on it as shown in Fig. A-1a and b. M_H is acting in the horizontal plane; it counteracts the moment caused by the air friction on the wing.

397

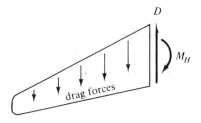

(a) top view of left wing.

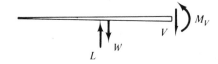

(b) rear view of left wing;
L = resultant of lifting forces.

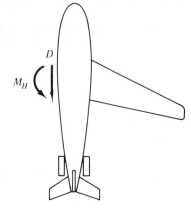

(c) top view of fuselage.

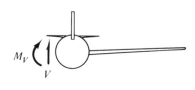

(d) rear view of fuselage.

Fig. A-1. Moments and forces on a wing and fuselage.

M_V is in the vertical plane; it counteracts the net moment caused by L and W. The same moments in magnitude but opposite in direction are acting on the fuselage, as shown in Fig. A-1c and d.

2. FREE-BODY DIAGRAMS

A free-body diagram is a sketch of a body (or a part of a body, or a combination of bodies) on which all significant forces and moments that act on the body are shown. Forces and moments internal to the body are not to be shown on a free-body diagram. They cancel out in pairs and, anyway, it would not be possible to show all of them. Judgment should be exercised in showing external forces and moments in the diagram. Real but insignificant forces should be left out because they clutter the picture, whereas the omission of a single important force makes the diagram incomplete and useless for the analysis of equilibrium.

EXAMPLES:

  (a) Somebody is trying to dislodge a rock with a lever, as shown in Fig. A-2a. The free-body diagrams of the rock, the lever, part of the lever, and the

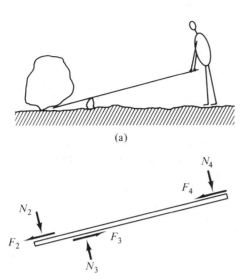

(a)

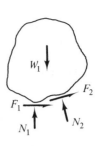

(b) free-body diagram
of the rock.

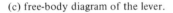

(c) free-body diagram of the lever.

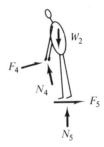

(d)   free-body diagram of a
segment of the lever (left
end).

(e) free-body diagram
of the worker.

Fig. A-2. Individual free-body diagrams from a system.

person are shown in the other parts of the same figure. Note that it is likely that the weight of the lever can be neglected compared to the other forces acting on it. The forces from the two hands are shown as if one hand held the lever; this is acceptable for most practical purposes. It is possible that the frictional forces F can be neglected in comparison with the normal forces N. In that case P is also negligible in (d).

Potentially significant unknown forces and moments should be shown. If at least the direction is known for any of these, the proper direction should be shown in the diagrams. Any direction may be drawn if it is not known for sure, but these should be consistent in related diagrams. A negative answer from the equilibrium equations (next section) indicates that the wrong direction was assumed in the free-body diagram.

(b)   Assume that a pole-vaulter has just left the ground in the beginning of his ascent. The free-body diagram of the pole at this instant is shown in

Fig. A-3.* There is a clockwise frictional moment at *A* because the pole is rotating counterclockwise, but this moment is small and can be neglected. The frictional forces at *B* and *C* are both drawn to the left because at the instant considered the vaulter is moving toward the left with a rapidly decreasing velocity; he leans on the pole fairly evenly with both hands to change his horizontal velocity. The directions of the normal forces at *B* and *C* are correct if the vaulter wants to apply an extra bending moment to the pole at the early stage of the jump. In this case N_B and N_C are not equal. If he only hangs on the pole, both N_B and N_C can be downward, and they may be similar in magnitude. The vaulter could also apply a distinct moment at *B* and another at *C*, but these would depend on the rotating strength of his wrists, which are at best quite small compared to his strength in pulling or pushing with the arms. The weight of the pole is also negligible compared to the other forces present.

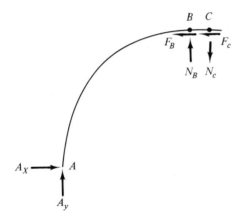

Fig. A-3. Free-body diagram of a pole-vaulting pole.

(c)   Consider a more complicated version of the airplane wing discussed in Section 1: There is a jet engine with weight W_E and thrust T hanging from each wing. In Fig. A-4 three mutually perpendicular views are drawn of a wing. For simplicity, the distributed forces L and D are drawn as resultant forces. The diagrams are complicated, and even people who are experienced with free-body diagrams may make mistakes in them. Note that a force or moment shown in one free-body diagram of an object may not appear in a different view of the same body (no components in that view). This finding is especially important because the reverse is also true: The absence of a force in one view does not mean the absence of a force on the body. For a general analysis three mutually perpendicular views of the body and the external loading on it must be drawn. The views may be selected for convenience.

---

*Practically all students do very poorly on this problem, perhaps because it is a complex dynamic situation.

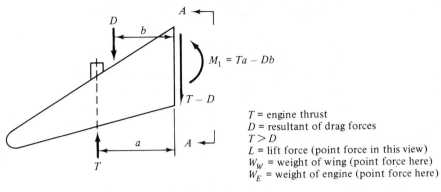

$M_1 = Ta - Db$

$T - D$

T = engine thrust
D = resultant of drag forces
$T > D$
L = lift force (point force in this view)
W_W = weight of wing (point force here)
W_E = weight of engine (point force here)

(a) top view of the left wing.

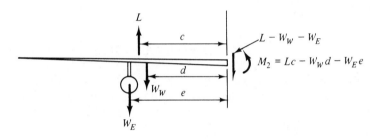

$L - W_W - W_E$

$M_2 = Lc - W_W d - W_E e$

(b) rear view of wing (T and D have no components here).

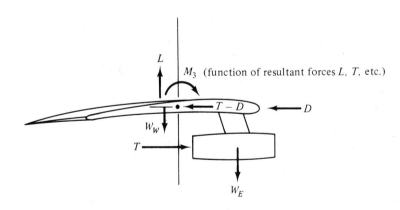

M_3 (function of resultant forces L, T, etc.)

$T - D$

D

(c) view A-A in Fig. A-5a.

Fig. A-4. Free-body diagrams of a wing.

3. EQUILIBRIUM EQUATIONS

If it is known that a body is in equilibrium, it is in equilibrium (stationary or moving at constant velocity) in any view of it. Naturally, a body may be in equilibrium in one direction and not in another, for example, an object falling vertically. Equilibrium equations for forces and moments can be written for any valid free-body diagram. The maximum number of independent equations for an object in perfect equilibrium under three-dimensional loading is six:

$$\sum F_x = 0, \qquad \sum M_x = 0$$
$$\sum F_y = 0, \qquad \sum M_y = 0 \qquad \qquad \text{(A-1)}$$
$$\sum F_z = 0, \qquad \sum M_z = 0$$

The *xyz* coordinates may be chosen arbitrarily to simplify the calculation of required unknown forces and moments.

4. CENTERS OF GRAVITY AND PRESSURE

The *center of gravity* of a body is the point where the resultant of the distributed weight forces is acting. For a homogeneous, regular geometric object the center of gravity is at the geometric center. For an object of complex geometry, the center of gravity is determined as follows.

The object is broken down into simple parts. For each part, it should be possible to determine the center of gravity by inspection. A convenient coordinate system is chosen for the location of the centers of gravity. For example, consider the flat piece of uniform material in Fig. A-5a. The location of the center of gravity can

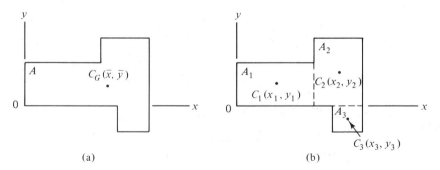

(a)   (b)

Fig. A-5. Example for calculating the center of gravity.

be guessed at this time. Its precise coordinates $\bar{x}$ and $\bar{y}$ are with respect to the arbitrary system. C_1, C_2, and C_3 are the centers of the simple parts chosen. The weight of each part is proportional to its area in this case.

The moment M_o of the whole piece about the origin (caused by gravity) is equal to $A\bar{x}$. The moment of a part is equal to its area times its central x coordinate (for example, $M_2 = A_2 x_2$). A given object can have only one moment about a given point, so

$$M_o = A\bar{x} = A_1 x_1 + A_2 x_2 + A_3 x_3$$

One could say that the moment about the origin was written in two different ways and these were equated. Thus,

$$\bar{x} = \frac{A_1 x_1 + A_2 x_2 + A_3 x_3}{A}$$

Similarly for the other coordinate, but noting that not all the individual areas are on the same side with respect to the x-axis,

$$\bar{y} = \frac{A_1 y_1 + A_2 y_2 - A_3 y_3}{A}$$

Because the object in the example is of uniform thickness, the center of gravity is at the middle of the plate in the thickness direction.

Using integration, the three coordinates of the center of gravity of an object of uniform material are

$$\bar{x} = \frac{\int x\,da}{\int da}, \qquad \bar{y} = \frac{\int y\,da}{\int da}, \qquad \bar{z} = \frac{\int z\,da}{\int da} \qquad \text{(A-2)}$$

The resultant force caused by a distributed pressure is often not at the geometric center of an area. The *center of pressure* is found in a way similar to that of the center of gravity. The moment of the resultant force about any point must equal the sum of the moments caused by infinitesimal forces generated by the pressure over the same total area.

5. AREA MOMENTS OF INERTIA

The moment of inertia (or second moment) of a small area dA about a point that is r distance away (Fig. A-6) is defined as

$$I = r^2 dA \qquad \text{(A-3)}$$

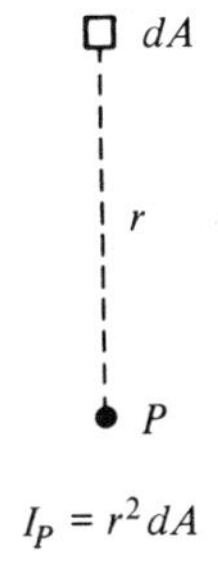

dA

r

P

Fig. A-6. Definition of moment of inertia.

$I_p = r^2\,dA$

This is an abstract concept. It is much easier to use in calculations than to understand the physical meaning of it. In fact, there is no standard description of its physical meaning. Some understanding may be obtained by considering rotational acceleration as an analogy. If the area dA is accelerated around point P, the rotational inertia (or resistance to change in velocity) depends on the mass associated with the area and on the square of the distance to the center of curvature of the path. The concept of rotational inertia is somewhat abstract, too, but one can rely on first-hand experiences to develop a "feel" for the correctness of the concept. In a simple analogy, the moment of inertia of a plane area is a way of describing the effort necessary to displace the area rotationally. Thus, one could say that I_x is the resistance of the whole area A to rotational displacement about the x-axis in Fig. A-7. I_y refers to rotational displacement about the y-axis. The polar moment of inertia, J_o, is the resistance to rotational displacement in the x-y-plane about the origin of the coordinates.

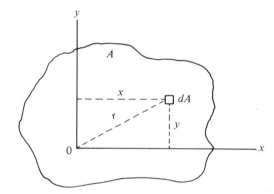

Fig. A-7. Sketch for calculating various moments of inertia.

The moments of inertia are additive. The area in Fig. A-5, for example, can be broken down to small parts for convenience. The sum of the moments of inertia of the individual parts is the moment of inertia of the whole area. In general, referring to Fig. A-7,

$$I_x = \int_A y^2 \, dA, \qquad I_y = \int_A x^2 \, dA, \qquad J_o = \int_A r^2 \, dA \qquad (\text{A-4})$$

Since $r^2 = x^2 + y^2$, $J_o = I_x + I_y$. Note that the moment of inertia is always a positive quantity regardless of where the area is with respect to the chosen coordinates.

A useful formula is provided by the *parallel-axis theorem*:

$$I' = I_c + Ad^2 \quad \text{or} \quad I_c = I' - Ad^2 \qquad (\text{A-5})$$

where I' = moment of inertia of area A about any line l

$\quad\;\; I_c$ = moment of inertia of area A about a line through the centroid of A and parallel to line l

$\quad\;\; d$ = distance between the two parallel lines

Examples:

(a) Find the moment of inertia of the rectangle about the x-axis shown in Fig. A-8. Here $da = bdy$, and

$$I_x = \int_{-h/2}^{h/2} y^2 bdy = b\left[\frac{y^3}{3}\right]_{-h/2}^{h/2} = \frac{1}{12}bh^3$$

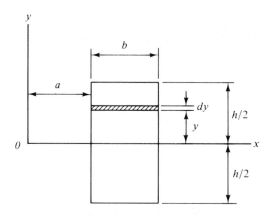

Fig. A-8. Example for calculating moments of inertia.

(b) Find the moment of inertia of the same rectangle about the y-axis. First, I_{y_c}, the moment of inertia about a line through the centroid and parallel to the y-axis must be found:

$$I_{y_c} = \frac{1}{12}hb^3$$

By the parallel-axis theorem,

$$I_y = \frac{1}{12}hb^3 + (bh)\left(a + \frac{b}{2}\right)^2$$

THE MODELING PROCESS

There are three goals in solving problems. The most important is to find specific answers to the questions posed. Another is to obtain the solutions efficiently and economically. This is often less important than the issue of correct answers, but there are situations when the answers become irrelevant if too much time or money is involved in providing them. The third goal is to learn and generalize from the process of solution of the problem. One may forget the specialized formulas and tricks used in certain areas, but the method of approach and the basic principles of problem-solving are remembered more easily. These can be transferred from one area to another.

The skill of an individual in solving problems can be improved gradually even without learning any new formulas for a while. Of course, some knowledge of facts is necessary. But intuition and experience are also important, especially in working on complex problems (whether they involve breaking new ground or not). The following items are of general interest in solving problems in strength of materials, and they are relevant to many other areas, as well.

1. ASSUMPTIONS AND SIMPLIFICATIONS

In many cases, it is best to start working with the simplest possible system that provides a chance of obtaining a reasonable answer. Any important assumptions made at this time (and sometimes there are many) must be stated, preferably in writing.

An interesting and occasionally productive approach is if the simplifications are preceded by a period of unbridled creative thinking. The idea is to write a list, as long as possible, of those items that appear to be relevant to the problem considered. At this stage no external or self-criticism of these items should be allowed (this is similar to the technique of organized brainstorming done by groups). Even a totally wild idea may be productive in that it may trigger another, perhaps practical, idea. On the other hand, any negative thoughts about any of the ideas impede the creative process. Eventually, a time comes when no new ideas are occurring. This is the time to review the list of ideas critically and to strike out those that appear unimportant or too farfetched. It is not shameful to strike out 90% or more of the list.

For an example of lists that can be produced during the expansive thinking period, consider a part of Problem X in Chapter 15: What could influence the strength of a single board? Loading rate, size, directions of fibers, kind of wood, age of wood, temperature, moisture content, humidity of air, surface finish, inner flaws, density, electric fields, magnetic fields, neutron radiation, prior loadings, degassing in vacuum, thin films of chemicals on surface, impregnation with certain chemicals, certain gases in the air, centrifuging, etc.

2. ANALYSIS OF ORDERS OF MAGNITUDE

One should be constantly alert that a useful first answer (and sometimes the final answer) can be obtained perhaps even if crude assumptions are made and if numbers are always rounded off liberally. The seeking of the right order of magnitude in the solution of a problem is particularly useful if one suspects gross inadequacies with respect to a problem. For example, what is the point of considering the stress concentration caused by a small hole or the residual stresses in a smooth shaft if the member will fail for sure anyway under the given loads?

In other cases the quantities that can be measured and the reproducibility of the data do not warrant working with precise numbers. For example, a stress of 41,286.17 psi or a fatigue life of 7.214×10^6 cycles are both given with unreasonable precision. It is difficult to measure stresses with an error not exceeding about 1%, and fatigue life data for identical specimens may show scatter with a factor of 2 or 3 or even more. Thus, the reasonable rounding off of these numbers gives $\sigma = 41$ ksi and $N_f = 7 \times 10^6$ cycles.

Stresses and strains may be rounded off within the nearest 10% or so of the values in most cases.

3. REFINEMENTS

The approximate answers may only be the initial steps in solving problems. It is often necessary to reexamine the original assumptions and refine the characteristics of the idealized system. New free-body diagrams may have to be drawn. Fundamental experiments or full-scale tests may have to be done to obtain more information about material properties and the behavior of the system under operating conditions. Refinements in the solution of problems tend to require more time and money. With the recent advances in the area of computers it has become possible to solve complex problems that could not be handled before. In some cases it could be tempting to reduce the burden of judgment and immediately solve the problem as precisely as possible. This can be encouraged only if the problem-solver still receives a reasonable experience in the basic process of modeling, solution, and evaluation of the results.

TABLES

Table 1

Conversion Factors for Units

U.S. customary	SI
atmosphere (760 mm Hg)	1.013×10^5 pascals (Pa = N/m²)
cycles per second (cps)	hertz (Hz)
degree Fahrenheit (°F)	degree Celsius (°C); $t_C = (t_F - 32)/1.8$; 0°C = 273.15°K
foot (ft)	0.3048* metre (m)
foot-pound force (ft-lbf; commonly, ft-lb)	1.356 newton-metre (N·m)
inch (in.)	0.0254* m = 2.54 cm = 25.4 mm
in²	6.45×10^{-4} m²
in³	1.64×10^{-5} m³
kip (1000 lbf)	4.448×10^3 newton (N) = 4.448 kN
kip/in² (ksi)	6.895×10^6 Pa $\cong$ 7 megapascals (MPa)
pound-force (lbf; commonly, lb)	4.448 N (1 kgf = 9.807 N)
pound (lb) (avoirdupois)	0.4536 kilogram (kg)
lbf/in² (psi)	6.895×10^3 Pa $\cong$ 7 kilopascals (kPa)
lb/in³	2.768×10^4 kg/m³
lb/ft³	16 kg/m³
ton (short, 2000 lb)	907.2 kg
	1 watt-hour (W·h) = 3.6 kilojoules (kJ)*

Multiplication factor	Prefix	Symbol
10^9	giga	G
10^6	mega	M
10^3	kilo	k
10^{-2}	centi	c
10^{-3}	milli	m
10^{-6}	micro	μ

*Exact. For further details, see ASTM Metric Practice, E380–76.

Table 2

Properties of Plane Areas

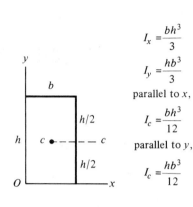

$$I_x = \frac{bh^3}{3}$$

$$I_y = \frac{hb^3}{3}$$

parallel to x,

$$I_c = \frac{bh^3}{12}$$

parallel to y,

$$I_c = \frac{hb^3}{12}$$

$$\overline{x} = b/3$$
$$\overline{y} = h/3$$

$$I_x = \frac{bh^3}{12} \qquad I_y = \frac{hb^3}{12}$$

parallel to x, $\quad I_c = \frac{bh^3}{36}$

parallel to y, $\quad I_c = \frac{hb^3}{36}$

$$I_x = \frac{5\pi r^4}{4}$$

$$I_c = \frac{\pi r^4}{4}$$

$$J_c = \frac{\pi r^4}{2}$$

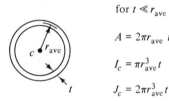

for $t \ll r_{ave}$

$$A = 2\pi r_{ave}\, t$$

$$I_c = \pi r_{ave}^3\, t$$

$$J_c = 2\pi r_{ave}^3\, t$$

semicircle

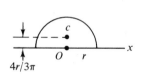

$$I_x = \frac{\pi r^4}{8}$$

parallel to x,

$$I_c = 0.035\pi r^4$$

$$J_O = \frac{\pi r^4}{4}$$

quarter ellipse

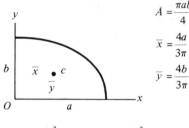

$$A = \frac{\pi ab}{4}$$

$$\overline{x} = \frac{4a}{3\pi}$$

$$\overline{y} = \frac{4b}{3\pi}$$

$$I_x = \frac{\pi ab^3}{16} \qquad I_y = \frac{\pi ba^3}{16}$$

parallel to x, $\quad I_c = 0.0175\pi ab^3$

parallel to y, $\quad I_c = 0.0175\pi ba^3$

Table 3

Slopes and Deflections of Uniform Beams

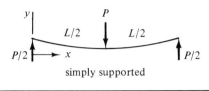

simply supported

$$|y'| = \frac{PL^2}{16EI} \quad \text{at } x = 0, L$$

$$y_{max} = -\frac{PL^3}{48EI} \quad \text{at } x = \frac{L}{2}$$

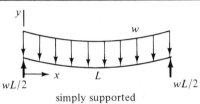

simply supported

$$|y'| = \frac{wL^3}{24EI} \quad \text{at } x = 0, L$$

$$y_{max} = -\frac{5wL^4}{384EI} \quad \text{at } x = \frac{L}{2}$$

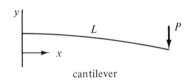

cantilever

$$y' = -\frac{PL^2}{2EI} \quad \text{at } x = L$$

$$y_{max} = -\frac{PL^3}{3EI} \quad \text{at } x = L$$

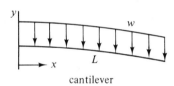

cantilever

$$y' = -\frac{wL^3}{6EI} \quad \text{at } x = L$$

$$y_{max} = -\frac{wL^4}{8EI} \quad \text{at } x = L$$

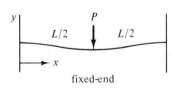

fixed-end

$$y' = 0 \quad \text{at } x = 0, \frac{L}{2}, L$$

$$y_{max} = -\frac{PL^3}{192EI} \quad \text{at } x = \frac{L}{2}$$

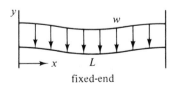

fixed-end

$$y' = 0 \quad \text{at } x = 0, \frac{L}{2}, L$$

$$y_{max} = -\frac{wL^4}{384EI} \quad \text{at } x = \frac{L}{2}$$

For other formulas, see Raymond J. Roark and Warren C. Young, *Formulas for Stress and Strain*, 5th ed. (New York: McGraw-Hill Book Company, 1975).
For properties of common structural shapes, see *Manual of Steel Construction*, 7th ed. (New York: American Institute of Steel Construction, 1970).

Table 4

Average Mechanical Properties of Materials

Material	E GPa/10^6 psi	G 10^6 psi	μ	σ_y/σ_y' ksi	σ_u MPa/ksi	% el. over / % RA	weight lb/in³	α 10^{-6} per °F	cyclic σ versus ϵ
Acrylics	3/0.4		0.36		55/8	5/	0.043	4.5	
		3.8	0.34						
Aluminums									
High purity, annealed	69/10			4/>4	83/12	40/	0.098	13.1	hardening
High purity, cold-worked	69/10			24/<24		6/	0.098	13.1	softening
1100 (26BHN)	69/10			14/9	110/16	10/88	0.098	13.1	softening
2014 T6 (155BHN)	72/10.5	4	0.33	67/60	510/74	13/25	0.1	12.8	approx. stable
2024 T4	72/10.5			44/64	475/69	20/35	0.1	12.9	hardening
5456 H311 (95BHN)	69/10			34/52	400/58	20/35	0.096	13.3	hardening
7075 T6	71/10.3			68/76	580/84	11/33	0.1	13.1	hardening
A108 (cast)	71/10.3			16/27	190/28	2/	0.1	11.9	hardening
Beryllium	290/42		0.1	12/	480/70	3/	0.067	6.4	
Boron	414/60		0.2	500/	3500/500		0.084	3	
Brass 6A (leaded yellow)	97/14	5.7	0.35	12/	240/35	40/30	0.306	11.5	
Bronze 385 (architectural)	97/14			20/	410/60	30/	0.306	11.6	
Cast irons									
Gray (class 30)	103/15	5.9	0.2	25/	210/30	nil	0.25	6	
Gray (class 60)	138/20		0.25	45/40	410/60	nil	0.27	6	softening
Concrete (compression)	17/2.5		0.15		20/3		0.087	6	
Coppers									
High purity, annealed	117/17	6.4	0.36	10/>10	230/33	50/	0.323	9.8	hardening
High purity, cold-worked	117/17	6.4	0.36	45/<45	350/50	10/	0.323	9.8	softening

Table 4 (Cont.)

Material	E GPa/10^6 psi	G 10^6 psi	μ	σ_y/σ_y' ksi	σ_u MPa/ksi	% el. over % RA	weight lb/in³	α 10^{-6} per °F	cyclic σ versus ϵ
Epoxy (cast, rigid)	3/0.5				70/10	4/	0.042	33	
Glass (plate, soda lime)	69/10		0.24		70/10	nil	0.15	8.5	
Graphite (general purpose)	7/1				7/1		0.06	1.5	
Kevlar 49 (fiber)	131/19				3600/525	2/	0.052		
Lead	14/2				17/2.5	45/	0.41	16.3	
Magnesium alloys (wrought)	45/6.5	2.4		35/	280/40	10/	0.065	14	
Molybdenum	324/47			82/	650/95	/60	0.37	2.7	
Nickel (annealed)	207/30			20/>20	380/55		0.321	7.4	hardening
Nylon 6 (cast)	4/0.55		0.4	13/	90/13	20/	0.04	44	
Polyethylene (medium density)					14/2	200/	0.033	120	
Polystyrene (molded)				5/	35/5	2–30/	0.039	40	
Polyvinyl chloride (rigid)	3/0.4				40/6	5–25/	0.05	30	
Rubbers Natural Neoprene			0.5 0.5		20/3 20/3	800/ 850/	0.034 0.045	400 350	
Steels SAE 1005 (hot-rolled, Low carbon, 90 BHN)	207/30	12	0.27	38/34	350/50	40/80	0.283	6.5	softening (low ϵ) hardening (high ϵ)

413

Table 4 (Cont.)

Material	E GPa/10^6 psi	G 10^6 psi	μ	σ_y/σ'_y ksi	σ_u MPa/ksi	% el. over % RA	weight lb/in³	α 10^{-6} per °F	cyclic σ versus ϵ
Steels (Cont.)									
SAE 1005 (cold-rolled, 125 BHN)	207/30	12	0.27	60/40	450/65	30/65	0.283	6.5	softening
SAE 1020 (normalized, 90 BHN)	207/30	12	0.27	35/38	450/65	30/65	0.283	6.5	softening (low ϵ) hardening (high ϵ)
Van-80 (225 BHN)	193/28	11.5	0.3	82/81	690/100	/68	0.283	6.5	approx. stable
RQC-100 (300 BHN)	203/29.5	12	0.27	112/75	830/120	/65	0.283	6.5	softening
SAE 1045 (225 BHN)	200/29	12	0.27	92/60	720/105	/65	0.283	6.5	softening
SAE 1045 (450 BHN)	207/30	12	0.27	220/140	1600/230	/55	0.283	6.5	softening
SAE 1045 (600 BHN)	207/30	12	0.27	270/250	2200/325	/40	0.283	6.5	approx. stable
AISI 4130 (365 BHN)	200/29	12	0.28	197/120	1400/207	/55	0.283	6.5	softening
SAE 4142 (670 BHN)	200/29	12	0.27	240/320	2400/355	/6	0.283	6.5	hardening
AISI 4340 (240 BHN)	193/28	12	0.32	92/66	830/120	/43	0.283	6.3	softening
AISI 4340 (410 BHN)	200/29	12	0.30	200/120	1500/215	/36	0.283	6.3	softening
SAE 5160 (430 BHN)	193/28	12		222/145	1700/242	/42	0.283	6.5	softening
AISI 52100 (520 BHN)	207/30	12	0.29	280/192	2000/295	/10	0.283	6.5	softening (low ϵ) hardening (high ϵ)
SAE 9262 (annealed, 260 BHN)	207/30	12	0.27	66/76	920/134	/14	0.283	6.5	hardening
SAE 9262 (410 BHN)	200/29	12	0.27	200/152	1600/227	/32	0.283	6.5	softening
H 11 (660 BHN)	207/30	12	0.27	300/340	2600/375	/33	0.281	7.4	hardening
AISI 304 Stainless (160 BHN)	186/27		0.27	37/104	740/108	/74	0.29	9.6	hardening
AM350 Stainless (annealed)	193/28		0.32	64/196	1300/191	/52	0.282	6.3	hardening
AM350 Stainless (496 BHN)	179/26		0.30	270/235	1900/276	/20	0.282	6.3	softening (low ϵ) hardening (high ϵ)
18 Ni Maraging (460 BHN)	186/27			260/195	1900/270	/56	0.29	6.3	softening

Table 4 (Cont.)

Material	E GPa/10^6 psi	G 10^6 psi	μ	σ_y/σ'_y ksi	σ_u MPa/ksi	% el. over % RA	weight lb/in³	α 10^{-6} per °F	cyclic σ versus ϵ
Teflon	0.4/0.06				20/3	300/	0.078	50	
Titanium	103/15	6.5		60/	480/70	25/	0.163	5.3	
Tungsten	379/55			220/	1500/220	nil	0.7	2.5	
Wood Douglas fir	14/2				55/8		0.02	3	
Oak	12/1.8				48/7		0.025	1.9	

NOTES:

1. $\sigma_y = 0.2\%$ offset yield strength in tension.
   $\sigma'_y = 0.2\%$ offset yield strength from cyclic σ-ϵ curve.

2. The severity of cyclic hardening or softening is indicated by σ'_y in comparison with σ_y.

3. The % elongation is from 2 in. gage length.

4. α = coefficient of thermal expansion.

5. Some of the data are from very few tests; others are averages of many tests where the scatter is not specified. All of these should be used with caution in critical design.

415

ANNOTATED BIBLIOGRAPHY

HISTORICAL INTEREST

1. DECAMP, L. SPRAGUE, *The Ancient Engineers*. New York: Ballantine Books, 1974. A comprehensive description of all aspects of engineering through the Middle Ages; puts the achievements into economic and social perspective; gives insights into the characters and life-styles of the people involved; a must for all engineers.

2. GALILEI, GALILEO, *Dialogues Concerning Two New Sciences*. Evanston, Ill.: Northwestern University Press, 1968. Describes the most extensive first attempts in fundamental analysis of structural members.

3. TATNALL, F. G., *Tatnall on Testing*. Metals Park, Ohio: American Society for Metals, 1966. Interesting accounts of the important people and their work in the area of mechanical testing; the invention and development of the strain gage and other devices up to but not including servo-controlled testing.

STATICS AND DYNAMICS

4. HIGDON, ARCHIE, ET AL., *Engineering Mechanics, Statics and Dynamics*, 2nd ed. Englewood Cliffs, N.J.: Prentice-Hall, Inc., 1976. An extensive work using vectors.

5. SHAMES, IRVING H., *Introduction to Statics*. Englewood Cliffs, N.J.: Prentice-Hall, Inc., 1971. A short, clear presentation of the most important topics in statics; uses vectors.

MATERIALS SCIENCE

6. DIETER, GEORGE E., JR., *Mechanical Metallurgy*. New York: McGraw-Hill Book Company, 1961. A comprehensive presentation of metallurgical fundamentals and

their applications to deformation and fracture of materials; materials testing; material behavior at elevated temperatures; residual stresses; statistics applied to materials testing; effects of various metal-forming processes.

7. SMITH, CHARLES O., *The Science of Engineering Materials*, 2nd ed. Englewood Cliffs, N.J.: Prentice-Hall, Inc., 1977. Covers metals, polymers, and ceramics; electrical, magnetic, thermal, and mechanical properties of materials; fabrication of materials; environmental effects; factors in selecting materials.

MECHANICS OF MATERIALS, INTERMEDIATE AND ADVANCED

8. POPOV, E. P., *Mechanics of Materials*, 2nd ed. Englewood Cliffs, N.J.: Prentice-Hall, Inc., 1976. Presents all fundamental principles of strength of materials, with most topics gradually extended to more complex problems; uses English and SI units equally.

9. SEELY, FRED B. AND JAMES O. SMITH, *Advanced Mechanics of Materials*, 2nd ed. New York: John Wiley & Sons, Inc., 1952. The major topics include unsymmetrical bending, curved beams, flat plates, noncircular torsion members, thick-walled cylinders, contact stresses, energy methods, and instability problems.

10. SHAMES, IRVING H., *Introduction to Solid Mechanics*. Englewood Cliffs, N.J.: Prentice-Hall, Inc., 1975. A rigorous treatment of mechanics of materials; advanced topics are interspersed with the more elementary ones; introduces tensor notation.

STRESS ANALYSIS

11. DOVE, RICHARD C. AND PAUL H. ADAMS, *Experimental Stress Analysis and Motion Measurement*. Columbus, Ohio: Charles E. Merrill Publishing Co., 1964. Discusses mechanical, electrical, and optical strain gages; grid techniques; the Moiré fringe method; brittle coatings; photoelasticity; and measurements of displacement, velocity, and acceleration.

12. DURELLI, A. J. AND V. J. PARKS, *Moiré Analysis of Strain*. Englewood Cliffs, N.J.: Prentice-Hall, Inc., 1970. An extensive presentation of methods for the determination of fields of displacement in moving or deforming bodies; includes holography.

13. DURELLI, A.J. AND W.F. RILEY, *Introduction to Photomechanics*. Englewood Cliffs, N.J.: Prentice-Hall, Inc., 1965. A simplified but complete presentation of the fundamentals of photoelasticity; detailed examples of applications are given in two-dimensional, three-dimensional, thermal, and dynamic problems.

14. JONES, ROBERT M., *Mechanics of Composite Materials*. New York: McGraw-Hill Book Company; Washington, D.C.: Scripta Book Company, 1975. Micromechanics and macromechanics of composites; particulate, fiber-reinforced, and laminated composites; uses matrices and tensors.

15. PETERSON, R. E., *Stress Concentration Design Factors*. New York: John Wiley & Sons, 1953. Charts and equations for various geometries and loading conditions.

16. ROARK, RAYMOND J. AND WARREN C. YOUNG, *Formulas for Stress and Strain*, 5th ed. New York: McGraw-Hill Book Company, 1975. Detailed discussions, formulas, and

examples for the mechanics analysis of numerous common structural elements; tables of numerical coefficients for maximum deformations and stresses; includes dynamic loading and temperature stresses.

STRUCTURAL MECHANICS

17. BEAUFAIT, FRED W., *Basic Concepts of Structural Analysis*. Englewood Cliffs, N.J.: Prentice-Hall, Inc., 1977. Classical and contemporary techniques for the analysis of determinate and indeterminate structures; includes techniques for computer programming.

18. CHAJES, ALEXANDER, *Principles of Structural Stability Theory*. Englewood Cliffs, N.J.: Prentice-Hall, Inc., 1974. A relatively simple treatment of elastic and inelastic buckling of common members and frames.

19. FERGUSON, PHIL M., *Reinforced Concrete Fundamentals*. New York: John Wiley & Sons, Inc., 1958. Conventional theory and ultimate strength theory for beams; frame analysis; prestressed concrete analysis.

20. WANG, CHU-KIA AND CHARLES G. SALMON, *Reinforced Concrete Design*, 2nd ed. New York: Intext Educational Publishers, 1973. Emphasizes the ultimate strength method; most of the discussion and examples are related to the 1971 ACI Code; practical aspects of proportioning beams and selecting materials; prestressed concrete analysis.

FAILURE ANALYSIS AND PREVENTION

21. ALMEN, J. O. AND P. H. BLACK, *Residual Stresses and Fatigue in Metals*. New York: McGraw-Hill Book Company, 1963. Discusses the nature of residual stresses and the practical aspects of creating such stresses for improving fatigue lives.

22. HERTZBERG, RICHARD W., *Deformation and Fracture Mechanics of Engineering Materials*. New York: John Wiley & Sons, 1976. Deformation mechanisms in metals and plastics; fracture toughness testing and design concepts; effects of temperature and chemical environment on the resistance to fracture; extensive failure analysis checklist.

23. PFLUGER, A. R. AND R. E. LEWIS, ed., *Weld Imperfections*. Reading, Mass.: Addison-Wesley Publishing Company, 1968. Describes a number of causes of weld imperfections in a variety of metals: destructive and nondestructive analyses of imperfections; includes the use of stress intensity factors and fracture toughness evaluations.

24. ROLFE, STANLEY T. AND JOHN BARSOM, *Fracture and Fatigue Control in Structures: Applications of Fracture Mechanics*. Englewood Cliffs, N.J.: Prentice-Hall, Inc., 1977. Emphasizes applications, engineering experience, economics, and recent developments in fracture criteria; samples of fracture control plans for aircraft, bridges, nuclear plants; provides considerable test data on structural materials.

25. SANDOR, BELA I., *Fundamentals of Cyclic Stress and Strain*. Madison, Wis.: The University of Wisconsin Press, 1972. Emphasizes plastic deformations in permutations of the signs of stress and strain, in cycle-dependent changes of mechanical properties, and in accumulation of fatigue damage.

26. TETELMAN, A. S. AND A. J. MCEVILY, JR., *Fracture of Structural Materials.* New York: John Wiley & Sons, Inc., 1967. Macroscopic and microscopic aspects of fracture; causes of embrittlement; fracture of metals and various nonmetallic materials.

SI UNITS

27. Standard for Metric Practice E380-76. Philadelphia: American Society for Testing and Materials, 1976.

ANSWERS TO SELECTED PROBLEMS

1-4. $P_{max} = 49 \times 10^5$ lb. **1-5.** $P_1 = 0.25$, $P_2 = 1$ N. **1-9.** $\sigma_{10} = 3$, $\tau_{10} = 17.1$, $\sigma_{20} = 11.7$, $\tau_{20} = 32.1$ ksi. **1-10.** $\sigma_{20} = 58.5$, $\tau_{20} = 160.5$, $\sigma_{40} = 206.6$, $\tau_{40} = 246.2$ MPa. **1-13.** $\sigma_5 = 2.3$, $\tau_5 = 26$, $\sigma_{45} = 150$, $\tau_{45} = 150$ ksi. **1-14.** $\sigma_{45} = \tau_{45} = 5$, $\sigma_{90} = 10$ MPa, $\tau_{90} = 0$.
1-20. $\sigma_{30} = 15$, $\tau_{30} = 8.7$ ksi. **1-21.** $\sigma_{30} = 34.5$, $\tau_{30} = 59.8$ MPa.
1-24. $\sigma_{30} = -52.5$, $\tau_{30} = 30.3$ ksi. **1-25.** $\sigma_{30} = -120.8$, $\tau_{30} = -209$ MPa.
1-27. $\sigma_{45} = -280$ MPa, $\tau_{45} = 0$. **1-30.** $\sigma_{45} = -25$, $\tau_{45} = 35$ ksi.
1-32. $\sigma_{30} = 154$, $\tau_{30} = 84.1$ ksi. **1-33.** $\sigma_{10} = -300$ MPa, $\tau_{10} = 0$.
1-35. $\sigma_{max} = 138$, $\sigma_{min} = 0$, $\tau_m = 69$ MPa. **1-38.** $\sigma_{max} = 0$, $\sigma_{min} = -70$, $\tau_m = 35$ ksi. **1-41.** $\sigma_{max} = \tau_m = -\sigma_{min} = 280$ MPa. **1-44.** $\sigma_{max} = 68$, $\sigma_{min} = -38$, $\tau_m = 53$ ksi. **1-47.** $\sigma_{max} = \sigma_{min} = -300$ MPa, $\tau_m = 0$.
1-50. $\tau_m = 2\sigma_1$. **1-53.** $\tau = 50$ MPa. **1-56.** $\sigma_y = 108$ ksi.
1-59. $\sigma_x = 120$, $\sigma_y = 30$ ksi. **1-63.** $\sigma_x = 250$ MPa.

2-1. $\epsilon = 2\%$. **2-5.** $e = 0.3$ in. **2-8.** $e = -0.25$ mm. **2-10.** $\epsilon_{circ.} = 0.01$.
2-15. $\epsilon_1 = 0.001$, $\epsilon_2 = 0$, $\gamma_m = 0.001$, $\epsilon_{30} = 0.00075$, $\gamma_{30} = -0.00088$.
2-19. $\epsilon_1 = -\epsilon_2 = 0.0035$, $\gamma_m = 0.007$, $\epsilon_{30} = 0.003$, $\gamma_{30} = 0.0035$.
2-22. $\epsilon_1 = 0$, $\epsilon_2 = -0.008$, $\gamma_m = 0.008$, $\epsilon_{30} = -0.006$, $\gamma_{30} = 0.0069$.
2-25. $D_{max} = 14$ m.

3-4. $E = 500$ GPa.　　**3-5.** $G = 50 \times 10^6$ psi.　　**3-10.** $\mu = 0.317$.
3-14. $\epsilon_{ax.} = 0.00286$.　　**3-18.** $\epsilon_z = -2.75 \times 10^{-4}$.　　**3-21.** $\epsilon_x = 7 \times 10^{-4}$.
3-24. $\sigma_y = 342$ MPa.　　**3-25.** $\sigma_z = 9.6$ ksi.　　**3-27.** $E_c = 30.8 \times 10^6$ psi.
3-30. $E_c = 173$ GPa.　　**3-32.** $V_f = 0.754$.

4-3. $\epsilon_t = 0.021$.　　**4-6.** $\epsilon_p = 0.0283$.　　**4-10.** $W = 167$ kPa.
4-30. $\sigma_f = 357$ ksi, $\sigma_y = 240$ ksi, $\% RA = 39.4$, $\epsilon_{tf} = 0.512$.
4-35. $\sigma_y = 85$, $\sigma'_y = 84$ ksi.　　**4-41.** $\sigma_y = 195$, $\sigma'_y = 120$ ksi.
4-46. $\sigma_y = 300$, $\sigma'_y = 165$ ksi.　　**4-49.** $\sigma_y = 310$, $\sigma'_y = 350$ ksi.
4-51. $\sigma_y = 68$, $\sigma'_y = 205$ ksi.　　**4-55.** $\sigma_y = 68$, $\sigma'_y = 75$ ksi.

5-2. $L = 2440$ ft, $e = 12.4$ in.　　**5-6.** $h = 0.95$ m.　　**5-10.** $P \cong 324$ kN.
5-14. $e = 10.53$ in.　　**5-17.** $\Delta P = 300$ kips.　　**5-23.** $F = 1500$ lb.
5-29. $D = 5$ mm, $\sigma_{max} = 125$ MPa.　　**5-32.** $D = 0.21$ in.
5-36. $P = 42.4$ kips.　　**5-39.** $A = 1.2$ mm.　　**5-43.** $a = 3.4$ cm.
5-48. $p = 208$ psi.　　**5-49.** $e_l = 0.04$ in., $D_1 = 48.07$ in.　　**5-51.** $D_1 = 3.007$ m,
$\Delta V = 0.099$ m³.　　**5-54.** $p = 0.006$ psi, $\sigma = 55$ psi.　　**5-56.** $t = 3 \times 10^{-5}$ m.

6-3. $V_1 = 5$, $M_1 = 5x$, $V_2 = 0$, $M_2 = 10$, $V_3 = -5$, $M_3 = -5x + 25$.
6-8. $V = 0$, $M = 1000$.　　**6-14.** $V = 300$, $M = -300x + 100$.
6-19. $V_1 = 450 - 600x$, $M_1 = 450x$, $V_2 = -150$, $M_2 = -150x + 300$.
6-23. $V_1 = 2000x$, $M_1 = -1000x^2$, $V_2 = 400$, $M_2 = -400(x - 0.1)$.
6-26. $V = 8x^2$, $M = -\frac{8}{3}x^3$.　　**6-29.** $V_1 = 1700 - 800x$,
$M_1 = 1700x - 400x^2$, $V_2 = 100$, $M_2 = 1700x - 1600(x - 1)$, $V_3 = -1900$,
$M_3 = 1700x - 1600(x - 1) - 2000(x - 3)$.　　**6-31.** $V = 83x^2$,
$M = 50 - \frac{83}{3}x^3$.

7-3. $M_a/M_b = b/a$.　　**7-6.** $\sigma_{max} = 26.7$ MPa.　　**7-9.** $\sigma_{max} = 25$ ksi,
$\sigma_A = 10$ ksi.　　**7-12.** $\sigma_{max} = 83$ MPa.　　**7-16.** $\sigma_{max} = 2.17$ MPa,
$\sigma_A = 1.08$ MPa.　　**7-21.** $\sigma_{max} = 8.3$ ksi, $\sigma_A = 330$ psi.　　**7-24.** $\sigma_{max} = 12.6$,
$\sigma_A = 4.5$ MPa.　　**7-29.** $\sigma_{max} = 4.9$, $\sigma_A = 2.45$ ksi.　　**7-33.** $\sigma_t = 13.9$,
$\sigma_c = 39.2$ ksi.　　**7-34.** $\sigma_t = 111$, $\sigma_c = 43$ MPa.　　**7-39.** $\sigma_{max} = 72$ psi.

7-42. $\sigma(x) = \dfrac{24x}{0.035x + 0.08}$ (x measured from free end).

7-45. $M_x = 82{,}550$ in.-lb, $M_y = 86{,}125$ in.-lb.   **7-48.** $\sigma_w = \pm 4.58$, $\sigma_{ge} = \pm 153.5$ MPa.   **7-50.** $\sigma_c = 12.9$, $\sigma_s = 283.5$ MPa.
7-52. $A_s = 0.004$ m².   **7-54.** $\sigma_c = 5.24$, $\sigma_t = 4.79$ ksi.   **7-57.** $e = 1.41$ in.
7-60. $e = 0.013$ m.   **7-62.** $\sigma_c = -78$, $\sigma_t = 66$ ksi.   **7-63.** $M_m = 412$ N·m.

CHAPTER 8

8-3. $M(x) = -300x^2$ in.-lb (x measured from right).   **8-7.** $x = 0.5$, $y' = -4.45 \times 10^{-4}$, $y = -1.48 \times 10^{-4}$ m (x measured from left).
8-10. $x = 1$ (from left), $y' = -0.015$, $y = -0.02$ m.   **8-13.** $x = 0$ (left), $y' = 1.24 \times 10^{-3}$, $y = -0.143$ in.   **8-15.** $x = 1$ (from left), $y' = 0.0163$, $y = 8.15 \times 10^{-3}$ m.   **8-18.** $x = 0$ (right), $y' = 8 \times 10^{-4}$, $y = -0.032$ in.
8-20. $y = 0.036$ m.   **8-22.** $y = 0.348$ m.   **8-24.** $y = 0.143$ in.
8-26. $y = 0.156$ in.

CHAPTER 9

9-2. $T = 14.8$ kN·m.   **9-5.** $T = 8.1 \times 10^5$ in.-lb.   **9-8.** $T = 5.7$ kN·m.
9-11. $T = 7.4 \times 10^5$ in.-lb.   **9-14.** a) $T_{e_{hollow}} = \frac{15}{16} T_{e_{solid}}$, b) $T_{fp_{hollow}} = \frac{7}{8} T_{fp_{solid}}$.
9-17. $T = G\pi R^3 \epsilon_{gage}$.   **9-20.** $\theta = 3.9 \times 10^{-3}$ rad.
9-23. $\theta_{BA} = 8.84 \times 10^{-3}$ rad.   **9-26.** $\theta_{BA} = 0.0158$ rad.
9-29. $\theta_{BA} = 0.0242$ rad.   **9-31.** $\theta_{AB} = 0.0193$ rad.   **9-34.** $R = 0.24$ in.
9-37. $N = 253$ rpm.   **9-40.** $T_1 = 0.747T_0$.   **9-43.** $T_1 = 0.26T_0$.

CHAPTER 10

10-1. $a = 24$ in.   **10-4.** $\sigma_1 = \sigma_2 = 167$ MPa.   **10-7.** $\sigma_1 = 20.7$, $\sigma_2 = 31$, $\sigma_3 = 41.4$ ksi.   **10-10.** $\sigma_1 = 356$, $\sigma_2 = 87$, $\sigma_1' = 116$, $\sigma_2' = 230$ MPa.
10-13. $\sigma_A = 376$, $\sigma_B = 332$, $\sigma_A' = 200$, $\sigma_B' = 510$ psi.   **10-15.** $\sigma = 28.3$ MPa.
10-18. $F = 930$, $V = 1070$ lb, $M_B = 30{,}700$ in.-lb.
10-20. $M_1 \approx M_2 = 12{,}500$ in.-lb.   **10-23.** $F = 200$, $V = 1800$ N, $M = -1870$ N·m.   **10-26.** $R = kd = 734$, $V = 266$ lb, $M = -19{,}150$ in.-lb.
10-29. $R = kd = 200$, $V = 2800$ N, $M = -1400$ N·m.
10-31. $\tau_m = 63.7$ MPa.   **10-34.** $\tau_m = 620$ psi.   **10-37.** $\tau_m = 44$ MPa.
10-39. $\tau_m = 85$ MPa.

CHAPTER 11

11-4. $0.5 < k < 1$.   **11-6.** $P_c = 30$ kN.   **11-9.** $x = 0.9$ in.
11-12. $R_0 = 0.0229$ m.   **11-16.** $R < 0.05$ m.   **11-18.** $P_c = 5$ kips.
11-19. $P_{c_7} = 6330$ lb $\simeq 24\, P_{c_1}$.   **11-21.** $P_c = 34.4$ kips.
11-24. $P_c = 940$ N.   **11-26.** $P_c = 2.4$ MN.

12-6. $\sigma_{\max} = 62$ ksi.   **12-9.** $\sigma_{\max} = 213$ MPa.   **12-11.** $\sigma_{\max} = 313$ MPa.
12-14. Equilibrium conditions must be satisfied.
12-22. $T_{\text{al}} > 470°$K, $T_{\text{st}} > 910°$K.
12-26. Loop is transformed using $\epsilon_p = \epsilon_t - \epsilon_e$.   **12-29.** $S = E\epsilon_e$;
$S = 75$ ksi, $N = 1$; $S = 28.5$ ksi, $N = 10^7$.   **12-34.** $IF = 5.58$,
$d_{\max} = 1.12$ in.   **12-37.** $k = 3 \times 10^6$ N/m.   **12-39.** $W = 52.3$ N.

13-1. $N_y = N_\tau = N_{\text{oct.}} = 2$, $N_u = 3.25$.   **13-4.** $N_y < 1$, $N_u = 1.33$, $N_\tau < 1$,
$N_{\text{oct.}} < 1$.   **13-7.** $N_y = 1.23$, $N_u = 2.16$, $N_\tau < 1$, $N_{\text{oct.}} = 1$.
13-10. $N_y = N_u = 1.8$.   **13-12.** $N_y = 2.65$, $N_u = 3.98$, $N_\tau = 2$, $N_{\text{oct.}} = 2.13$.
13-15. $\sigma = 2.9$ ksi.   **13-18.** No yielding, no fracture.
13-21. Ductile fracture possible.   **13-24.** $\sigma_u > 500$, or $\tau_{ys} < 240$.
13-27. $\sigma_u > 100$, or $\tau_{ys} < 49.5$.   **13-30.** $2c = 3.2$ cm.
13-33. $\sigma < 319$ ksi.   **13-36.** $\sigma < 830$ MPa.

14-1. $\sigma_{\max} = 7.6$ ksi.   **14-4.** $\sigma_{\max_1} = -2$, $\sigma_{\max_2} = -668$ kPa.
14-7. $\sigma_{\max} = 324$ MPa.   **14-9.** $\tau_m = 626$ psi.   **14-12.** $\tau_m = 1.1$ kPa.
14-14. $\sigma_{\max} = 4.14$ ksi.   **14-15.** $\sigma_{\max} = 41.6$ MPa.   **14-18.** $\sigma_{\max} = 11.46$ ksi.
14-20. $\sigma_{\max} = 34$ MPa.   **14-21.** $\sigma_{\max} = 8$ ksi.   **14-24.** $\sigma_{\max} = 4.7$ MPa.

INDEX

Average Mechanical Properties of Materials

See pp. 412-415 for the complete Table

Material	E GPa/10^6 psi	G 10^6 psi	μ	σ_y/σ_y' ksi	σ_u MPa/ksi	% el. over % RA	weight lb/in³	α 10^{-6} per °F	cyclic σ versus ϵ
Aluminums									
High purity, annealed	69/10	3.8	0.34	4/>4	83/12	40/	0.098	13.1	hardening
High purity, cold-worked	69/10			24/<24	110/16	6/	0.098	13.1	softening
1100 (26BHN)	69/10			14/9		10/88	0.098	13.1	softening
2014 T6 (155BHN)	72/10.5	4	0.33	67/60	510/74	13/25	0.1	12.8	approx. stable
2024 T4	72/10.5			44/64	475/69	20/35	0.1	12.9	hardening
5456 H311 (95BHN)	69/10			34/52	400/58	20/35	0.096	13.3	hardening
7075 T6	71/10.3			68/76	580/84	11/33	0.1	13.1	hardening
A108 (cast)	71/10.3			16/27	190/28	2/	0.1	11.9	hardening
Cast irons									
Gray (class 30)	103/15	5.9	0.2	25/	210/30	nil	0.25	6	
Gray (class 60)	138/20		0.25	45/40	410/60	nil	0.27	6	softening
Concrete (compression)	17/2.5		0.15		20/3		0.087	6	
Coppers									
High purity, annealed	117/17	6.4	0.36	10/>10	230/33	50/	0.323	9.8	hardening
High purity, cold-worked	117/17	6.4	0.36	45/<45	350/50	10/	0.323	9.8	softening
Epoxy (cast, rigid)	3/0.5				70/10	4/	0.042	33	
Glass (plate, soda lime)	69/10		0.24		70/10	nil	0.15	8.5	
Nylon 6 (cast)	4/0.55		0.4	13/	90/13	20/	0.04	44	
Polyethylene (medium density)					14/2	200/	0.033	120	
Polystyrene (molded)				5/	35/5	2–30/	0.039	40	
Polyvinyl chloride (rigid)	3/0.4				40/6	5–25/	0.05	30	
Rubbers									
Natural			0.5		20/3	800/	0.034	400	
Neoprene			0.5		20/3	850/	0.045	350	